AF779838

Plant Protein Processing: Technological Innovation and New Application Scenarios

Plant Protein Processing: Technological Innovation and New Application Scenarios

Guest Editors

Zhongjiang Wang
Zengwang Guo
Zhaoxian Huang

MDPI

Basel • Beijing • Wuhan • Barcelona • Belgrade • Novi Sad • Cluj • Manchester

Guest Editors

Zhongjiang Wang
College of Food Science
Northeast Agricultural
University
Harbin
China

Zengwang Guo
College of Food Science
Northeast Agricultural
University
Harbin
China

Zhaoxian Huang
School of Food Science
and Engineering
Hainan University
Haikou
China

Editorial Office
MDPI AG
Grosspeteranlage 5
4052 Basel, Switzerland

This is a reprint of the Special Issue, published open access by the journal *Foods* (ISSN 2304-8158), freely accessible at: https://www.mdpi.com/journal/foods/special_issues/UYYUD96DC9.

For citation purposes, cite each article independently as indicated on the article page online and as indicated below:

Lastname, A.A.; Lastname, B.B. Article Title. *Journal Name* **Year**, *Volume Number*, Page Range.

ISBN 978-3-7258-7675-4 (Hbk)
ISBN 978-3-7258-7676-1 (PDF)
https://doi.org/10.3390/books978-3-7258-7676-1

© 2026 by the authors. Articles in this reprint are Open Access and distributed under the Creative Commons Attribution (CC BY) license. The reprint as a whole is distributed by MDPI under the terms and conditions of the Creative Commons Attribution-NonCommercial-NoDerivs (CC BY-NC-ND) license (https://creativecommons.org/licenses/by-nc-nd/4.0/).

Contents

About the Editors

Zhongjiang Wang

Zhongjiang Wang is a professor at the College of Food Science, Northeast Agricultural University. His research interests primarily focus on theoretical research, common key technological innovation, and industrialization demonstration in the field of soybean processing. He has presided over and completed 22 national and provincial-level projects, including for the National Key Research and Development Program and the National Natural Science Foundation of China. A highly cited researcher listed in the global top 2% of scientists, he has published extensively in top-tier journals, with an H-index of 34. A total of 125 patents have been authorized to him; as the first principal investigator, he has received the first-class award for natural sciences in Heilongjiang Province, the second-class award for scientific and technological progress in Hainan Province, the second prize of the 13th Dabonong Science and Technology Award, and other provincial and ministerial-level awards such as the Young Science and Technology Award of Heilongjiang Province, totaling 7 awards; as a main contributor, he has received 12 provincial and ministerial-level awards such as the first-class award for technological invention of Heilongjiang Province.

Zengwang Guo

Zengwang Guo is a professor at the College of Food Science, Northeast Agricultural University, where he also serves as the Deputy Director of the Grain Engineering Department. His research focuses on the texture improvement and flavor regulation of plant-based meat analogs, the development of plant-based foods, and the high-value processing and utilization of plant proteins. He has hosted four provincial and ministerial-level scientific research projects, such as sub-projects of the Key Research and Development Program of Shandong Province. He has published numerous papers in top-tier journals, and holds 11 authorized invention patents. His innovative research has led to the development of five new functional soybean milk powders and four new soybean protein products, generating significant economic benefits with an added sales revenue of 260 million yuan for enterprises.

Zhaoxian Huang

Zhaoxian Huang is a researcher and doctoral supervisor recognized as a High-Level Talent in Hainan Province. He currently serves as the Deputy Director of the International Joint Research Center for High-Value Processing of Tropical Characteristic Protein Resources in Hainan. His research interests center on the high-value processing of plant proteins and the development of functional lipids. He has presided over 20 provincial and ministerial-level projects, including for the National Natural Science Foundation of China and the Key Research and Development Program of Hainan Province. He has published over 50 high-level academic papers, including more than 30 SCI papers published in top journals.

foods

MDPI

Article

The Impact of Varying Enzymatic Pretreatment Durations of Wheat Gluten on the Flavour Characteristics of High-Moisture Plant-Based Extrudates

Xiaodong Li [1], Huihui Dai [2], Boning Mao [1], Hongzhou An [2,*], Yanhong Bai [1,2,*] and Lovedeep Kaur [3]

1 College of Food and Bioengineering, Zhengzhou University of Light Industry, Zhengzhou 450001, China; xdli0610@163.com (X.L.); b.mao@zzuli.edu.cn (B.M.)
2 College of Food Science and Technology, Henan University of Technology, Zhengzhou 450001, China; daihuihui920@haut.edu.cn
3 School of Food and Advanced Technology, Massey University, Palmerston North 4442, New Zealand; l.kaur@massey.ac.nz
* Correspondence: anhz@haut.edu.cn (H.A.); baiyanhong212@163.com (Y.B.)

Abstract

This study examined the effects of varying enzymatic pretreatment durations (0–80 min) of wheat gluten on flavour characteristics of high-moisture plant-based extrudates (HMPEs). Through a comprehensive analysis involving sensory evaluation, electronic tongue, free amino acid (FAA) profiling, electronic nose, and headspace solid-phase microextraction-gas chromatography-mass spectrometer (HS-SPME-GC-MS) analysis of volatile odour compounds, it was found that HMPEs with moderate enzymatic pretreatment (40 min) achieved the highest overall sensory score. Electronic tongue and FAA results confirmed a significant enhancement in umami and sweetness, while electronic nose effectively discriminated differences in odour profiles. Extending pretreatment durations gradually reduced beany off-flavours substances (hexanal reduced by up to 174.7 μg/kg) and encouraged the formation of meaty aroma compounds (furans and pyrazines). However, excessive pretreatment (>40 min) reduced acceptance due to burnt odour caused by the excessive accumulation of pyrazines, particularly 2,3-diethyl-5-methylpyrazine. Six key volatile odour compounds were identified by integrating the analysis of variable importance projection (VIP $\geq$ 1) and relative odour activity value (ROAV $\geq$ 1), offering a foundation for targeted flavour regulation in HMPEs.

Keywords: wheat gluten; enzymatic pretreatment; high-moisture plant-based extrudates; flavour characteristics; sensory evaluation; taste; odour

1. Introduction

With the global population steadily increasing and consumers becoming more focused on health, environmental protection, and sustainable diets, plant protein-based meat analogues are rapidly expanding in the market as a viable alternative to traditional animal meat [1]. Compared to conventional meat, plant protein-based meat analogues offer several benefits, such as reduced cholesterol intake, lower carbon emissions, and enhanced resource efficiency [2]. High-moisture extrusion technology is the leading technique for creating meat-like fibrous structures from plant proteins, producing products with textures and tastes similar to animal muscle, while also offering improved safety and nutritional benefits [3]. However, as consumer expectations evolve, modern consumers are concerned

 https://doi.org/10.3390/foods15050912

not only with nutritional value but also with the overall sensory experience, particularly the flavour profile, which has become a crucial factor influencing market acceptance [4,5].

The flavour quality of plant-based meat analogues remains a notable challenge. Those meat analogues, often produced by soybean or pea proteins due to their high yield and rich essential amino acids, are usually associated with undesirable off-flavours, particularly a pronounced beany off-flavour, and their overall flavour profile differs markedly from authentic meat [6,7]. Current research primarily aims to eliminate these off-flavours and develop desirable meat-like flavours. In terms of eliminating off-flavours from raw materials, heat treatment, including conventional heating, microwave [8], radio frequency [9], and other techniques, can directly remove off-flavours at the source or indirectly reduce the formation of off-flavours precursors by inhibiting lipoxygenase activity. Additionally, fermentation [10], enzymatic hydrolysis [11], and physical embedding technologies [12] can convert or mask volatile off-flavours components. During the processing of extruded plant-based meat analogues, proteins, lipids, polysaccharides, and other components in the raw materials undergo complex chemical reactions, such as the Maillard reaction, lipid oxidation, Strecker degradation, caramelization, and thiamine degradation, under high-temperature and high-pressure conditions. These reactions generate numerous volatile odour components that can effectively eliminate or suppress off-flavours and impart a pleasant aroma [4]. Furthermore, adding small amounts of exogenous flavour additives, including protein hydrolysates, yeast extracts, Maillard reaction precursors (reducing sugars, amino acids, nucleotides, and thiamine), natural spices, and vegetable oils, can effectively mask off-flavours, develop a meat-like flavour, and enhance the taste of extruded plant-based meat analogs [12–16].

Enzymatic hydrolysis is a highly effective method for modifying plant proteins, widely applied to enhance their functional properties [17]. Recent studies indicate that controlled enzymatic hydrolysis of proteins not only releases taste-active peptides and free amino acids, which contribute directly to desirable tastes such as umami and sweetness, but also supplies abundant precursors for the Maillard reaction, facilitating the generation of meaty aroma compounds during subsequent processing [18]. However, most existing studies focused on using protein enzymatic hydrolysates to prepare meaty Maillard flavor bases via thermal reaction with reducing sugars [14,18], rather than being used as a high-proportion raw material directly for high-moisture extrusion, and its comprehensive impact on the flavour quality of extruded plant-based meat analogs remains insufficiently explored. In our previous research, wheat gluten with varying durations of enzymatic pretreatment (0–80 min, degree of hydrolysis (DH) 0–22%) was blended with soy protein isolate at a 4:6 ratio to prepare high-moisture plant-based extrudates (HMPEs). We found that only moderate enzymatic pretreatment (40 min, DH 12.36%) of wheat gluten effectively addressed the challenge of screw conveying caused by high viscosity during high-moisture extrusion, and also regulated the fibrous structure of HMPEs [19]. Unlike the conventional exogenous addition of small amounts of flavour additives [12–16], this endogenous modification strategy involved a high proportion of wheat gluten enzymatic hydrolysates as raw materials to prepare HMPEs, which will provide abundant peptides and free amino acids as flavor precursors. Meanwhile, this approach has been proven to reduce mechanical energy input and alter extrusion processing conditions [19]. These factors will synergistically influence flavor development together, particularly through modulation of the Maillard reaction pathways. However, the composition and content of these precursors is closely linked to the enzymatic pretreatment time, and insufficient or excessive pretreatment may fail to improve or even deteriorate the flavour. Yet, relevant flavor research has not been systematically investigated.

https://doi.org/10.3390/foods15050912

Therefore, this study aims to systematically elucidate the effect of varying enzymatic pretreatment durations (0 min, 20 min, 40 min, 60 min, and 80 min) of wheat gluten on the overall flavour quality of HMPEs. We will thoroughly examine how enzymatic pretreatment of wheat gluten influences the taste and odour characteristics of HMPEs through a combination of sensory evaluation, electronic tongue analysis, free amino acid analysis, electronic nose analysis, and headspace solid-phase microextraction-gas chromatography-mass spectrometer (HS-SPME-GC-MS). Additionally, key odour components responsible for changes in the odour characteristics of HMPEs will be identified using cluster analysis, partial least squares-discriminant analysis, variable importance projection (VIP) analysis, and relative odour activity value (ROAV) analysis of volatile odour components in HMPEs. This research investigates the application of enzymatic pretreatment technology on wheat gluten, aiming to provide a theoretical foundation and technical support for developing the next generation of plant-based meat analogs with both ideal texture and superior flavour.

2. Materials and Methods

2.1. Materials

Soy protein isolate and wheat gluten were sourced from Yihai Kerry (Shanghai, China). Neutrase (50,000 U/g, BR grade) and flavourzyme (30,000 U/g, BR grade) were supplied by Soleibo Technology Co., Ltd. (Beijing, China) and used directly without further purification. Cyclohexanone was obtained from McLean Biochemical Technology Co., Ltd. (Shanghai, China). Sulfonylsalicylic acid was acquired from Kemio Chemical Reagent Co., Ltd. (Tianjin, China).

2.2. Enzymatic Pretreatment of Wheat Gluten

The enzymatic pretreatment of wheat gluten was conducted following our previous research [20]. Briefly, wheat gluten was dissolved in water at a concentration of 35% (w/v). The pH-value of the mixture was adjusted to 7.0, and the temperature was maintained at 30 °C. The reaction commenced with the simultaneous addition of neutral protease and flavourzyme with specific activities of 250 U/g and 150 U/g to the substrate, respectively. The reaction times were set at 0, 20, 40, 60, and 80 min. Subsequently, the mixture was placed in a boiling water bath for 10 min to deactivate the enzymes. Finally, the obtained wheat gluten hydrolysates were freeze-dried and ground into powder for further use. The DH of wheat gluten hydrolysates for the respective time points were as follows: 20 min (8.91%), 40 min (12.11%), 60 min (13.39%), and 80 min (14.43%), meanwhile, the molecular weight distribution showed a corresponding shift toward lower molecular weight peptides, which have been all determined in our previous research [20].

2.3. Preparation of High-Moisture Plant-Based Extrudates (HMPEs)

The HMPEs were produced using a co-rotating twin-screw extruder (CLEXTRAL Ev025, Clestero Co., Ltd., Firminy, France), featuring a screw diameter of 25 mm and a length-to-diameter ratio of 24:1, in accordance with the methodology outlined in our previous study [19]. The raw material blend, consisting of soy protein isolate and wheat gluten hydrolysates at a dry basis ratio of 6:4, was processed at a screw speed of 280 rpm and a feed rate of 4.6 kg/h, with the mixture's moisture content maintained at approximately 57%. The extruder's barrel was divided into six zones, with the temperature profile set at 30, 90, 120, 140, 150, and 160 °C from zone I to VI, respectively. A cooling die with dimensions of 30 mm × 4 mm was affixed to the extruder outlet to reduce the product's temperature to 60 °C.

https://doi.org/10.3390/foods15050912

2.4. Sensory Evaluation

The sensory evaluation of HMPE was conducted in the sensory evaluation labortory of the College of Food Science and Technology at Henan University of Technology, and obtained approval from its Institutional Ethics Committee (Approval No. HAUTEC-2025-32). The panel comprised 12 assessors, all postgraduates in food science who had undergone professional sensory training. The HMPEs were uniformly cut into pieces measuring 5 cm × 5 cm and were randomly assigned three-digit codes for the five samples. Assessors were provided with disposable plates, gloves, and mouthwash, and were required to rinse their mouths thoroughly between sample evaluations to minimise cross-over effects. The samples were scored according to the criteria outlined in Table S1, with attributes such as fibrous structure, texture, colour, taste, and odour being assessed.

2.5. Electronic Tongue Analysis

The HMPEs were ground following rapid freezing with liquid nitrogen. Subsequently, 10.00 g of the powdered HMPE was combined with 40 mL of deionized water using a vortex mixer, then shaken for 30 min at room temperature for extraction. The mixture underwent centrifugation at 10,000× g for 10 min, and the supernatant was collected. The extraction process was repeated by adding another 40 mL of deionised water to the residue. The combined supernatants were filtered and adjusted to a final volume of 100 mL. This prepared solution was transferred into a dedicated electronic tongue beaker for taste profiling using an electronic tongue system (MOS ATREE, Alpha MOS Co., Ltd., Toulouse, France). The measurement conditions were set as follows: 120 s for signal acquisition and 10 s for sensor cleaning, with four replicated determinations per sample [21].

2.6. Determination of Free Amino Acids in HMPEs

Free amino acids were extracted and determined following the method outlined by Luo et al. [22]. Precisely 0.5 g of powdered HMPE was mixed with 5 mL of distilled water. The mixture underwent vortexing and was then shaken for 1 h. After centrifugation at 6000× g for 20 min, 400 μL of the supernatant was combined with 100 μL of 10% sulfosalicylic acid. This mixture was left at 4 °C for 60 min to ensure complete protein precipitation. It was subsequently centrifuged at 12,000× g for 10 min. The supernatant obtained was centrifuged again under identical conditions to achieve clarity. Finally, the supernatant was diluted to the required concentration, filtered through a 0.45 μm aqueous phase membrane, and analysed using an automatic amino acid analyser (S-433D, Sykam GmbH, Munich, Germany).

2.7. Electronic Nose Analysis

The odour profiles of the HMPEs were examined using an electronic nose (Fox 4000, Alpha M.O.S., Toulouse, France). For each measurement, precisely 2.00 g of the powdered HMPE was weighed and placed into a 20 mL headspace vial. The analysis was conducted under these conditions: the headspace vial was heated to 60 °C with an equilibration time of 5 min, after which 2.5 mL of the headspace gas was injected into the system. The sensor response was recorded for 120 s. To prevent cross-contamination between samples, a system cleaning step lasting 200 s was carried out after each analysis [23].

2.8. Analysis of Volatile Odour Compounds in HMPEs by HS-SPME-GC-MS

Volatile odour compounds in HMPEs were extracted using HS-SPME, following a modified method from the literature [24]. Briefly, 2.00 g of powdered HMPE was placed in a 20 mL headspace vial, to which 10 μL of the internal standard (cyclohexanone) was added, and the vial was sealed. The vial was equilibrated at 60 °C before a pre-conditioned 50/30 μm

https://doi.org/10.3390/foods15050912

DVB/CAR/PDMS fibre (conditioned at 250 °C for 30 min, SUPELCO Analytical, Inc., Bellefonte, PA, USA) was exposed to the sample for 30 min to allow adsorption. Subsequently, the fibre was inserted into the GC inlet and desorbed at 250 °C for 5 min. GC-MS (A 7890 GC equipped with a 7000D MS detector, Agilent Technologies, Inc., Santa Clara, CA, USA) analysis was conducted using an HP-5 MS capillary column (30 m × 250 μm × 0.25 μm. Agilent Technologies, Inc.) with helium as the carrier gas at a flow rate of 1.8 mL/min in splitless mode. The oven temperature program was as follows: initially held at 40 °C for 5 min, then increased to 130 °C at 2 °C/min with a 5 min hold, and finally raised to 250 °C at 10 °C/min with another 5 min hold. MS detection utilised electron impact ionisation at 70 eV, with ion source and quadrupole temperatures set at 230 °C and 150 °C, respectively, scanning m/z 33–400 in full scan mode.

Volatile odour compounds were identified through the NIST database and retention index, which was calculated from n-alkanes. Their relative contents were quantified using the internal standard method, as per the following formula:

$$C = \frac{A \times C_S \times V_S}{A_{S \times m}} \quad (1)$$

Note: C, relative contents of volatile odour compounds; C_S, concentration of the internal standard; A, peak area of the sample; A_S, peak area of the internal standard; V_S, volume of the internal standard; m, weight of the sample.

Further analysis employed cluster analysis, orthogonal partial least squares-discriminant analysis (OPLS-DA), and relative odour activity value (ROAV) on the relative contents of the odour compounds to identify and screen key differential compounds. The ROAV was calculated as follows:

$$\text{ROAV} \approx 100 \times \frac{C}{T} \times \frac{T_S}{C_S} \quad (2)$$

Note: C and T, relative contents and sensory threshold of volatile odour compounds; C_S and T_S, relative contents and sensory threshold of normalized reference volatile odour compounds [25].

2.9. Statistical Analysis

Data are presented as mean ± standard deviation. Statistical analysis was performed using SPSS statistical software (IBM, Version 20, New York, NY, USA). All quantitative data were tested for normality and homogeneity of variance. For data that did not meet the assumptions for parametric tests, the non-parametric Kruskal-Wallis H test was applied, followed by Dunn's test for post hoc multiple comparisons ($p < 0.05$). Descriptive sensory data were analyzed using a two-way mixed-model ANOVA with samples as the fixed factor and assessors as the random factor. Multivariate statistical analysis of Electronic tongue and Electronic nose data were analyzed by their built-in software. Orthogonal OPLS-DA of colatile odour compounds was performed using the online MetaboAnalyst 6.0 platform (https://www.metaboanalyst.ca/, accessed on 15 December 2025).

3. Results and Discussion

3.1. Sensory Evaluation

A thorough sensory analysis based on subjective evaluation was conducted to assess the effects of different enzymatic pretreatment durations of wheat gluten on the quality of HMPEs. This section initially performed sensory evaluation of the fibrous structure, texture and colour of HMPEs, attempting to objectively verify the finding of our previous studies that moderate enzymatic pretreatment achieved the optimal texture structure of

https://doi.org/10.3390/foods15050912

HMPE [19]. Subsequently, sensory analysis was carried out to evaluating the taste and odour characteristics of HMPEs, which was the core research focus of this study.

As illustrated in Figure 1A, the sensory scores for the fibrous structure of HMPEs initially increased and then decreased with the enzymatic pretreatment durations of wheat gluten extended from 0 to 80 min. This pattern was largely due to the varied effects of enzymatic hydrolysis on the protein network structure. Moderate enzymatic pretreatment of wheat gluten effectively mitigated the excessively dense fibrous structure caused by the high texturization degree in HMPEs without such pretreatment. Conversely, excessive enzymatic pretreatment could severely impair wheat gluten's ability to form a three-dimensional network, significantly diminishing the viscoelasticity of the protein melt during extrusion. This led to a very low degree of texturization and an overly loose fibrous structure, ultimately obstructing the formation of an ordered fibrous structure [19]. Similarly, the texture scores of HMPEs also showed an initial rise followed by a decline with increasing enzymatic pretreatment time, peaking at 40 min. The HMPE without enzymatic pretreatment exhibited excessive hardness and poor elasticity, which was mainly caused by the overly dense and rigid fibrous structure resulting from the highest texturization degree. In contrast, moderate enzymatic pretreatment (e.g., 40 min) achieved balanced texturization, effectively disrupting excessive protein cross-linking and enhancing hardness and elasticity with a desirable fibrous structure. However, excessive enzymatic pretreatment severely degraded wheat gluten proteins, considerably weakening the protein network integrity and significantly reducing the viscoelasticity of the protein melt during extrusion. Consequently, the material failed to form well-organised fibrous, resulting in insufficient texturization, an overly soft texture, and notably reduced elasticity [19,26]. The results of fiber structure and texture objectively confirmed that moderate enzymatic pretreatment achieved the optimal texture structure of HMPE. The colour sensory score of HMPEs declined progressively with longer enzymatic pretreatment of wheat gluten. This change resulted mainly from the enzymatic hydrolysis of wheat gluten, producing abundant free amino acids that underwent Maillard reactions during extrusion, darkening the product. This interpretation was supported by our earlier observations that L^* values decreased while a^* and b^* values increased of HMPEs [19].

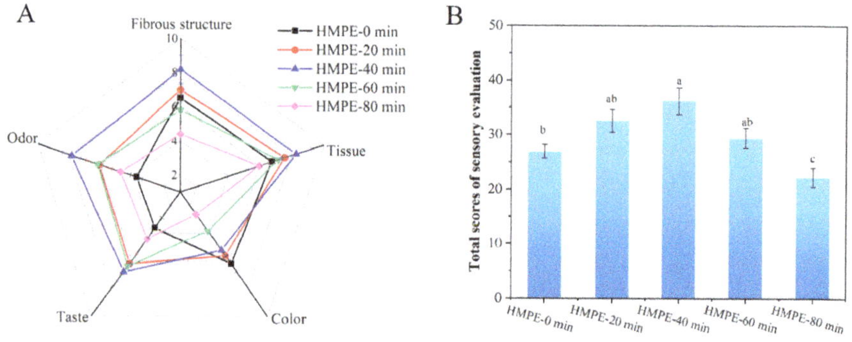

Figure 1. Sensory evaluation radar chart (**A**) and overall sensory scores (**B**) of the HMPEs obtained with varying enzymatic pretreatment durations of wheat gluten. Different lowercase letters in figure (**B**) mean significant differences ($p < 0.05$).

Regarding the sensory evaluation of flavour characteristics of HMPEs, both the sensory scores for odour and taste initially increased and then decreased as the enzymatic pretreatment duration for wheat gluten extended, peaking at 40 min. This pattern might be attributed to the production of small peptides and free amino acids during the enzymatic pretreatment of wheat gluten. These compounds not only directly enhanced desirable

 https://doi.org/10.3390/foods15050912

taste components, such as umami and sweetness, but also participated in Maillard reactions during extrusion, producing roasted aroma substances that effectively masked beany off-flavours and improved odour acceptability [18]. However, excessive enzymatic pretreatment led to an over-accumulation of amino acids and peptides, which in turn promoted overly intense Maillard reactions. This resulted in burnt or bitter substances that adversely affected both the odour and taste characteristics of HMPEs [4].

In Figure 1B, the overall sensory evaluation score revealed that the HMPE obtained with 40 min of wheat gluten enzymatic pretreatment achieved the highest score. This indicated that moderate enzymatic pretreatment of wheat gluten could significantly enhance the texture and flavour of HMPE. This was essentially attributed to the moderate DH (12.11%) of wheat gluten and the shifted molecular weight distribution toward low-molecular-weight peptides [20], which avoided the drawbacks of insufficient hydrolysis (0–8.91% DH) and excessive hydrolysis (13.39–14.43% DH), while optimizing the protein network for desirable texture and providing abundant taste-active peptides and flavor precursors for improved taste and odour. However, given the subjective nature of sensory evaluations and the fact that changes in flavour characteristics are inherently linked to alterations in taste and odour substances, further investigation is warranted. The underlying material basis for changes in the taste and odour properties of HMPEs resulting from varying durations of wheat gluten enzymatic pretreatment will be thoroughly explored using electronic tongue analysis, free amino acid content determination, electronic nose analysis, and quantification of volatile odour compounds. Notably, this descriptive sensory analysis result only revealed the impact of various enzymatic pretreatment durations of wheat gluten on the objective sensory characteristics of the HMEPs. Future research should recruit a larger consumer group for preference testing based on this foundation, in order to directly evaluate the market acceptance potential of products and establish a correlation model between descriptive attributes and consumer preference.

3.2. Electronic Tongue Analysis

An electronic tongue was utilised to objectively assess the impact of varying enzymatic pretreatment durations of wheat gluten on the taste properties of HMPEs. The electronic tongue mimics human taste perception by generating specific responses to taste compounds through its five fundamental taste sensors: sourness, sweetness, bitterness, saltiness, and umami [27]. The collected sensor response values of HMPEs were analyzed using principal components analysis (PCA) to systematically evaluate the differences in taste properties among HMPEs obtained with varying enzymatic pretreatment duration of wheat gluten. Subsequently, a radar chart was then plotted to depict the taste profiles of the HMPEs based on the response values, with the results displayed in Figure 2.

The PCA results depicted in Figure 2A indicated that the first two principal components of PC1 and PC2 explained 66.081% and 33.204% of the total variance, respectively, with a cumulative contribution rate of 99.285% (>85%). This suggested that these components collectively capture the majority of the core taste-related information of HMPEs subjected to varying durations of wheat gluten enzymatic pretreatment [22]. In the PCA plot, the taste response distribution areas for the four groups of HMPEs, each subjected to different enzymatic pretreatment times, were markedly distant from the non-enzymatic pretreatment group, indicating a significant difference in taste properties of HMPEs with or without wheat gluten enzymatic pretreatment. Additionally, among the four enzymatic pretreatment (20 min, 40 min, 60 min and 80 min) groups, the taste response distribution areas of HMPEs were arranged systematically along the negative axes of PC1 and PC2, with relatively small distances between. This phenomenon indicated that varying the duration of enzymatic pretreatment of wheat gluten significantly affected the taste properties of

https://doi.org/10.3390/foods15050912

HMPEs, albeit less so than the presence or absence of enzymatic pretreatment. Meanwhile, only a partial overlap was observed in the taste response distribution areas between the HMPEs obtained with the time of enzymatic pretreatment at 40 min and 60 min, indicating a certain similarity in taste properties.

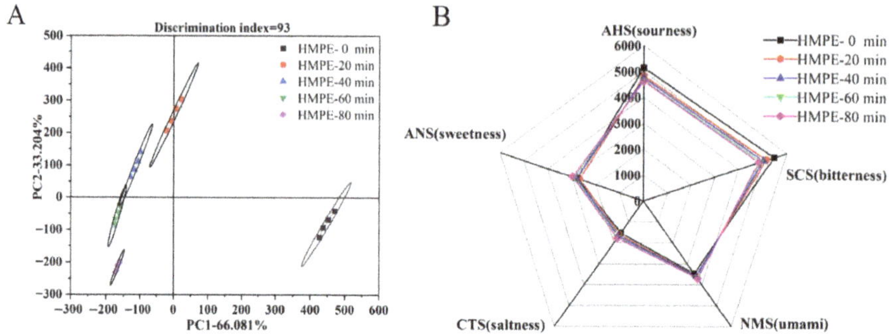

Figure 2. PCA (**A**) and radar fingerprint (**B**) of electronic tongue data for HMPEs obtained with varying enzymatic pretreatment durations of wheat gluten.

Radar chart analysis (Figure 2B) further revealed that while the five groups of HMPEs exhibited generally similar taste profiles, distinct differences emerged in sensor response values across the samples. Notably, as the enzymatic pretreatment time of wheat gluten increased, the sensor response values of umami, saltness, and sweetness of HMPEs gradually increased, whereas the sourness values decreased consistently, and bitterness values initially fell before rising again. The enhanced umami and sweetness response values improved the taste characteristics of HMPEs, accounting for the higher taste scores observed in sensory evaluations. This improvement might be attributed to the abundant peptides and amino acids generated by enzymatic hydrolysis of wheat gluten using complex enzymes (neutral protease and flavourzyme). These enzymes not only boosted the levels of positive taste components, such as sweet and umami amino acids, but also effectively reduced the formation of negative taste components, such as bitter peptides [28]. However, excessive enzymatic hydrolysis of wheat gluten (beyond 40 min) might inadvertently increase the concentration of bitter compounds. Additionally, the large amounts of small peptides and free amino acids can undergo Maillard reactions under the high-temperature and high-pressure conditions of subsequent high-moisture extrusion, leading to the formation of new bitter substances [4]. This results in a resurgence of bitter response values and taste sensory scores. The specific alterations in these taste-active compounds, particularly free amino acids, are examined in the following section.

3.3. Free Amino Acids Analysis

Free amino acids (FAAs) play a significant role in shaping the overall flavour profile of food, not only through their inherent taste properties and synergistic interactions with other flavour compounds, but also by serving as essential flavour precursors that undergo transformations during food processing [22]. Building on the previous sensory evaluation and electronic tongue analysis of HMPEs, this study further investigated the composition and content of FAAs in HMPEs, aiming to elucidate the molecular basis for taste changes in HMPEs subjected to varying durations of wheat gluten enzymatic pretreatment. The results are summarized in Table 1.

https://doi.org/10.3390/foods15050912

Table 1. Free amino acid composition (mg/100 g) in HMPEs obtained with varying enzymatic pretreatment durations of wheat gluten.

Amino Acid	HMPE-0 min	HMPE-20 min	HMPE-40 min	HMPE-60 min	HMPE-80 min
Asp	11.25 ± 0.56 [b]	14.89 ± 0.34 [a]	13.41 ± 0.55 [a]	13.45 ± 0.43 [a]	10.54 ± 0.45 [b]
Glu	3.40 ± 0.12 [c]	14.40 ± 0.07 [b]	17.65 ± 0.66 [b]	20.08 ± 0.27 [a]	20.58 ± 0.23 [a]
Asn	14.22 ± 0.50 [d]	31.25 ± 0.27 [c]	35.17 ± 0.28 [b]	38.00 ± 0.27 [b]	40.04 ± 1.12 [a]
∑UAA	28.87	60.54	66.23	71.53	71.16
Thr	2.01 ± 0.20 [d]	21.27 ± 0.56 [c]	23.66 ± 1.52 [bc]	28.09 ± 0.04 [a]	27.12 ± 0.37 [ab]
Ser	3.78 ± 0.13 [b]	4.79 ± 0.08 [a]	3.88 ± 0.48 [c]	5.40 ± 0.55 [a]	6.42 ± 0.17 [a]
Gly	3.68 ± 0.18 [c]	11.77 ± 0.16 [b]	13.41 ± 0.14 [ab]	14.62 ± 0.71 [a]	15.76 ± 0.53 [a]
Ala	0.11 ± 0.01 [d]	14.51 ± 0.56 [c]	17.90 ± 0.56 [b]	21.54 ± 0.18 [a]	22.47 ± 0.49 [a]
Pro	4.82 ± 1.45 [b]	22.35 ± 1.61 [a]	28.51 ± 4.23 [a]	29.97 ± 0.45 [a]	33.01 ± 0.19 [a]
∑SAA	14.4	74.69	87.36	99.62	104.78
His	2.36 ± 0.29 [c]	14.10 ± 0.31 [a]	15.39 ± 0.21 [a]	15.59 ± 0.05 [a]	14.54 ± 0.26 [a]
Arg	20.07 ± 2.25 [a]	18.15 ± 0.93 [a]	17.94 ± 2.47 [a]	18.78 ± 1.02 [a]	21.08 ± 1.34 [a]
Tyr	3.71 ± 0.12 [d]	29.74 ± 0.14 [c]	37.07 ± 0.04 [b]	42.90 ± 0.51 [a]	46.95 ± 1.10 [a]
Val	1.05 ± 0.22 [d]	15.77 ± 0.16 [c]	16.15 ± 0.13 [bc]	17.49 ± 1.10 [ab]	18.28 ± 0.42 [a]
Met	0.70 ± 0.20 [d]	15.37 ± 0.10 [c]	19.68 ± 0.21 [b]	23.02 ± 0.05 [a]	25.63 ± 0.06 [a]
Phe	1.89 ± 0.12 [d]	4.40 ± 0.52 [c]	10.70 ± 0.23 [b]	11.57 ± 0.04 [b]	12.18 ± 0.16 [a]
Ile	1.58 ± 0.16 [d]	117.99 ± 0.62 [c]	142.7 ± 0.06 [b]	161.21 ± 0.34 [a]	172.2 ± 1.94 [a]
Leu	1.90 ± 0.01 [d]	6.97 ± 0.12 [c]	8.52 ± 0.09 [b]	9.45 ± 0.04 [a]	9.27 ± 0.27 [a]
Lys	332.76 ± 5.96 [b]	359.72 ± 2.38 [a]	348.2 ± 4.34 [a]	346.55 ± 1.63 [a]	348.64 ± 3.74 [a]
∑BAA	366.02	582.21	616.35	646.56	668.77
∑TAA	409.27	717.4	769.9	817.67	844.69

Notes: UAA represents the total amount of fresh amino acids; SAA represents the total amount of sweet amino acids; BAA represents the total amount of bitter amino acids; TAA represents the total amount of amino acids; Different letters in the same line mean significant differences ($p < 0.05$).

The FAAs in HMPEs produced without enzymatic pretreatment of wheat gluten predominantly comprised Lys, Arg, Asn, and Asp, in that order. However, following enzymatic pretreatment, the main amino acids were Lys, Ile, Asn, and Tyr. This shift might be primarily attributed to the specific cleavage of the compound proteases (neutral protease and flavourzyme), which preferentially hydrolyzed peptide bonds containing hydrophobic and aromatic amino acids [29]. Meanwhile, as the enzymatic pretreatment durations increased from 0 to 80 min, the total free amino acid contents in the HMPEs increased significantly from 409.27 mg/100 g to 844.69 mg/100 g. This marked increase indicated effective protein degradation and the release of FAAs induced by enzymatic pretreatment, enriching the taste components of HMPEs, which aligned with the enhanced taste intensity detected by the electronic tongue. From the perspective of taste attributes, both umami and sweetness amino acids showed a continuous increase with prolonged enzymatic pretreatment time, directly supporting the rising umami and sweetness sensor response values observed in the electronic tongue. Notably, although the total content of bitter amino acids also exhibited an upward trend, sensory evaluation and electronic tongue analysis revealed a decrease followed by an increase in bitter taste perception. This phenomenon could be attributed to the effective control of bitter peptide generation by complex enzymes during early hydrolysis, the masking effect of abundant umami and sweet amino acids, and the fact that bitter perception was a complex integration of multiple taste signals [18]. However, with excessive enzymatic pretreatment of wheat gluten, the accumulated bitter amino acids and peptides, as well as new bitter substances formed during the process of extrusion, might eventually exceed the masking threshold, leading to increased bitterness perception [4]. It should be noted that, in addition to free amino acids,

https://doi.org/10.3390/foods15050912

taste peptides generated during enzymatic hydrolysis and their significant synergistic interaction with free amino acids also played a crucial role in influencing the overall taste profile [30]. However, due to the complexity of peptide composition in the enzymatic hydrolysates and the further intricate alterations these peptides undergo under the high-temperature and high-pressure conditions of extrusion processing, a detailed analysis of their specific composition and taste contribution was beyond the scope of this study.

3.4. Electronic Nose Analysis

In this study, an electronic nose equipped with 17 specific odour sensors was utilised to objectively assess the impact of varying enzymatic treatment durations on the odour properties of HMPEs. The electronic nose mimics human olfactory perception by generating specific responses to odour compounds through its sensors, thereby reflecting the overall odour characteristics of samples and avoiding the biases inherent in human evaluation [24]. Sensor response values for odour components in HMPEs were collected and analysed using discriminant factor analysis (DFA) to systematically evaluate the differences in odour properties among HMPEs derived from wheat gluten enzymatic pretreatment at different stages. A radar chart was then plotted to illustrate the odour profiles of the HMPEs based on these response values, with the results presented in Figure 3.

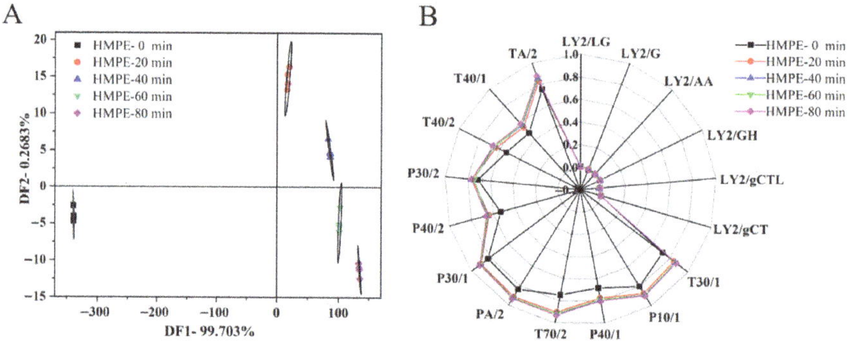

Figure 3. DFA (**A**) and radar fingerprint (**B**) of electronic nose data for HMPEs obtained with varying enzymatic pretreatment durations of wheat gluten.

DFA results in Figure 3A revealed that the differentiation indexes of DF1 and DF2 were 99.703% and 0.268%, respectively, and the cumulative contribution rate of DF1 and DF2 reached 99.971% (>85%), which indicated that these two factors adequately captured the core odour characteristics of HMPEs subjected to varying durations of wheat gluten enzymatic pretreatment [31]. In the DFA plot, the odour response distribution areas for the five groups of HMPEs were distinctly separated, with no overlap, highlighting significant differences in odour properties based on the pretreatment duration. Notably, a clear distinction was evident between the four enzymatic pretreatment groups and the non-enzymatic group, confirming that enzymatic pretreatment of wheat gluten significantly modified HMPEs. Furthermore, among the four enzymatic pretreatment groups (20 min, 40 min, 60 min, and 80 min), the odour response distribution areas were arranged systematically along the positive DF1 and negative DF2 axes. Although these areas did not overlap, they were relatively close to each other, suggesting that the duration of enzymatic pretreatment also affected the odour properties of HMPE. However, the presence or absence of enzymatic pretreatment remained the primary factor, aligning with taste variation results from electronic tongue analysis.

https://doi.org/10.3390/foods15050912

As shown in the radar chart (Figure 3B), the six L-series odour sensors exhibited no significant response to the odour substances in HMPEs, whereas the P- and T-series sensors showed marked responses. Notably, the odour profiles of HMPEs varied significantly with different durations of wheat gluten enzymatic pretreatment. The odour profile ranges for the four enzymatic pretreatment groups extended beyond that of the non-enzymatic pretreatment group, indicating that enzymatic pretreatment of wheat gluten enhanced the overall odour intensity of HMPEs. However, as the enzymatic pretreatment duration increased from 20 to 80 min, the expansion of the odour profile range was limited, suggesting that the differences in odour intensity due to varying pretreatment times were less pronounced than those between enzymatic and non-enzymatic pretreatment. This observation aligned with sensory evaluation results. Differences in odour profiles between enzymatically and non-enzymatically pretreated HMPEs were evident in the response values of all odour sensors except the L-series. In contrast, variations among the four enzymatic pretreatment groups were primarily reflected in the response values of sensors such as TA/2, T40/1, T40/2, P40/2, T70/2, PA/2, and T30/1, indicating significant changes in the content of organic, polar, and aromatic compounds. These changes were likely due to small peptides and free amino acids produced during the enzymatic pretreatment of wheat gluten, which underwent Maillard reactions under high-temperature and high-pressure conditions during extrusion, generating more odour substances. It is important to note that while the electronic nose effectively identified changes in the odour profile, it could not precisely quantify specific odour compounds or their correlation with odour property changes. Therefore, GC-MS should be utilised in future analyses to elucidate the detailed composition of volatile odour compounds and uncover the reasons for changes in odour properties of HMPEs due to varying enzymatic pretreatment durations of wheat gluten.

3.5. Qualitative and Quantitative Analysis of Volatile Odour Compounds

A qualitative and quantitative analysis of volatile odour compounds in HMPEs, derived from wheat gluten with varying enzymatic pretreatment durations, was performed using SPME-GC-MS. A total of 32 volatile odour compounds were identified and quantified in different HMPEs, including 6 aldehydes, 3 ketones, 4 alcohols, 7 furans, 3 thiophenes, and 9 pyrazines, as summarized in Table 2.

Aldehydes, as key aroma compounds in foods, are primarily derived from lipid oxidation and the Maillard reaction. Given their typically low odour thresholds, even minimal aldehyde concentrations can significantly impact a product's overall odour profile [32]. The results in Table 2 demonstrated significant changes in the relative content of aldehydes in HMPEs with prolonged enzymatic pretreatment durations of wheat gluten. Notably, the content of fatty aldehydes, represented by the grassy-smelling hexanal, exhibited a consistent downward trend. Hexanal, the main contributor to the beany off-flavour [6], showed the most substantial reduction, with a maximum decrease of 174.7 μg/kg, effectively diminishing the beany off-flavours in HMPEs. Conversely, the levels of aromatic aldehydes like benzaldehyde and phenylacetaldehyde, primarily formed through the oxidative degradation of proteins or amino acids [33], increased significantly. These compounds impart roasted almond and rose aromas, respectively, enhancing the odour properties of HMPEs. This compositional shift can be attributed to several factors. Firstly, the extended enzymatic pretreatment of wheat gluten gradually meant a gradual increase in DH (0–14.43%) and the aggregation of molecular weight distribution towards low molecular weight peptides, which reduced the viscosity of material during extrusion, decreasing mechanical energy input and mitigating thermal effects, thereby suppressing lipid oxidation [19,20]. Secondly, polypeptide compounds released during the enzymatic pretreatment exhibited certain antioxidant activity, further inhibiting lipid oxidation [34].

 https://doi.org/10.3390/foods15050912

Additionally, the increased availability of aromatic free amino acids, such as phenylalanine in the hydrolysates, provides abundant precursors for Strecker degradation, promoting the formation of aromatic aldehydes.

Table 2. Relative contents (μg/kg) of volatile odour compounds in HMPEs obtained with varying enzymatic pretreatment durations of wheat gluten.

Compounds [1]	RI [2]	Threshold [3] μg/kg	HMPE-0 min	HMPE-20 min	HMPE-40 min	HMPE-60 min	HMPE-80 min
Aldehydes							
Hexanal	780	5	605.64 ± 6.50 [a]	522.77 ± 4.30 [b]	478.12 ± 6.39 [b]	447.92 ± 6.23 [c]	430.94 ± 1.88 [c]
Heptanal	902	2.8	89.88 ± 0.48 [a]	72.36 ± 0.99 [b]	63.60 ± 1.01 [c]	59.42 ± 0.77 [c]	54.80 ± 0.95 [c]
Benzaldehyde	960	750	199.14 ± 7.11 [d]	441.27 ± 4.80 [c]	489.46 ± 7.99 [bc]	543.54 ± 4.20 [b]	631.20 ± 4.64 [a]
Benzeneacetaldehyde	1044	6.3	-	42.39 ± 1.08 [c]	53.84 ± 1.58 [bc]	61.42 ± 0.63 [b]	74.52 ± 2.30 [a]
Nonanal	1102	1.1	74.37 ± 0.11 [a]	72.66 ± 1.46 [a]	68.14 ± 2.85 [a]	66.74 ± 0.77 [ab]	62.89 ± 0.38 [b]
Decanal	1205	3	29.42 ± 1.08 [a]	25.28 ± 1.59 [b]	25.10 ± 1.70 [b]	28.06 ± 3.35 [a]	26.86 ± 0.88 [b]
Ketones							
2-Methyl-3-Octanone	985	21	24.05 ± 1.07 [a]	-	-	-	-
2-Nonanone	1093	41	46.47 ± 0.90 [a]	45.03 ± 2.03 [a]	45.19 ± 0.74 [a]	43.54 ± 1.05 [a]	32.99 ± 1.18 [b]
2-Decanone	1192	8.3	34.29 ± 4.77 [a]	25.80 ± 1.66 [b]	27.09 ± 2.25 [b]	27.35 ± 0.85 [b]	26.85 ± 1.97 [b]
Alcohols							
3-Methyl-butanol	730	460	14.41 ± 0.20 [d]	187.82 ± 3.35 [c]	212.12 ± 2.99 [bc]	223.05 ± 3.55 [b]	240.15 ± 3.76 [a]
1-Hexanol	868	5.6	20.04 ± 0.66 [b]	35.38 ± 1.58 [a]	23.58 ± 1.90 [b]	26.14 ± 1.29 [b]	24.44 ± 1.76 [b]
1-Octen-3-ol	978	1.5	50.23 ± 0.72 [a]	47.23 ± 0.27 [a]	42.48 ± 1.25 [b]	40.14 ± 0.56 [b]	37.78 ± 0.48 [b]
3,5-Octadien-2-ol	1037	-	29.44 ± 0.63 [a]	29.38 ± 2.12 [a]	26.55 ± 1.12 [a]	25.57 ± 2.28 [a]	24.47 ± 2.28 [a]
Furans							
2-Ethylfuran	691	8000	168.53 ± 4.81 [c]	188.98 ± 6.88 [bc]	199.82 ± 0.92 [b]	211.12 ± 7.33 [ab]	221.90 ± 1.58 [a]
Furfural	836	9.56	-	13.15 ± 0.36 [c]	19.23 ± 0.83 [c]	34.95 ± 1.57 [b]	41.11 ± 1.09 [a]
2-Furanmethanol	852	1900	231.90 ± 3.41 [e]	855.95 ± 6.43 [d]	1066.98 ± 44.52 [c]	1298.75 ± 13.83 [b]	1605.62 ± 27.61 [a]
2-Pentylfuran	994	5.8	5625.80 ± 63.75 [c]	5954.76 ± 23.9 [bc]	6111.08 ± 23.21 [b]	6262.06 ± 16.03 [a]	6384.57 ± 69.71 [a]
Maltol	1118	1.24	221.51 ± 5.74 [d]	353.24 ± 7.12 [c]	415.42 ± 8.46 [b]	472.85 ± 8.06 [a]	508.70 ± 6.93 [a]
3-Hydroxy-2,3-dihydromaltol	1134	-	15.15 ± 0.27 [d]	62.76 ± 1.77 [c]	71.75 ± 1.45 [b]	80.19 ± 1.38 [a]	91.20 ± 2.26 [a]
3-Phenylfuran	1228	-	-	18.61 ± 1.01 [c]	23.19 ± 0.66 [b c]	32.55 ± 1.81 [b]	39.62 ± 1.16 [a]
Thiophenes							
2-Butylthiophene	1070	-	38.87 ± 0.80 [b]	38.62 ± 1.47 [b]	47.73 ± 1.10 [a]	50.65 ± 2.29 [b]	55.06 ± 1.90 [a]
2-Pentylthiophene	1164	-	134.60 ± 2.84 [d]	144.47 ± 2.18 [c]	161.48 ± 0.74 [b]	168.95 ± 1.88 [b]	182.80 ± 3.52 [a]
2-Hexylthiophene	1274	-	30.65 ± 0.75 [d]	42.66 ± 2.16 [c]	50.95 ± 1.13 [b]	58.20 ± 1.95 [a]	66.17 ± 1.16 [a]
Pyrazines							
Pyrazine	737	2	78.57 ± 0.54 [c]	102.82 ± 6.41 [b]	116.00 ± 0.67 [a]	125.85 ± 3.03 [a]	132.71 ± 1.08 [a]
2,5-Dimethylpyrazine	916	1.75	134.30 ± 4.36 [e]	376.42 ± 10.04 [d]	477.71 ± 2.54 [c]	536.83 ± 5.25 [b]	615.53 ± 2.92 [a]
2-Ethyl-3-methylpyrazine	1005	500	60.48 ± 0.49 [e]	144.04 ± 8.52 [d]	180.91 ± 2.18 [c]	208.29 ± 5.83 [a]	249.62 ± 6.77 [a]
2-Ethenyl-6-methylpyrazine	1017	40	11.41 ± 1.72 [d]	45.02 ± 2.08 [c]	59.21 ± 1.95 [b]	73.87 ± 2.13 [a]	86.78 ± 1.45 [a]
2-Acetylpyrazine	1022	60	17.08 ± 0.60 [a]	-	-	-	-
3-Ethyl-2,5-dimethylpyrazine	1081	8.6	132.58 ± 3.33 [d]	174.50 ± 2.29 [c]	191.96 ± 3.23 [b]	218.04 ± 1.42 [a]	233.47 ± 5.60 [a]
2,3-Diethyl-5-methylpyrazine	1200	0.0031	18.86 ± 0.12 [d]	21.87 ± 0.65 [c]	32.02 ± 0.23 [b]	37.23 ± 0.77 [a]	42.18 ± 1.04 [a]
n-Pentylpyrazine	1216	1	8.17 ± 0.88 [a]	-	-	-	-
2,5-Dimethyl-3-(3-methylbutyl)-Pyrazine	1308	600	34.60 ± 0.72 [e]	198.75 ± 1.47 [d]	254.67 ± 11.24 [c]	295.12 ± 7.15 [b]	322.96 ± 8.69 [a]

[1] Volatile odour compounds were identified based on retention index (RI) and mess information in NIST 20 database; [2] RI, retention index was calculated on HP-5 column using N-ketones (C6–C22); [3] Odour thresholds were obtained by reviewing the literature of Sohail et al. [32]. Different letters in the same line mean significant differences ($p < 0.05$).

Ketone compounds, such as the typical lipid oxidation products 2-nonanone and 2-decanone, were often reported to impart buttery and cheesy notes [35], potentially enhancing the meat-like aroma of products. However, the content of these ketone compounds decreased with extended enzymatic pretreatment durations of wheat gluten, and their high sensory thresholds limited their overall contribution to odour properties. In terms of alcohol compounds, the content of saturated alcohols like 3-methyl-butanol increased significantly, rising by up to approximately 16 times. Despite this increase, the influence of 3-methylbutanol on odour properties was restricted by its generally high odour thresholds. Notably, the content of unsaturated alcohols, particularly 1-octen-3-ol, which was known for its mushroom aroma and significant role in beany off-flavours [6], decreased gradually with prolonged enzymatic pretreatment of wheat gluten. Additionally, the odour threshold of 1-octen-3-ol was relatively low (1.5 μg/kg), which would have a positive significance in weakening the beany off-flavours of the HMPEs.

https://doi.org/10.3390/foods15050912

The analysis of furan compounds indicated that 2-pentylfuran, with its buttery and meaty aroma, predominated among the seven furan compounds in HMPEs, followed by maltol, which has a caramel-like odour. Notably, both 2-pentylfuran and maltol possess relatively low thresholds, significantly enhancing the odour characteristics of HMPEs [36]. Furthermore, the relative content of all furan compounds increased to varying extents as the enzymatic pretreatment duration for wheat gluten was extended, thereby improving the odour characteristics of HMPEs. Furan compounds typically arise from glycolysis, Maillard reactions, and the cyclization of unsaturated aldehydes, with branched furan compounds primarily forming through the Maillard reaction pathway [35]. Prolonging the enzymatic pretreatment of wheat gluten enriched the flavour precursors, such as peptides and free amino acids, available for the Maillard reaction during extrusion, thus fostering the formation of furan compounds and significantly enhancing the grilled meat aroma of the HMPEs.

Thiophene compounds are known for their characteristic sulphurous and meaty aroma, primarily resulting from the oxidative degradation of sulphur-containing amino acids, the Maillard reaction, and interactions with lipid oxidation products [37]. As shown in Table 2, the relative content of three thiophene compounds in HMPEs increased progressively with the extending enzymatic pretreatment durations of wheat gluten, enhancing the odour characteristics of HMPEs. This increase was mainly due to the rise in sulphur-containing free amino acids (such as methionine) in the enzymatic products of wheat gluten, which served as rich precursors for the formation of these thiophene compounds.

Pyrazine compounds primarily arise from the condensation of α-amino ketones, which are produced through the oxidative degradation of amino acids (especially lysine with 2 amino groups) during the Maillard reaction. These compounds are crucial for imparting nutty, grilled, or roasted flavours to food [35]. As shown in Table 2, the relative content of nine pyrazine compounds in HMPEs increased to varying extents with the extension of the enzymatic pretreatment durations of wheat gluten. Notably, the relative content of pyrazine, 2,5-dimethylpyrazine, 3-ethyl-2,5-dimethylpyrazine, and 2,3-diethyl-5-methylpyrazine, which had relatively low sensory thresholds, increased significantly. This increase was largely due to the intensified Maillard reaction, driven by the abundant free amino acids—especially lysine, the most abundant FAA (Table 1) and a highly efficient precursor for pyrazine formation—released during the enzymatic hydrolysis of wheat gluten [35]. Particularly significant is 2,3-diethyl-5-methylpyrazine, with an exceptionally low sensory threshold of 0.0031 μg/kg, enabling it to produce a strong olfactory stimulus even at minimal concentrations, thereby playing a vital role in developing roasted flavours in foods [35]. However, it is important to note that excessively high levels of low-threshold pyrazine compounds can lead to an unpleasant burnt odour. This is likely the primary reason for the decline in sensory scores for odour when the enzymatic pretreatment duration of wheat gluten exceeds 40 min, highlighting more nuanced differences in the odour properties of HMPEs.

In summary, the extent of enzymatic pretreatment of wheat gluten markedly altered the composition and relative content of volatile odour compounds, thereby significantly influencing the odour characteristics of HMPEs. A moderate pretreatment duration of 40 min could enhance the presence of compounds imparting roasted and meaty aromas, such as aromatic aldehydes, furans, and pyrazines, while effectively reducing those responsible for beany off-flavours, like hexanal and 1-octen-3-ol. Conversely, excessive enzymatic pretreatment might lead to an increase in compounds associated with a burnt odour, particularly excessive pyrazines. This shift originated from the enzymatic degradation of proteins, which systematically altered the composition and concentration of the flavour precursors available for Maillard and Strecker reactions. Notably, although the contents of most

https://doi.org/10.3390/foods15050912

precursors (such as free amino acids) increased continuously from 0 to 80 min, providing a richer substrate for flavour-forming reactions, sensory scores declined beyond 40 min. This indicated that the final flavour quality was not linearly correlated with precursor abundance and odorant content but was governed by a balance of reaction pathways and intensities. Furthermore, our previous study showed that extended hydrolysis of wheat gluten reduced torque, pressure, and specific mechanical energy during extrusion, which would generally suppress the Maillard reaction [19]. However, the continued increase in Maillard products (e.g., furans and pyrazines) observed here suggested that the substantial rise in flavour precursors (peptides and amino acids) at higher hydrolysis degrees might exert a dominant positive influence on reaction progression.

3.6. Cluster Heat Map Analysis of Volatile Odour Compounds

The cluster analysis heatmap of volatile odour compounds can visualize the subtle differences in the odour characteristics of HMPEs obtained with varying enzymatic pretreatment durations of wheat gluten [38]. From Figure 4, all samples are distinctly separated into two primary clusters: one comprising HMPEs without enzymatic pretreatment and the other including those with pretreatment times ranging from 20 to 80 min. This clearly demonstrated that enzymatic pretreatment was the primary factor influencing the odour characteristics of HMPEs, aligning with the odour profile changes identified by the electronic nose. Notably, sub-clustering within the enzymatic pretreatment groups indicated that the duration of pretreatment was a secondary yet significant factor. Specifically, the odour characteristics of HMPEs with 20-min and 40-min pretreatments were more similar to each other, while the 60-min and 80-min groups formed another adjacent sub-cluster. This pattern suggested a distinct phase shift in the volatile odour compounds of HWPEs when the pretreatment time exceeded 40 min, marked by a significant accumulation of Maillard reaction products, particularly furan and pyrazine compounds.

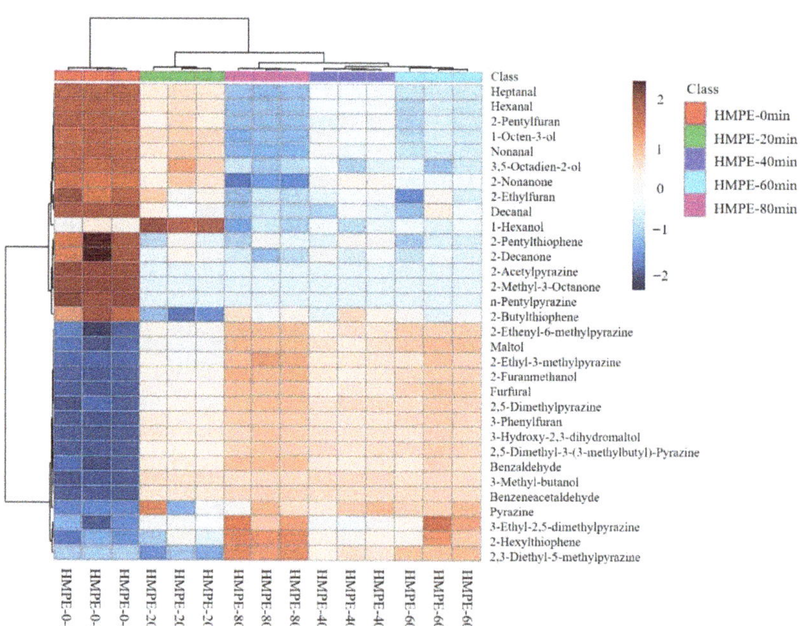

Figure 4. Cluster heat map analysis of volatile odour compounds in HMPEs obtained with varying enzymatic pretreatment durations of wheat gluten.

https://doi.org/10.3390/foods15050912

Primary and secondary clusters analysis through the cluster heat map, combined with the significant accumulation of Maillard reaction products (particularly pyrazines) in the 60 min and 80 min enzymatic groups, provides chemical substantiation for the highest odour score for the HMPE with 40 min enzymatic pretreatment of wheat gluten in sensory evaluation. While these Maillard reaction products at appropriate levels contributed desirable roasted and nutty notes for HMPEs, their excessive accumulation resulting from an over-intense Maillard reaction with rich free amino acids as precursor substances might lead to an overly intense aroma profile and even undesirable burnt off-flavours, thereby reducing overall odour acceptability.

3.7. OPLS-DA Analysis of Volatile Odour Compounds

The OPLS-DA, effective for distinguishing sample groups and constructing discriminant models [25], was used to analyse the relative abundances of 32 volatile odour compounds in HMPEs (5 pretreatment durations, $n = 3$ per group, 15 samples in total) and to identify key odourants responsible for odour differences. The 5-fold cross-validation results of the 2-component OPLS-DA model (Figure 5A) exhibited excellent performance, with the goodness-of-fit (R^2) and prediction ability (Q^2) values of 0.9698 and 0.9618, respectively. Furthermore, a permutation test with 100 permutations yielded $p < 0.01$ (Figure 5B), confirming model reliability and excluding overfitting [39]. These statistical metrics collectively indicated strong explanatory power and robust predictive accuracy for identifying discriminatory compounds. The 2D score plot for the two-component model (Figure 5C) showed the HMPE without enzymatic pretreatment of wheat gluten isolated in the second quadrant, while HMPEs with 20 min enzymatic pretreatment occupied the lower part of the third quadrant and those with 40, 60, and 80 min pretreatment lay in the first quadrant. Moreover, the 40, 60, and 80 min samples clustered closely. Thus, the model fully separated the volatile profiles of the HMPEs, which was consistent with the distribution of HMPEs in the electronic nose results.

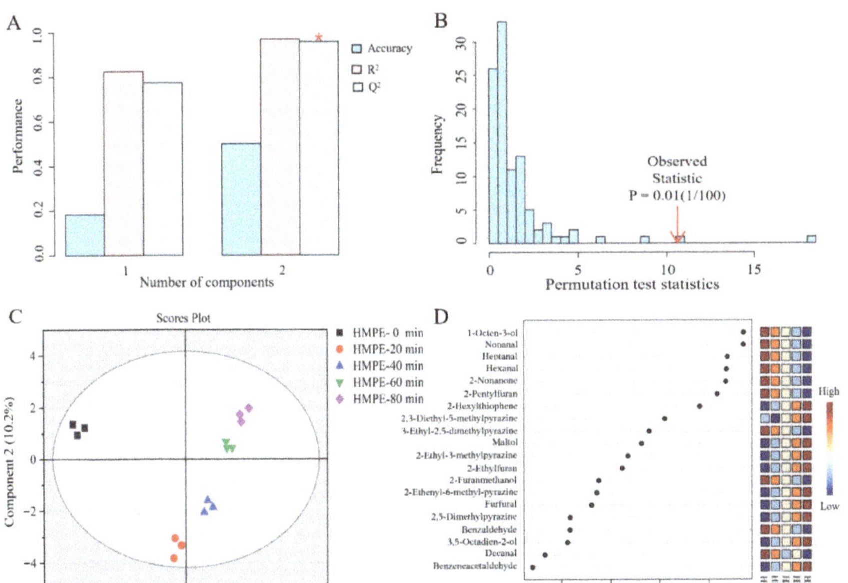

Figure 5. OPLS-DA analysis of volatile odour compounds in HMPEs obtained with varying enzymatic pretreatment durations of wheat gluten. (**A**) The goodness of fit (R^2) and prediction (Q^2) of the OPLS-DA

https://doi.org/10.3390/foods15050912

model, the asterisk (*) indicates the selected model with two components, which exhibited optimal performance; (**B**) The significant permutation test of the OPLS-DA model; (**C**) 2D scores plot with two components in the model; (**D**) Variable importance projection (VIP) value of volatile odour compounds in HMPEs.

The variable importance in projection (VIP) value was used to assess each variable's contribution in the OPLS-DA model. Variables with VIP $\geq$ 1 are generally regarded as significant contributors to the model [40]. As shown in Figure 5D, 18 volatile odour compounds had VIP values $\geq$ 1: 1-octen-3-ol, nonanal, heptanal, hexanal, 2-nonanone, 2-pentylfuran, 2-hexylthiophene, 2,3-diethyl-5-methylpyrazine, maltol, 3-ethyl-2,5-dimethylpyrazine, 2-ethyl-3-methylpyrazine, 2-ethylfuran, 2-furanmethanol, 2-ethyl-6-methylpyrazine, furfural, 2,5-dimethylpyrazine, benzaldehyde, and 3,5-octadien-2-ol. These compounds were identified as the statistically significant differential volatile compounds driving the odour profile variations of HMPEs. However, VIP analysis alone cannot directly indicate each compound's actual odor contribution, as it overlooks sensory thresholds essential for odor perception—only exceeding the threshold can be perceived.

3.8. Relative Odour Activity Value (ROAV) Analysis of Volatile Odour Compounds

The contribution of volatile odour compounds to HMPEs depends not only on their contents but also on their detection thresholds, the ratio of content to threshold is therefore the key index for assessing their odour impact [41]. Therefore, the ROAV method was applied to evaluate the actual influence of each compound on the odour properties of HMPEs. In this study, 2,3-diethyl-5-methylpyrazine served as the normalised reference for ROAV calculations. Compounds with ROAV $\geq$ 1 were deemed essential volatile odour compounds, while those with 0.1 $\leq$ ROAV $\leq$ 1 were classified as modified volatile odour compounds [42]. The results are given in Table 3. Six essential volatile odour compounds (ROAV $\geq$ 1) were identified in HMPEs: hexanal, nonanal, 2-pentylfuran, maltol, 2,5-dimethylpyrazine and 2,3-diethyl-5-methylpyrazine, all of which had VIP $\geq$ 1. Five modified volatile odour compounds were also identified: heptanal, 2-methyl-3-octanone, 1-octen-3-ol, pyrazine and 3-ethyl-2,5-dimethylpyrazine. Among these, 2-pentylfuran (ham flavour), maltol (caramel flavour), 2,5-dimethylpyrazine (popcorn flavour) and 2,3-diethyl-5-methylpyrazine (roasted flavour) are likely the principal contributors to an ideal meat-like odour profile [35]. By contrast, hexanal (grass flavour), nonanal (citrus flavour) and 1-octen-3-ol (mushroom flavour), as well as excessive 2,3-diethyl-5-methylpyrazine (roasted flavour), are likely associated with undesirable beany off-flavours and burnt notes, respectively [6,35].

Table 3. Relative odour activity value (ROAV) of volatile odour compounds in HMPEs obtained with varying enzymatic pretreatment durations of wheat gluten.

Compounds	Threshold ug/Kg	HMPE-0 min	HMPE-20 min	HMPE-40 min	HMPE-60 min	HMPE-80 min	Odour Describes
Hexanal	5	1.99	1.48	0.93	0.75	0.63	grass, green
Heptanal	2.8	0.53	0.37	0.22	0.18	0.14	fat, citrus, rancid
Benzaldehyde	750	<0.01	<0.01	<0.01	<0.01	<0.01	almond, burnt sugar
Benzeneacetaldehyde	6.3	<0.01	<0.01	<0.01	<0.01	<0.01	rosy
Nonanal	1.1	1.04	0.86	0.60	0.55	0.50	fat, citrus, green
Decanal	3	0.11	0.12	<0.01	<0.01	<0.01	fatty, rancid, burnt
2-Methyl-3-Octanone	21	<0.01	-	-	-	-	fruity, nutty
2-Nonanone	41	<0.01	<0.01	<0.01	<0.01	<0.01	flower petal, floral
2-Decanone	8.3	<0.01	<0.01	<0.01	<0.01	<0.01	heavy, sweet
3-Methyl-butanol	460	<0.01	<0.01	<0.01	<0.01	<0.01	alcohol, fruity
1-Hexanol	5.6	<0.01	<0.01	<0.01	<0.01	<0.01	minty, flower, green
1-Octen-3-ol	1.5	0.55	0.45	0.27	0.22	0.19	mushroom
2-Ethylfuran	8000	<0.01	<0.01	<0.01	<0.01	<0.01	fruity, floral
Furfural	9.56	<0.01	<0.01	<0.01	<0.01	<0.01	almond, woody

https://doi.org/10.3390/foods15050912

Table 3. *Cont.*

Compounds	Threshold ug/Kg	HMPE-0 min	HMPE-20 min	HMPE-40 min	HMPE-60 min	HMPE-80 min	Odour Describes
2-Furanmethanol	1900	<0.01	<0.01	<0.01	<0.01	<0.01	caramel
2-Pentylfuran	5.8	15.94	14.55	10.20	8.99	8.09	butter, caramel
Maltol	1.24	2.94	4.04	3.24	3.18	3.02	butter, caramel
Pyrazine	2	0.65	0.73	0.56	0.52	0.49	nutty
2,5-Dimethylpyrazine	1.75	1.26	3.05	2.64	2.55	2.59	popcorn, roasted
2-Ethyl-3-methylpyrazine	500	<0.01	<0.01	<0.01	<0.01	<0.01	roasted
2-Ethenyl-6-methyl-pyrazine	40	<0.01	<0.01	<0.01	<0.01	<0.01	roasted, earthy
2-Acetylpyrazine	60	<0.01	<0.01	<0.01	<0.01	<0.01	roasted
3-Ethyl-2,5-dimethylpyrazine	8.6	0.25	0.29	0.22	0.21	0.20	nutty, roasted
2,3-Diethyl-5-methylpyrazine *	0.0031	100	100	100	100	100	roasted, burnt
n-Pentylpyrazine	1	<0.01	<0.01	<0.01	<0.01	<0.01	roasted
2,5-Dimethyl-3-(3-methylbutyl)-pyrazine	600	<0.01	<0.01	<0.01	<0.01	<0.01	roasted

* 2,3-Diethyl-5-methylpyrazine was employed as the normalized reference substance to calculate the relative odour activity value (ROAV) for volatile odour compounds in HMPEs.

3.9. Screening of Key Volatile Flavour Compounds Based on VIP and ROAV

By integrating the previous analyses for VIP ≥ 1 and ROAV ≥ 1 [40], this study ultimately identified six key volatile odour compounds that exhibited both statistically significant changes and substantive contributions to the overall odour of HMPEs obtained with varying enzymatic pretreatment durations of wheat gluten: hexanal, nonanal, 2-pentylfuran, maltol, 2,5-dimethylpyrazine, and 2,3-diethyl-5-methylpyrazine. These compounds can be categorized into two functional groups: (1) Positive and synergistic contributors (2-pentylfuran, maltol, 2,5-dimethylpyrazine, and 2,3-diethyl-5-methylpyrazine) that provide desirable meaty, caramel, and roasted notes; (2) Negative contributors (hexanal, nonanal, and excess 2,3-diethyl-5-methylpyrazine) that cause beany off-flavour and burn odours. The shifting balance among these six compounds accounts for the decrease in beany off-flavour, the initial enhancement and followed by deterioration of the meat-like aroma with increasing enzymatic pretreatment. Notably, 2,3-diethyl-5-methylpyrazine, which has a roasted odour, shows an exceptionally high ROAV owing to its extremely low threshold and is therefore a likely driver of the burnt odour observed after excessive enzymatic hydrolysis.

3.10. Correlation Analysis Between Key Volatile Compounds and Odour Sensory Scores

Pearson correlation analysis was performed on the mean contents of the six key volatile compounds and odour sensory scores. The results showed negative correlations for hexanal ($r = -0.47$) and nonanal ($r = -0.21$), and positive correlations for 2,5-dimethylpyrazine ($r = 0.47$), maltol ($r = 0.42$), 2-pentylfuran ($r = 0.38$) and 2,3-diethyl-5-methylpyrazine ($r = 0.24$). Among negative contributors, hexanal with the highest absolute r-value was the core beany off-flavour compound, whose reduction improved odour acceptability; nonanal was a secondary off-flavour substance with weak negative correlation. For positive contributors, 2,5-dimethylpyrazine and maltol were the main sources of meaty and caramel aromas with higher positive r-values; 2,3-diethyl-5-methylpyrazine had the lowest r-value, which might be due to its excessive accumulation after 40 min caused burnt odour and offset its roasted aroma contribution.

4. Conclusions

This study systematically demonstrated that the enzymatic pretreatment duration of wheat gluten serves as thecritical determinant for the overall quality of HMPEs. Enzymatic pretreatment of wheat gluten for 40 min achieved an optimal balance between texture

https://doi.org/10.3390/foods15050912

and flavour characteristics of HMPEs. Moderate enzymatic pretreatment (≤40 min) of wheat gluten enhanced umami and sweetness perception of HMPEs (through the release of more free amino acids), promoted meaty aromas (via increased furans and pyrazines) and suppressed beany off-flavour (e.g., the decreased hexanal and nonanal). However, excessive enzymatic pretreatment (>40 min) of wheat gluten led to the production of bitter and burnt off-flavors of HMPEs. Through integrated VIP ≥ 1, ROAV ≥ 1 and correlation analyses, six key odour compounds were identified: hexanal and nonanal as negative contributors, 2-pentylfuran, maltol, 2,5-dimethylpyrazine, and 2,3-diethyl-5-methylpyrazine as positive contributors. And, the excessive accumulation of 2,3-diethyl-5-methylpyrazine might be fundamentally linked to the development of burnt off flavour.

In summary, the novelty of this work lies in employing enzymatic hydrolysis of wheat gluten as an endogenous strategy to intrinsically regulate flavor formation in HMPEs by modulating flavour precursors supply and extrusion processing, which breaks through the limitation of conventional exogenous addition of minor flavour additives, providing theoretical insights and practical methods to enhance the flavour quality of alternative proteins. Future work should quantify taste-active peptides in enzymatic hydrolysates and HMPEs, and validate the identified key odorants via absolute quantification and recombination-omission tests to facilitate precise flavor tailoring for plant-based meat analogs. Additionally, larger-scale consumer preference testing is needed to directly link descriptive flavour attributes to market acceptance.

Supplementary Materials: The following supporting information can be downloaded at: https://www.mdpi.com/article/10.3390/foods15050912/s1, Table S1: Criteria for sensory evaluation.

Author Contributions: X.L.: Methodology, Investigation, Data curation, Writing—original draft; H.D.: Methodology, Data curation; B.M.: Methodology, Data curation, Writing—review & editing; H.A.: Project administration, Funding acquisition, Conceptualization; Y.B.: Supervision, Funding acquisition, Conceptualization; L.K.: Visualization, Writing—review & editing, Conceptualization. All authors have read and agreed to the published version of the manuscript.

Funding: This work was financially supported by the Research and Development Project of Henan Province (No. 251111111200); and the Zhengzhou University of Light Industry Postdoctoral Research Start-up Fund (No. 13501050200).

Institutional Review Board Statement: The sensory evaluation analysis was conducted in accordance with the Declaration of Helsinki, and the protocol was approved by the Institutional Ethics Committee of Henan University of Technology Henan University of Technology (Approval No. HAUTEC-2025-32, Approval Date: 7 November 2025).

Informed Consent Statement: Informed consent was obtained from all subjects involved in the study.

Data Availability Statement: The original contributions presented in this study are included in the article/Supplementary Material. Further inquiries can be directed to the corresponding authors.

Conflicts of Interest: The authors declare no conflicts of interest.

References

1. Sha, L.; Xiong, Y. Plant protein-based alternatives of reconstructed meat: Science, technology, and challenges. *Trends Food Sci. Technol.* **2020**, *102*, 51–61. [CrossRef]
2. Mosibo, O.K.; Ferrentino, G.; Alam, M.R.; Morozova, K.; Scampicchio, M. Extrusion cooking of protein-based products: Potentials and challenges. *Crit. Rev. Food Sci. Nutr.* **2022**, *62*, 2526–2547. [CrossRef] [PubMed]
3. Zhang, J.; Chen, Q.; Kaplan, D.L.; Wang, Q. High-moisture extruded protein fiber formation toward plant-based meat substitutes applications: Science, technology, and prospect. *Trends Food Sci. Technol.* **2022**, *128*, 202–216. [CrossRef]
4. Wang, Y.; Tuccillo, F.; Lampi, A.-M.; Knaapila, A.; Pulkkinen, M.; Kariluoto, S.; Coda, R.; Edelmann, M.; Jouppila, K.; Sandell, M.; et al. Flavor challenges in extruded plant-based meat alternatives: A review. *Compr. Rev. Food Sci. Food Saf.* **2022**, *21*, 2898–2929. [CrossRef]

 https://doi.org/10.3390/foods15050912

5. Jiang, W.; Yang, X.; Li, L. Flavor of extruded meat analogs: A review on composition, influencing factors, and analytical techniques. *Curr. Res. Food Sci.* **2024**, *8*, 100747. [CrossRef]
6. Wang, B.; Zhang, Q.; Zhang, N.; Bak, K.H.; Soladoye, O.P.; Aluko, R.E.; Fu, Y.; Zhang, Y. Insights into formation, detection and removal of the beany flavor in soybean protein. *Trends Food Sci. Technol.* **2021**, *112*, 336–347. [CrossRef]
7. Chen, Y.; Feng, X.; Blank, I.; Liu, Y. Strategies to improve meat-like properties of meat analogs meeting consumers' expectations. *Biomaterials* **2022**, *287*, 121648. [CrossRef]
8. Jiang, Z.; Pulkkinen, M.; Wang, Y.; Lampi, A.; Stoddard, F.; Salovaara, H.; Piironen, V.; Strohm, T. Faba bean flavour and technological property improvement by thermal pre-treatments. *LWT-Food Sci. Technol.* **2016**, *68*, 295–305. [CrossRef]
9. Jiang, Y.; Wang, S.; He, F.; Fan, Q.; Ma, Y.; Yan, W.; Tang, Y.; Yang, R.; Zhao, W. Inactivation of lipoxygenase in soybean by radio frequency treatment. *Int. J. Food Sci. Technol.* **2018**, *53*, 2738–2747. [CrossRef]
10. Nedele, A.K.; Gross, S.; Rigling, M.; Zhang, Y. Reduction of green off-flavor compounds: Comparison of key odorants during fermentation of soy drink with *Lycoperdon pyriforme*. *Food Chem.* **2021**, *334*, 127591. [CrossRef]
11. Arteaga, V.G.; Guardia, M.A.; Muranyi, I.; Eisner, P.; Schweiggert-Weisz, U. Effect of enzymatic hydrolysis on molecular weight distribution, techno-functional properties and sensory perception of pea protein isolates. *Innov. Food Sci. Emerg. Technol.* **2020**, *65*, 102449. [CrossRef]
12. Guo, Z.; Teng, F.; Huang, Z.; Lv, B.; Lv, X.; Babich, O.; Yu, W.; Li, Y.; Wang, Z.; Jiang, L. Effects of material characteristics on the structural characteristics and flavor substances retention of meat analogs. *Food Hydrocoll.* **2020**, *105*, 105752. [CrossRef]
13. Kyriakopoulou, K.; Keppler, J.K.; van der Goot, A.J. Functionality of ingredients and additives in plant-based meat analogues. *Foods* **2021**, *10*, 600. [CrossRef]
14. Chiang, J.; Hardacre, A.; Parker, M. Effects of Maillard-reacted beef bone hydrolysate on the physicochemical properties of extruded meat alternatives. *J. Food Sci.* **2020**, *85*, 567–575. [CrossRef]
15. Dai, H.; An, H. Effects of cysteine on physicochemical properties of high-moisture extrudates prepared from plant protein. *Foods* **2022**, *11*, 3109. [CrossRef] [PubMed]
16. Milani, T.; Conti, A. Textured soy protein with meat odor as an ingredient for improving the sensory quality of meat analog and soy burger. *J. Food Sci. Technol.* **2024**, *61*, 743–752. [CrossRef] [PubMed]
17. Bekiroglu, H.; Acar, Z.D.; Sagdic, O. Sustainable plant-based protein hydrolysates: Utilization of waste proteins modified by enzymatic hydrolysis in techno-functional applications. *Int. J. Biol. Macromol.* **2025**, *333*, 148823. [CrossRef]
18. Huang, X.; Wang, P.; Xue, W.; Cheng, J.; Yang, F.; Yu, D.; Shi, Y. Preparation of meaty flavor additive from soybean meal through the Maillard reaction. *Food Chem. X* **2023**, *19*, 100780. [CrossRef]
19. Dai, H.; An, H. Effect of protease hydrolysis pretreatment on extruder response and the structural characteristics of high-moisture plant-protein extrudates. *J. Food Eng.* **2024**, *376*, 112062. [CrossRef]
20. Dai, H.; An, H. Study on the structural characteristics of wheat gluten enzymatic hydrolysates and their effect on the texturization degree of high-moisture plant-protein extrudates. *J. Cereal Sci.* **2024**, *119*, 103974. [CrossRef]
21. Wang, L.; Xie, J.; Wang, Q.; Hu, J.; Jiang, Y.; Wang, J.; Tong, H.; Yuan, H.; Yang, Y. Evaluation of the quality grade of Congou black tea by the fusion of GC-E-Nose, E-tongue, and E-eye. *Food Chem. X* **2024**, *23*, 101519. [CrossRef]
22. Luo, X.; Xiao, S.; Ruan, Q.; Gao, Q.; An, Y.; Hu, Y.; Xiong, S. Differences in flavor characteristics of frozen surimi products reheated by microwave, water boiling, steaming, and frying. *Food Chem.* **2022**, *372*, 131260. [CrossRef]
23. Luo, X.; Ren, G.; Xiong, S.; An, Y.; Huang, K.; Hu, Y. Mechanistic insights into volatile odor changes in surimi gels with typical cross-linking degrees during frozen storage based on lipidomics analysis. *Food Chem. X* **2025**, *29*, 102630. [CrossRef]
24. Xu, M.; Jin, Z.; Gu, Z.; Rao, J.; Chen, B. Changes in odor characteristics of pulse protein isolates from germinated chickpea, lentil, and yellow pea: Role of lipoxygenase and free radicals. *Food Chem.* **2020**, *314*, 126184. [CrossRef]
25. Zhao, G.; Zhang, J.; Wang, S.; Yu, X.; Zhang, Q.; Zhu, C. Influence of heating temperatures and storage on the odor of duck meat and identification of characteristic odorous smell compounds. *Food Chem. X* **2024**, *21*, 101242. [CrossRef]
26. Dai, H.H.; An, H.Z.; Ma, Y.X.; Guo, Y.T.; Du, Y.; Zhu, X.Q.; Luo, Q. Effects of lysine on the physiochemical properties of plant-protein high-moisture extrudates. *LWT-Food Sci. Technol.* **2023**, *173*, 114410. [CrossRef]
27. Weng, Z.; Sun, L.; Wang, F.; Sui, X.; Fang, Y.; Tang, X.; Shen, X. Assessment the flavor of soybean meal hydrolyzed with Alcalase enzyme under different hydrolysis conditions by E-nose, E-tongue and HS-SPME-GC-MS. *Food Chem. X* **2021**, *12*, 100141. [CrossRef] [PubMed]
28. Schlegel, K.; Sontheimer, K.; Hickisch, A.; Wani, A.A.; Eisner, P.; Schweiggert-Weisz, U. Enzymatic hydrolysis of lupin protein isolates-Changes in the molecular weight distribution, technofunctional characteristics, and sensory attributes. *Food Sci. Nutr.* **2019**, *7*, 2747–2759. [CrossRef] [PubMed]
29. Wei, Y.; Liu, B.; Zhang, H.; Yan, K. Isolation and identification of bitter peptides during sequential hydrolysis of wheat gluten by enzyme preparations with endo-and exo-activities. *Food Chem.* **2024**, *460*, 140491. [CrossRef]
30. Zhu, J.; Liang, L.; Kang, Z.; Chen, J.; Zhu, Q.; Sun, B.; Zhang, Y. Umami-enhancing peptides from *Acetes chinensis*: Insights into taste mechanism through multiligand docking with T1R1/T1R3. *Food Chem.* **2025**, *495*, 146538. [CrossRef]

31. Chen, X.; Chen, H.; Xiao, J.; Liu, J.; Tang, N.; Zhou, A. Variations of volatile flavour compounds in finger citron (*Citrus medica* L. var. *sarcodactylis*) pickling process revealed by E-nose, HS-SPME-GC-MS and HS-GC-IMS. *Food Res. Int.* **2020**, *138*, 109717. [CrossRef]
32. An, Y.; Wen, L.; Li, W.; Zhang, X.; Hu, Y.; Xiong, S. Insight into the evolution of aroma compounds during thermal processing of surimi gel from silver carp (*Hypophthalmichthys molitrix*). *Food Chem.* **2022**, *374*, 131762. [CrossRef]
33. Olivares, A.; Luis Navarro, J.; Flores, M. Effect of fat content on aroma generation during processing of dry fermented sausages. *Meat Sci.* **2011**, *87*, 264–273. [CrossRef]
34. Zheng, L.; Xie, S.; Zhuang, Z.; Liu, Y.; Tian, L.; Niu, J. Effects of yeast and yeast extract on growth performance, antioxidant ability and intestinal microbiota of juvenile Pacific white shrimp (*Litopenaeus vannamei*). *Aquaculture* **2021**, *530*, 735941. [CrossRef]
35. Bassam, S.M.; Noleto-Dias, C.; Farag, M.A. Dissecting grilled red and white meat flavor: Its characteristics, production mechanisms, influencing factors and chemical hazards. *Food Chem.* **2022**, *371*, 131139. [CrossRef] [PubMed]
36. Sohail, A.; Al-Dalali, S.; Wang, J.; Xie, J.; Shakoor, A.; Asimi, S.; Shah, H.; Patil, P. Aroma compounds identified in cooked meat: A review. *Food Res. Int.* **2022**, *157*, 111385. [CrossRef] [PubMed]
37. Wang, T.; Zhen, D.; Tan, J.; Xie, J.; Cheng, J.; Zhao, J. Characterization of initial reaction intermediates in heated model systems of glucose, glutathione, and aliphatic aldehydes. *Food Chem.* **2020**, *305*, 125482. [CrossRef] [PubMed]
38. Lu, K.; Liu, L.; Zi, J.; Song, L.; Xie, W. New insights from flavoromics on different heating methods of traditional fermented shrimp paste: The volatile components and metabolic pathways. *LWT-Food Sci. Technol.* **2022**, *168*, 113880. [CrossRef]
39. Zhang, T.; Dai, S.; Niu, B.; Shen, C.; Chen, H.; Liu, R.; Chen, H.; Wu, W.; Gao, H. Investigating the mechanism behind warmed-over flavor in prepared pork products during their shelf life using GC-IMS. *Food Chem. X* **2025**, *32*, 103258. [CrossRef]
40. Li, B.; Mu, Q.; Guo, H.; Jia, W.; Zhang, M.; Liu, Y.; Jin, X. Effects of different roasting temperatures on the flavor characteristics of jujube wine: Analysis based on HPLC, HS-GC-IMS, and HS-SPME-GC-MS. *Food Chem. X* **2025**, *30*, 102891. [CrossRef]
41. Zheng, Y.; Oellig, C.; Zhang, Y.; Liu, Y.; Chen, Y.; Zhang, Y. Characterization of the key odourants in goji wines in three levels of sweetness by applications of sensomics approach. *Food Chem.* **2024**, *461*, 140803. [CrossRef] [PubMed]
42. Wang, M.; Zhang, J.; Chen, J.; Jing, B.; Zhang, L.; Yu, X. Characterization of differences in flavor in virgin rapeseed oils by using gas chromatography-mass spectrometry, electronic nose, and sensory analysis. *Eur. J. Lipid Sci. Technol.* **2020**, *122*, 1900205. [CrossRef]

Disclaimer/Publisher's Note: The statements, opinions and data contained in all publications are solely those of the individual author(s) and contributor(s) and not of MDPI and/or the editor(s). MDPI and/or the editor(s) disclaim responsibility for any injury to people or property resulting from any ideas, methods, instructions or products referred to in the content.

 https://doi.org/10.3390/foods15050912

foods

MDPI

Article

Effects of Treatment Methods on the Formation, Structure, and Functional Properties of Soy Protein Amyloid Fibrils

Qian Zhang [1], Yanmei Deng [1], Yanling Lu [1], Long Han [1], Qian Ma [1], Lei Guo [1,2] and Fangyu Fan [1,2,3,*]

1 College of Biological and Food Engineering, Southwest Forestry University, Kunming 650224, China; 13619690455@163.com (Q.Z.); 14736649465@163.com (Y.D.); luyanling20220222@163.com (Y.L.); longhan24yn@163.com (L.H.); maqian0224@163.com (Q.M.); guolei@swfu.edu.cn (L.G.)
2 Forest Resources Exploitation and Utilization Engineering Research Center for Grand Health of Yunnan Provincial Universities, Southwest Forestry University, Kunming 650224, China
3 Key Laboratory of Forest Disaster Warning and Control of Yunnan Province, Southwest Forestry University, Kunming 650224, China
* Correspondence: ffy118@163.com

Abstract: To investigate the effects of different treatment methods on soybean protein amyloid fibrils (SPAF), this study examined the effects of ultrasonication, cold plasma, heat, and NaCl treatment on the formation, structure, and functional properties of SPAF. SPAF structural analyses indicated that all treatments promoted SPAF assembly, with the order of effectiveness being: heat treatment > ultrasonication > cold plasma treatment > NaCl treatment. Regarding functional properties, the heat treatment group also demonstrated superior overall performance, including the highest solubility (88.15%), optimal emulsifying activity (79.63 m^2/g) and foaming capacity (169.44%), and the highest thermal denaturation temperature (107.49 °C). Conversely, ultrasonication and cold plasma treatments, which generated shorter fibrils, offered moderate functional improvements. In contrast, NaCl treatment provided limited functional enhancement due to the formation of coarse aggregates. Consequently, heat treatment was identified as the most effective approach to promote SPAF formation and enhance functional properties. These findings provide a theoretical basis for the process optimization of SPAF in the food industry.

Keywords: soy protein isolate; amyloid fibrils; protein modification; treatment methods; structure and functional properties; food functionality

1. Introduction

Soy protein isolate (SPI), a major component of defatted soybean meal, is extensively available, cost-effective, and non-toxic. It is rich in nutritional value, with an amino acid composition that meets human nutritional requirements, and exhibits higher digestibility than most plant proteins [1]. In addition, SPI contains various bioactive components, including bioactive peptides and isoflavones. These compounds have demonstrated protective and therapeutic effects against multiple chronic metabolic diseases [2]. Nevertheless, the functionality of SPI is constrained by its molecular structure. SPI primarily contains two storage protein components, 7S (β-conglycinin) and 11S (glycinin), whose native globular structures are stabilized by hydrogen bonding and disulfide linkages. These interactions stabilize a compact spherical conformation, resulting in complex tertiary and quaternary conformations, which contribute to the poor SPI flexibility and interface properties, as well as a large apparent molecular size [3]. This limits the application of SPI in food systems. Thus, modifications are commonly applied to SPI to improve its functional properties. Among various modification methods, amyloid fibrils have attracted considerable attention

https://doi.org/10.3390/foods14223835

because of their biocompatibility and safety, their capacity to markedly enhance protein functional properties, and their broad application potential in the food industry [4]. Amyloid fibrils are a special class of protein aggregate [5]. This fibrillar structure, characterized by its high aspect ratio, exceptional flexibility, and intertwined network, offers unique advantages for food applications owing to its biocompatibility and harmless nature [6]. Amyloid fibrils have been demonstrated to enhance both the stability and delivery efficiency of bioactive molecules [7]. Chen et al. [8] demonstrated that whey protein isolate amyloid fibrils can bind β-carotene. This not only enhances β-carotene's water solubility and stability but also enables controlled release during in vitro gastrointestinal digestion. Ji et al. [9] reported that curcumin binds to soybean protein amyloid fibrils (SPAF), enhancing curcumin's stability, antioxidant capacity, and sustained-release properties. In addition to being excellent carriers for bioactive molecules, amyloid fibrils can enhance protein functional properties. Liu et al. [10] demonstrated that the emulsifying properties of ovalbumin amyloid fibrils are enhanced substantially compared to native ovalbumin. Similarly, Xie et al. [11] also confirmed that the emulsifying activity index (increased from 11.72 to 19.48 m^2/g) and foaming capacity (increased from 85% to 128%) of the fava bean protein amyloid fibrils were substantially enhanced compared to natural fava bean protein. Moreover, different treatment methods yield amyloid fibrils with diverse structure and functional properties.

Heat treatment ranks among the most prevalent techniques for protein modification. This process induces protein denaturation, aggregation, and structural rearrangement. During heat treatment, stabilizing interactions within proteins—notably noncovalent interactions (e.g., hydrogen bonding, hydrophobic and ionic interactions) and disulfide bonds—are disrupted. Subsequently, the protein undergoes ordered self-assembly by intermolecular interactions, forming amyloid fibrils [12]. Ultrasonication treatment, a non-thermal processing technology, induces cavitation, generating intense shear forces and microjets that disrupt non-covalent interactions within proteins. This phenomenon induces the conformational unfolding of proteins, exposing their hydrophobic cores and creating favorable conditions for amyloid fibril formation [13]. Cold plasma treatment, a physical modification, generates reactive oxygen and nitrogen species (RONS) along with high-energy electrons. These agents induce protein unfolding and initiate oxidative cross-linking [10]. Under specific conditions, this interaction facilitates the self-assembly of amyloid fibrils. In addition, NaCl treatment affects the assembly of amyloid fibrils primarily through electrostatic screening. This effect reduces interprotein charge repulsion, accelerates fibril aggregation, and stabilizes the fibril structure through enhanced hydrophobic interactions [14]. While studies have examined the effects of individual treatment methods on protein fibrillation, such as ultrasonication and heat treatments, systematic comparisons of heat, non-heat, physical, and chemical methods under identical conditions remain scarce.

Therefore, this study methodically analyzed the influence of four methods: ultrasonication, cold plasma, heat, and NaCl treatment on the formation, structure, and functional properties of SPAF. The extent of fibrillation and alterations in secondary structure were measured through Thioflavin T (ThT) fluorescence spectroscopy and Fourier transform infrared (FTIR) spectroscopy, respectively. The morphology of SPAF was examined using scanning electron microscopy (SEM). Furthermore, the structure under different treatments was comprehensively characterized using the surface hydrophobicity (H_0) and X-ray diffraction (XRD). Functional properties were evaluated by determining the emulsifying activity index (EAI), emulsifying stability index (ESI), thermal stability, solubility, foaming capacity (FC), and foam stability (FS). This study systematically compared the effects of different treatments on the formation, structure, and functional properties of SPAF, aiming to provide a foundation for the high-value utilization of SPAF in the food industry.

2. Materials and Methods

2.1. Materials

The source of the soybean meal was China Grain Reserves Corporation Zhenjiang Grain and Oil Co., Ltd. (Zhenjiang, China). 1-Amino-8-naphthalenesulfonic acid (ANS) and ThT were purchased from Shanghai Yuanye Biotechnology Co., Ltd. (Shanghai, China). All chemicals were analytically pure.

2.2. Preparation of SPAF

SPI was acquired by alkaline solubilization and acid precipitation [15]. SPI was dispersed in deionized water at 3% (*w*/*v*). This solution was stirred for 20 min, adjusted to pH 2.0 with 1 M HCl, and centrifuged for 10 min at 8000 rpm. After collection, the supernatant was kept at 4 °C for 24 h. All treatments were conducted using the same batch of 3% (*w*/*v*) SPI dispersion at pH 2.0 as the raw material. For heat treatment, the samples were exposed to 80 °C for 10 h [16]. For ultrasonication treatment, the samples were ultrasonicated at 200 W and 25 °C for 30 min [17]. For cold plasma treatment, the samples were subjected to treatment in the reactor for 60 s at 70 V and 1.00 A [18]. For NaCl treatment, an accurately measured mass of NaCl was added to the samples based on their initial volumes to achieve a final concentration of 160 mM, followed by 20 min of magnetic stirring at room temperature [14]. All processed samples were stored at 4 °C. Samples were labeled as follows: USPAF (ultrasonication treatment), CSPAF (cold plasma treatment), HSPAF (heat treatment), and NSPAF (NaCl treatment).

2.3. Characteristics of SPAF

2.3.1. FTIR

The freeze-dried sample was ground with KBr (1:100, m/m) and pressed into an infrared-transparent pellet. Subsequently, FTIR spectra were recorded on an FTIR spectrometer (IRPrestige-21, Shimadzu, Kyoto, Japan) from 400 to 4000 cm^{-1} at 4 cm^{-1} resolution with 16 scans. OMNIC 9.2 and Peakfit 4.12 were used to process data and calculate protein secondary structure [19].

2.3.2. ThT Fluorescence

A stock solution was prepared by mixing ThT with phosphate-buffered salt solution (PBS, 150 mM NaCl, pH 7.0) at a 4:5 (*w*/*v*) ratio and sterilized by filtration through a 0.22 μm membrane. A 40 μL aliquot of the sample (10 mg/mL) was combined with 5 mL of the stock solution that had been diluted 50-fold with PBS and was incubated for 2 min at 25 °C. Fluorescence intensity was measured with excitation at 460 nm and emission at 490 nm (slit widths 5 nm for both) using a fluorescence microplate reader (Molecular Devices, San Jose, CA, USA) [14].

2.3.3. Fluorescence Spectroscopy

The sample solutions were prepared at 0.06 mg/mL using deionized water that was acidified to pH 2.0 with 1 M HCl. Fluorescence spectra were acquired using a fluorescence microplate reader at an excitation wavelength of 290 nm (5 nm slit width) and by scanning the emission from 300 to 400 nm [20].

2.3.4. SEM

The freeze-dried sample was analyzed using SEM (Mira LMS, Tescan Orsay Holding, Brno, Czech Republic) after gold sputtering, at 200 times magnification and an acceleration voltage of 5 kV [21].

2.3.5. H_0

A series of sample dilutions (0.1–0.5 mg/mL) was prepared in deionized water acidified to pH 2.0 using 1 M HCl. Following the addition of 50 μL of 8 mM ANS to 2 mL of the sample, the solution was incubated for 2 min in the dark. Fluorescence intensity was measured using a fluorescence microplate reader at excitation 370 nm and emission 470 nm (both slit widths 5 nm) [9].

2.3.6. XRD

The XRD pattern was acquired with an Ultima IV X-ray diffractometer (Rigaku, Tokyo, Japan) using Cu-Kα radiation (λ = 1.5418 Å). The data were gathered under operating conditions of 40 kV and 40 mA over a 2θ range of 4–75° (step size: 0.013°; scanning rate: 0.056°/s) [15].

2.3.7. Dityrosine

The sample solutions were prepared at 1.0 mg/mL with deionized water acidified to pH 2.0 using 1 M HCl. Fluorescence intensity was measured using a fluorescence microplate reader at excitation 325 nm and emission 420 nm (slit widths 5 nm for both) [10].

2.3.8. Thermal Stability

Thermal stability was analyzed using Differential Scanning Calorimetry (DSC, 3500 Sirius, Netzsch, Selb, Germany). Samples (5.0 mg) were loaded into aluminum crucibles and heated from 30 to 200 °C at 10 °C/min under an N_2 flow of 20 mL/min [22].

2.3.9. Solubility

The sample solutions were prepared at 10.0 mg/mL with deionized water acidified to pH 2.0 using 1 M HCl. The solutions were then centrifuged for 10 min at 8000 rpm. Subsequently, the supernatant was analyzed for protein solubility using an automated Kjeldahl nitrogen analyzer (K1100, Hanon, Beijing, China).

2.3.10. Turbidity

The sample solutions were prepared at 10.0 mg/mL in deionized water acidified to pH 2.0 with 1 M HCl. Absorbance was measured at 500 nm using a spectrophotometer (UV-2600, Shimadzu, Kyoto, Japan).

2.3.11. Free Sulfhydryl Groups (Free-SH), Total Sulfhydryl (Total-SH), and Disulfide Bonds (S–S)

Free-SH, Total-SH, and S–S contents were measured according to the method of Liu et al. [10], with minor modifications:

Free-SH: 0.5 mL of sample was mixed with 0.02 mL of Ellman's reagent and 2.0 mL of an 8 M urea in Tris-glycine buffer. After being vortexed for 1 min, the mixture was incubated in the dark for 30 min. The UV-2600 spectrophotometer was employed to determine the absorbance at a wavelength of 412 nm.

Total-SH: 0.5 mL of sample was mixed with 1 mL of 6 M guanidine hydrochloride and 0.025 mL of β-mercaptoethanol (β-Me), followed by incubation for 1 h. The mixture was combined with 2.5 mL of 12% (*w*/*v*) trichloroacetic acid (TCA) solution and left to stand for 1 h. The precipitate was collected by centrifugation (10,000 rpm, 10 min), washed with 12% (*w*/*v*) TCA, and centrifuged again under identical conditions, repeating this process three times to remove β-Me. The final pellet was redissolved in 2.0 mL of a solution of 8 M urea in Tris-glycine buffer, and the subsequent steps were performed following the procedure for Free-SH determination. Equation (1) was used to determine the concentrations of Free-SH (μmol/g) and Total-SH (μmol/g), and Equation (2) was used to assess the S–S (μmol/g) content:

$$SH(\mu mol/g) = \frac{73.53 \times A_{412} \times D}{C} \quad (1)$$

$$S\text{–}S(\mu mol/g) = \frac{SH_{total} - SH_{free\ sulfhydryl}}{2} \quad (2)$$

C represents protein concentration (mg/mL), D denotes the dilution factor, and 73.53 is derived from $10^6/(1.36 \times 10^4)$.

2.3.12. EAI and ESI

The sample solutions were prepared at 10.0 mg/mL with deionized water acidified to pH 2.0 using 1 M HCl. Then, the sample solutions were centrifuged (8000 rpm, 10 min), after which the supernatant (15 mL) and soybean oil (5 mL) were mixed. Next, the mixture was emulsified with a high-speed disperser (FJ200-SH, Lu xi, Liaocheng, China) for 1 min at 10,000 rpm. Immediately after emulsification, a 50 μL aliquot of the emulsion was mixed with 10 mL of 0.1% (*w*/*v*) SDS solution. The absorbance at 500 nm (A_0) was determined with a UV-2600 spectrophotometer. After 10 min, a further 50 μL of the aliquot was mixed with 10 mL of a 0.1% (*w*/*v*) SDS solution, and the absorbance at 500 nm (A_{10}) was determined again using the UV-2600 spectrophotometer [23]. Equations (3) and (4) were used to calculate the EAI and ESI, respectively:

$$EAI(m^2/g) = \frac{2 \times 2.303 \times A_0 \times N}{C \times (1 - \theta) \times 10^4} \quad (3)$$

$$ESI(\%) = \frac{A_0}{A_0 - A_{10}} \quad (4)$$

where N denotes the dilution factor, θ represents the oil volume fraction (%), C is the solution concentration (g/mL), and A_0 and A_{10} signify the absorbance at 0 min and 10 min.

2.3.13. FC and FS

The sample solutions were prepared at 10.0 mg/mL with deionized water acidified to pH 2.0 using 1 M HCl. The sample (V_1) was homogenized for 1 min at 10,000 rpm using a high-speed disperser. The volume immediately after homogenization was recorded as V_2. After standing for 10 min, the volume was recorded as V_3. Equations (5) and (6) were used to calculate the FC and FS, respectively:

$$FC(\%) = \frac{V_2 - V_1}{V_1} \times 100\% \quad (5)$$

$$FS(\%) = \frac{V_3 - V_1}{V_1} \times 100\% \quad (6)$$

V_1, V_2, and V_3 are the sample volumes before homogenization, immediately after homogenization, and after standing for 10 min, respectively.

2.4. Statistical Analysis

Data were presented as the mean ± standard deviation (SD) of three independent experiments (n = 3). Data were analyzed using IBM SPSS Statistics 27. Specifically, one-way ANOVA followed by Waller–Duncan post hoc tests was applied to compare groups. Graphs were generated using Origin 2021 software. Statistical significance is defined as $p < 0.05$.

3. Results and Discussion

3.1. FTIR Analysis

FTIR can assess changes in protein functional groups and chemical bonds. Figure 1A shows that the 3000–3500 cm^{-1} range peak is associated with O-H and N-H stretching vibrations, with peak shifts indicating altered hydrogen-bonding interactions [24]. Notably, the redshift observed for NSPAF implies the formation of new hydrogen bonds during fibrillation. In contrast, the blueshift in USPAF, CSPAF, and HSPAF indicates that the extent of hydrogen bond destruction during fibrillation outweighed the formation of intermolecular hydrogen bonds, which is consistent with Mo et al.'s observations during whey protein fibrillation [25]. Additionally, the peak at 2960 cm^{-1}, which can be used to analyze hydrophobic interactions, is attributed to C-H stretching vibrations in CH_3 and CH_2 groups [26]. Compared to SPI, the redshift in all samples reflects increased hydrophobicity induced by hydrolysis and fibrillation [27]. Given the heightened sensitivity of the amide I region (1600–1700 cm^{-1}) to secondary structure perturbations, it is widely employed to monitor protein conformational transitions [28].

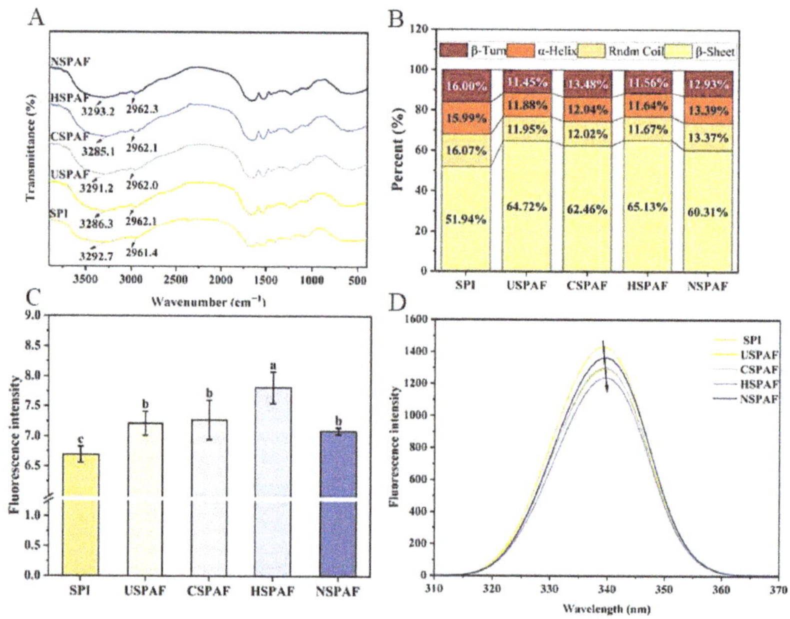

Figure 1. FTIR spectra (**A**), quantitative secondary structure analysis (**B**), ThT fluorescence intensity (**C**), and intrinsic fluorescence spectra (**D**) of SPI, USPAF, CSPAF, HSPAF, and NSPAF. Data in (**C**) is presented as mean ± SD (n = 3). Different lowercase letters above the bars indicate significant differences ($p < 0.05$).

As shown in Figure 1B, the β-sheet content for SPI, USPAF, CSPAF, HSPAF, and NSPAF was 51.94%, 64.72%, 62.46%, 65.13%, and 60.31%, respectively. Following fibrillation, all four treatments exhibited an increase in the β-sheet content compared to SPI. These results are in line with the findings of Oboroceanu et al. [29], who through peak area fitting of the amide I region, confirmed that the formation of the β-sheet structure is a key characteristic of the fibrillation process. Among these, HSPAF exhibited the highest β-sheet content, with 65.13%, owing to the 10 h heat treatment, providing sustained energy to unfold the protein structure. Then, loose peptide chains undergo directional arrangement, ultimately forming a regular structure dominated by β-sheets [21]. Feng et al. [30] showed that the formation of protein amyloid fibrils is essentially a molecular self-assembly process. The increase in the β-sheet content of USPAF is primarily due to cavitation caused by ultrasonication, which breaks non-covalent bonds in proteins, resulting in structural changes that promote amyloid fibril formation [31]. The main reason for the increase in CSPAF β-sheet content is that cold plasma treatment generates energetic particles, UV radiation, and RONS, such as ozone and peroxides. These reactive agents induce protein unfolding and initiate oxidative cross-linking, thereby driving amyloid fibrillation [18]. An increase in NSPAF β-sheet content was observed, possibly due to NaCl-induced modulation of protein charge distribution and intermolecular interactions. This modulation promotes amyloid fibril formation. Notably, the treatments could be ranked in the following order based on their effectiveness in promoting β-sheet formation: heat > ultrasonication > cold plasma > NaCl. Thus, heat treatment most effectively induced β-sheet formation and amyloid fibril self-assembly.

3.2. ThT Fluorescence Analysis

ThT fluorescence intensity is widely used to monitor amyloid fibril formation. ThT dye specifically recognizes and intercalates into the β-sheets, exhibiting fluorescence enhancement that correlates with the content of protein amyloid fibrils [15]. Figure 1C illustrates that the fluorescence intensities of SPI, USPAF, CSPAF, HSPAF, and NSPAF were 6.69, 7.21, 7.27, 7.81, and 7.09 a.u., respectively. All treated samples exhibited significantly higher fluorescence intensities than SPI ($p < 0.05$). These results confirm the formation of amyloid fibrils featuring β-sheets in the samples [32]. The low fluorescence intensity of SPI may be due to the absence of extensive cross-β amyloid structures in its native state. HSPAF exhibited the highest fluorescence intensity, likely due to its degree of fibrillation and greatest β-sheet content [33]. The results indicate that the effectiveness of treatments in promoting SPAF formation follows the order: heat treatment > ultrasonication treatment > cold plasma treatment > NaCl treatment. Among them, heat treatment is the most effective in promoting amyloid fibril formation.

3.3. Fluorescence Spectroscopy Analysis

Shifts in the fluorescence spectrum are a result of altered polarity around tyrosine (Tyr) and tryptophan (Trp) residues. Consequently, perturbations to protein tertiary structure can be monitored by observing these concomitant spectral shifts [34]. The hydrophobic regions of SPI contain Trp and Tyr, which promote β-sheet formation and facilitate π-π stacking interactions [31]. Figure 1D illustrates that, compared with SPI, USPAF, CSPAF, HSPAF, and NSPAF all exhibited reduced maximum fluorescence intensities. This decrease may be attributed to protein hydrolysis, molecular rearrangement, and encapsulation of Trp and Tyr within the newly formed protein aggregates. HSPAF exhibited the lowest fluorescence intensity among all samples, likely due to its highest degree of fibrillation (as evidenced by ThT fluorescence), which leads to greater burial of Trp residues. Additionally, Xia et al. [20] found that Trp residues exhibit sensitivity to heat-induced oxidative effects, resulting in lowered fluorescence emission intensity. The maximum emission wavelengths of USPAF, CSPAF, HSPAF, and NSPAF exhibited red shifts in the fluorescence spectra. This red shift indicates that the microenvironment around the Trp and Tyr residues becomes more polar. These indicate that, during the fibrillation process, SPI undergoes structural changes involving partial unfolding/disaggregation and self-assembly. According to Zhang et al. [35], during fibrillation, the fluorescence of whey protein amyloid fibrils decreased and underwent a red shift over time.

3.4. SEM Analysis

SEM was employed to characterize the morphology and macroscopic aggregate structure of the freeze-dried samples. In Figure 2, HSPAF exhibits continuous, well-defined aggregates, attributable to the 10 h heat treatment, which induces protein unfolding and promotes the ordered self-assembly of β-sheets, forming a highly ordered structure [21]. In addition, the enhanced cross-linking of intermolecular disulfide bonds during the heat treatment process further promotes the longitudinal extension of aggregates [36]. USPAF exhibited short aggregates, likely due to localized high pressure and microjets generated by ultrasonic cavitation, which not only induces protein fragmentation but also disrupts noncovalent bonds [37]. CSPAF exhibited short aggregates, likely because cold plasma-active species cause random chemical modifications of proteins [38]. NSPAF showed the highest surface roughness. This can be explained by Na^+, a weak salting-out ion, which partially disrupts the protein hydration layer and, via electrostatic screening, weakens intermolecular repulsion, thereby enhancing hydrophobic interactions and promoting localized aggregation [39]. An excessively high aggregation rate can lead to non-uniform aggregates. In summary, the treatments yielded distinct aggregate morphologies, reflecting their different mechanisms of action.

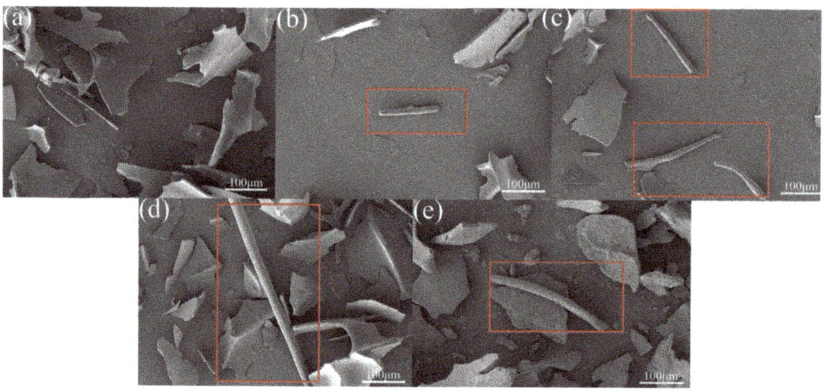

Figure 2. SEM of SPI (**a**), USPAF (**b**), CSPAF (**c**), HSPAF (**d**), and NSPAF (**e**). The red squares highlight aggregation structures.

3.5. H_0 Analysis

ANS specifically binds to exposed hydrophobic protein surfaces. When proteins undergo conformational changes or denaturation, buried hydrophobic amino acid residues become solvent-exposed. The binding of ANS to these exposed residues leads to an increase in fluorescence intensity [40]. Figure 3A illustrates that the H_0 values of USPAF, CSPAF, HSPAF, and NSPAF were significantly increased compared to those of SPI ($p < 0.05$). During protein amyloid fibril formation, proteins are hydrolyzed into low-molecular-weight peptides, causing the hydrophobic groups within the natural proteins to become exposed, increasing H_0 [41]. Among them, the highest H_0 in HSPAF could be attributed to the heat treatment (80 °C, 10 h), which provides a continuous and uniform energy input, resulting in protein denaturation and revealing additional hydrophobic groups. These revealed hydrophobic groups promote protein aggregation. It was reported by Green et al. [42] that, during fibrillation, hydrophobic interactions promote the formation of β-sheets. The differences in H_0 values reflect the varying contributions of hydrophobic interactions during protein fibrillation. The highest H_0 value was found in the heat treatment, suggesting that hydrophobic interactions contribute to the fibrillation process under this condition.

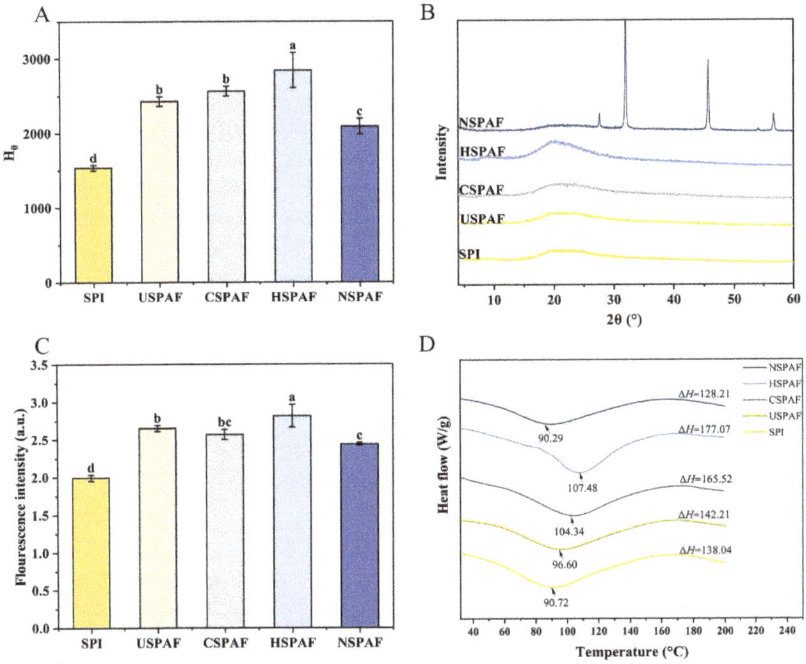

Figure 3. H_0 (**A**), XRD spectra (**B**), dityrosine content (**C**), and DSC thermograms (**D**) of SPI, USPAF, CSPAF, HSPAF, and NSPAF. Data in (**A**,**C**) are presented as mean ± SD (n = 3). Different lowercase letters above the bars indicate significant differences ($p < 0.05$).

3.6. XRD Analysis

XRD is employed for crystal structure analysis [43]. By analyzing the displacement of diffraction angles (2θ) and intensity alterations in diffraction patterns, the structural orderliness of molecular arrangements in protein aggregates can be evaluated [44]. Figure 3B shows the XRD patterns of the samples. All samples showed characteristic diffraction peaks of β-sheets at 2θ = 18.9° [15]. Compared to SPI, peak intensities were enhanced in USPAF, CSPAF, and HSPAF, indicating that these treatments effectively promoted the formation of β-sheet. Among them, HSPAF exhibited the most pronounced diffraction enhancement effect, consistent with the degree of fibrillation indicated by ThT fluorescence and the secondary-structure results from FTIR. This is attributed to the highest degree of fibrillation and the highest β-sheet content in HSPAF [16]. However, the characteristic peak intensity at 2θ = 18.9° in NSPAF was weakened. This may be attributed to Na^+-induced protein aggregation, which leads to the formation of partially irregular structures [45]. Additionally, the sharp peaks observed in the XRD spectrum after NaCl treatment were attributed to the overlap between the crystalline diffraction peaks of NaCl and the characteristic peaks of protein amyloid fibrils, according to the ICDD standards (PDF 00-005-0628) [46].

3.7. Dityrosine Analysis

Dityrosine is integral to the stability and architecture of amyloid fibrils and exhibits fluorescent properties with a maximum emission wavelength of 420 nm [47]. In Figure 3C, the dityrosine contents of SPI, USPAF, CSPAF, HSPAF, and NSPAF were 1.99, 2.65, 2.57, 2.81, and 2.44 a.u., respectively. The dityrosine levels in USPAF, CSPAF, HSPAF, and NSPAF were higher than those of SPI, likely owing to oxidative cross-linking of tyrosine residues during fibrillation. When SPI undergoes depolymerization and reassembly, the tyrosine residues in its structure become exposed from the hydrophobic core, making them susceptible to oxidation to form dityrosine [10]. Among them, HSPAF had the highest dityrosine content, probably because the heat treatment lasting 10 h drives tyrosine residues, originally concealed within the hydrophobic core, into exposure to a polar microenvironment. This subsequently triggers oxidative modification to form dityrosine. The results indicate that when protein self-assembles to form amyloid fibrils, dityrosine cross-linking also occurs. Liu et al. [10] obtained similar results, demonstrating that the driving force behind ovalbumin fibrillation involves not only hydrogen bonds and hydrophobic interactions but also dityrosine covalent bonds.

3.8. Thermal Stability Analysis

The endothermic peak temperature in the DSC curve corresponds to the denaturation temperature (T_d) of proteins. The enthalpy change (ΔH) reflects the energy required for protein denaturation [48]. Figure 3D shows that T_d and ΔH of SPI were 90.72 °C and 138.05 J/g, respectively. The T_d values of USPAF, CSPAF, and HSPAF increased to 96.59 °C, 104.33 °C, and 107.49 °C, respectively, with their corresponding ΔH also rising to 142.21 J/g, 165.52 J/g, and 177.07 J/g. Among these, HSPAF showed the highest thermal stability. This can be attributed to a two-step mechanism: first, the structural unfolding induced by pre-denaturation with heat, which exposes buried hydrophobic regions and active sites; second, the subsequent continuous heating that provides the necessary energy for the self-assembly of peptide chains into well-ordered amyloid fibrils dominated by β-sheet structures, thereby forming more amyloid fibrils [16]. NSPAF exhibited lower T_d and ΔH, at 90.29 °C and 128.21 J/g, respectively, than SPI. This reduction is likely due to NaCl treatment disrupting electrostatic repulsion among proteins, which promotes protein refolding and the formation of partially disordered aggregates, reducing thermal stability. The increase in disordered structures typically leads to a lower ΔH, which is consistent with the findings of Jiang et al. [48]. Combined analysis of fluorescence spectroscopy and secondary structure content demonstrated that the treatment groups with enhanced thermal stability (including heat, cold plasma, and ultrasound) consistently exhibited higher β-sheet content and lower tryptophan fluorescence intensity compared to those of SPI. This suggests that the ordered assembly of β-sheet enriched amyloid fibrils drives enhanced thermal stability. The enhanced thermal stability resulting from protein fibrillation aligns with earlier observations by Bolder et al. [49].

3.9. Solubility Analysis

Protein solubility affects numerous key functional properties, such as emulsification, gelation, and foaming [50]. Figure 4A shows that the solubility of SPI was 70.79%. After treatments, the solubilities of USPAF, CSPAF, HSPAF, and NSPAF were 78.99%, 80.05%, 88.13%, and 74.19%, respectively. USPAF, CSPAF, and HSPAF experienced a significant enhancement in solubility ($p < 0.05$). During fibrillation, this increase is due to structural changes that expose buried hydrophilic amino acid residues, as well as hydrolysis, which produces low-molecular-weight soluble peptides. These changes collectively enhance solubility. The hydrolysis products then undergo ordered self-assembly via hydrophobic interactions and β-sheet formation, yielding SPAF with excellent surface characteristics. Moreover, during fibrillation, protein hydrolysis and self-assembly maintain a dynamic equilibrium. This balance ensures SPAF formation without excessive aggregation, further increasing sample solubility [48]. Among all samples, HSPAF exhibited the highest solubility, likely due to the formation of well-structured amyloid fibrils through heat-induced protein self-assembly. Wang et al. [16] demonstrated that proteins heated for 8–10 h at pH 2.0 and 85 °C develop an optimal fibrous structure. Nevertheless, the solubility values of SPI and NSPAF were found to be virtually identical, with no significant difference ($p > 0.05$). This result is likely due to the ionic effect during fibrillation, and Na^+ may accelerate aggregation, thereby disrupting the dynamic equilibrium maintained between proteolysis and self-assembly.

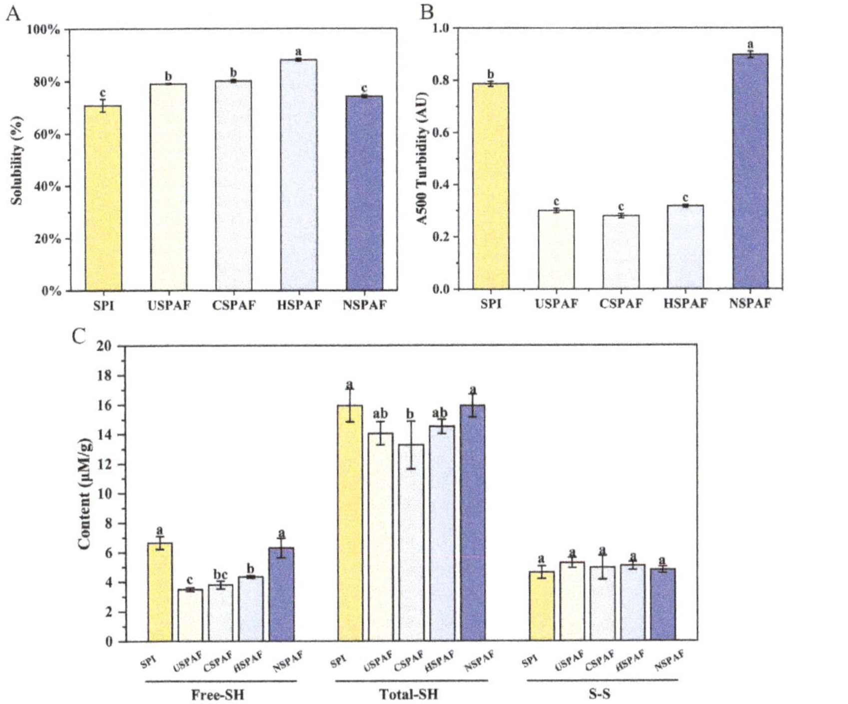

Figure 4. Solubility (**A**), turbidity (**B**), and contents of Free-SH, Total-SH, and S–S of SPI, USPAF, CSPAF, HSPAF, and NSPAF. Data in (**A–C**) are presented as mean ± SD (n = 3). Different lowercase letters above the bars indicate significant differences ($p < 0.05$).

3.10. Turbidity Analysis

Turbidity is directly related to the aggregation state and particle size of the sample and thus can indirectly reflect solubility [51]. Figure 4B illustrates that the turbidity of SPI was 0.78 AU, which may be due to the presence of numerous insoluble proteins. The turbidity of USPAF, CSPAF, and HSPAF decreased significantly to 0.30 AU, 0.28 AU, and 0.32 AU, respectively ($p < 0.05$), compared with SPI. The decrease is likely due to the mechanical effect of ultrasound, modification of proteins by reactive oxygen species in plasma, and thermotropic denaturation, respectively. These treatments induce the expansion and restructuring of the proteins, which promote the generation of soluble SPAF. The turbidity value of the NSPAF was 0.89 AU, likely due to NaCl-induced formation of larger protein aggregates. Ji et al. [45] also demonstrated that samples with 160 mM NaCl exhibited larger particle sizes. This finding is consistent with the results regarding solubility.

3.11. Free-SH, Total-SH, and S–S Analysis

During amyloid fibril formation, protein rearrangements lead to the exposure of buried sulfhydryl groups. These exposed sulfhydryl groups are consumed during self-assembly through disulfide bond formation and sulfhydryl-disulfide exchange reactions, which are critical for stabilizing the fibril structure [52]. As shown in Figure 4C, the Free-SH and Total-SH contents of SPI were 6.65 and 15.95 μmol/g, respectively. After treatments, the Free-SH contents of USPAF, CSPAF, HSPAF, and NSPAF were 3.48, 3.79, 4.34, and 6.30 μmol/g, respectively. The corresponding Total-SH contents were 14.05, 13.26, 14.53, and 15.93 μmol/g, respectively. The observed decline is due to the structural unfolding of SPI during fibrillation, which promotes the oxidation of exposed -SH groups into S–S bonds [53]. Among them, CSPAF showed the lowest Free-SH and Total-SH content, likely due to over-oxidation of cysteine residues and disulfides by RONS generated by cold plasma, converting a fraction into sulfinic and sulfonic acids [54]. The Free-SH and Total-SH contents showed no significant difference between NSPAF and SPI ($p > 0.05$), suggesting that NaCl treatment may promote SPAF formation through other mechanisms. As shown in Figure 4C, the S–S content remained statistically unchanged across all treatment groups ($p > 0.05$), indicating that it was not the primary driver of SPAF formation under these conditions.

3.12. FC and FS Analysis

FC reflects the protein's ability to rapidly adsorb at the air–water interface and generate foam upon shear. FS indicates the protein's ability to maintain foam structure and delay foam collapse [55]. Figure 5A illustrates that the FC and FS of SPI were 102.22% and 48.64%, respectively. The FC values for USPAF, CSPAF, HSPAF, and NSPAF reached 151.11%, 165.83%, 169.44%, and 151.67%, with corresponding FS values of 51.33%, 57.99%, 57.06%, and 51.77%, respectively. The significant improvements ($p < 0.05$) in FC and FS for all treated samples (USPAF, CSPAF, HSPAF, NSPAF) compared to SPI were closely associated with the increased formation of amyloid fibrils. During fibrillation, protein unfolding exposes hydrophobic groups and enhances molecular flexibility. These changes promote more rapid and efficient formation of a viscoelastic protein film at the interface, thereby enhancing FC [56]. The FS increase results from the formation of SPAF, which enhances its adsorption capacity at the interface. This process effectively reduces interfacial tension and strengthens the viscoelasticity of the interfacial film, thereby delaying foam rupture [57]. Among all groups, HSPAF exhibited the highest FC and FS, while no significant difference was found ($p > 0.05$). Based on ThT fluorescence results, heat treatment appears to be the most effective means for inducing protein fibril formation. Subjected to a 10 h heat treatment, the sample underwent more extensive molecular cross-linking, resulting in the formation of a greater quantity of stable SPAF. Cold plasma may act on the surface of proteins through reactive species generated, increasing the surface properties of the proteins and thereby enhancing their adsorption kinetics and interfacial spreading ability at the air–water interface, ultimately resulting in improvements in FS and FC. Oboroceanu et al. [55], who also reported that protein fibrillation can improve both FC and FS, observed similar phenomena in their study.

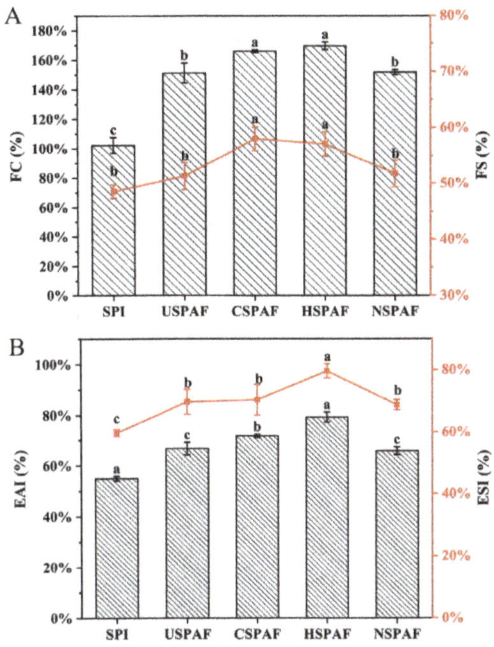

Figure 5. FC and FS (**A**) and EAI and ESI (**B**) of SPI, USPAF, CSPAF, HSPAF, and NSPAF (FS and ESI are shown in red). Data in (**A**,**B**) are presented as mean ± SD (n = 3). Different lowercase letters above the bars indicate significant differences ($p < 0.05$).

3.13. EAI and ESI Analysis

EAI quantifies protein adsorption at the oil–water interface, whereas ESI assesses emulsion stability over time [58]. Figure 5B illustrates that, compared with SPI, the EAI and ESI of USPAF, CSPAF, HSPAF, and NSPAF were significantly higher ($p < 0.05$). This can be attributed to SPAF having a greater specific surface area and higher interfacial activity, which facilitates faster diffusion to the interface and the formation of stable interfacial films. Additionally, protein unfolding during fibrillation increases H_0. The increased H_0 enhances protein-oil interactions, leading to greater adsorption at the oil–water interface [30]. Man et al. [51] also confirmed that fibril aggregates, due to their unique surface characteristics and structure, could serve as highly promising emulsifiers in food processing. Among all samples, HSPAF showed the highest EAI (79.63 m^2/g) and ESI (79.26%). Based on the H_0 and ThT fluorescence analysis, this phenomenon may be because HSPAF has a higher H_0 value than other groups. Additionally, heat induced the formation of more SPAF, which possesses superior surface properties. However, NSPAF showed the lowest EAI (65.75 m^2/g) and ESI (68.73%) among all treated samples, likely due to the ionic environment favoring the formation of larger, irregular aggregates. The increase in these aggregates reduced the effective coverage of emulsion droplet surfaces. Therefore, the unique structural characteristics induced by different treatments impart specific functional advantages to the samples. Heat treatment induces the formation of highly ordered aggregates, and the samples exhibit optimal functional properties, making them suitable for high-stability emulsions and gels. Ultrasonication treatment produces short aggregates conducive to rapid adsorption, making them ideal for foaming systems. Cold plasma treatment enhances interfacial activity through surface modification, which is applicable in fine emulsions and composite gels. Conversely, NaCl treatment forms coarse aggregates, resulting in limited improvement in functional properties.

3.14. Correlation Analysis

Based on Pearson correlation analysis (Figure 6), the relationship between the degree of SPAF formation and its structures and functional properties was investigated. The positive correlation between ThT fluorescence intensity and β-sheet content, H_0, and dityrosine content confirms that SPAF formation is fundamentally an ordered self-assembly process dominated by β-sheets [30], and indicates that during fibrillation, proteins unfold and expose hydrophobic regions and tyrosine residues, which provide the key driving force for fibril assembly. In addition, the negative correlation between ThT fluorescence intensity and Free-SH and Total-SH, coupled with its positive correlation with S–S content, indicates that the oxidation of -SH to form S-S further promotes intermolecular cross-linking, thereby stabilizing the fibrillar structure [52]. Further analysis indicates that the positive correlation between ThT fluorescence intensity and T_d, solubility, EAI, and FC demonstrates that the SPAF exhibits excellent thermal stability, solubility, and interfacial functionality. Moreover, H_0, dityrosine content, and β-sheet content positively correlate with T_d, solubility, EAI, and FC. This indicates that enhanced hydrophobic interactions, dityrosine formation, and orderly β-sheet dominated structural stacking act synergistically during fibrillation. Together, they promote SPAF formation, thereby driving improvements in surface activity, thermal stability, and solubility. Notably, the positive correlation between H_0 and solubility likely stems from the protein conformational transitions and ordered assembly during fibrillation. Protein unfolding increases H_0, while hydrolysis exposes hydrophilic groups and generates small soluble peptides. Furthermore, the ordered rearrangement of peptide chains forms structurally regular soluble fibers. In this process, the enhancement of solubility plays a dominant role, resulting in the overall observed trend of their simultaneous increase. The aforementioned conclusions are corroborated by the experimental results presented in Sections 3.1–3.14 of this paper. In summary, the formation process of SPAF involves multi-level structural evolution and the synergistic interaction of multiple molecular forces, with its structural order being closely associated with enhanced functional properties.

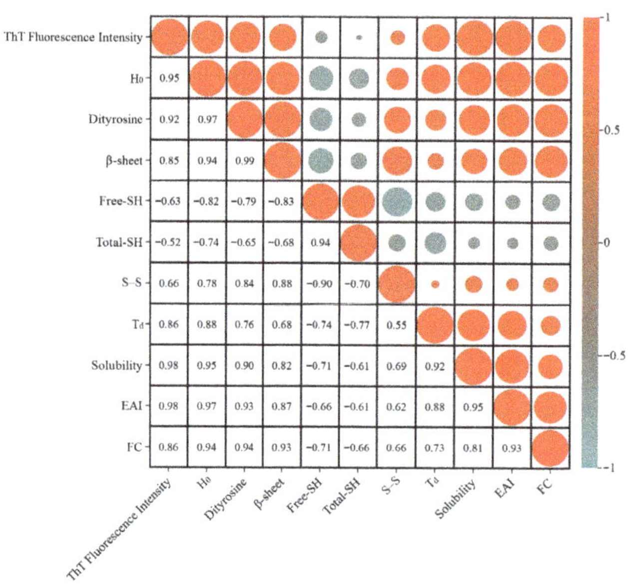

Figure 6. Statistical heat map correlation matrix of the relationship between different variables.

4. Conclusions

This study examined the impact of ultrasonication, cold plasma, heat, and NaCl treatment on the formation, structure, and functional properties of SPAF. The results indicated that all four treatments promoted β-sheet formation and SPAF self-assembly through the major driving forces of hydrogen bonding, hydrophobic interactions, and oxidative cross-linking, with efficacy in the order: heat > ultrasonication > cold plasma > NaCl treatment. Specifically, the heat treatment group exhibited the most outstanding comprehensive performance, featuring the highest β-sheet content, the most distinct fibrillar morphology, and optimal solubility, thermal stability, and interfacial properties. Although ultrasonication and cold plasma treatments differ in their mechanisms of action, both effectively modify the interfacial behavior of proteins and exhibit comparable performance in functional properties. In contrast, NaCl treatment formed coarse aggregates, resulting in limited improvement of functional properties. Consequently, heat treatment is the most effective method for preparing high-performance SPAF, and its operation is convenient and straightforward, possessing excellent potential for industrial application. These findings provide a basis for the strategic, high-value utilization of proteins in the food industry. Future research should further investigate multimethod processing synergies and evaluate the performance of SPAF within complex food matrices, including its digestive characteristics, long-term stability, and efficacy as a novel emulsifier or encapsulation system for bioactive substances, thereby advancing its application in functional foods.

Author Contributions: Q.Z.: writing—review and editing, writing—original draft, visualization, software, methodology, investigation, formal analysis. Y.D.: data curation, conceptualization, validation, software, methodology. Y.L.: formal analysis, data curation, methodology. L.H.: investigation, formal analysis. Q.M.: validation, methodology. L.G.: investigation, formal analysis. F.F.: writing—review and editing, validation, resources, supervision, project administration, methodology, funding acquisition, data curation, conceptualization. All authors have read and agreed to the published version of the manuscript.

Funding: The authors thank the Basic Research Projects of Free Exploration Category of Central Guided Local Science and Technology Development Funds (grant NO. 202407AA110007), the Yunnan Fundamental Research Projects (grant NO. 202401AS070012).

Institutional Review Board Statement: Not applicable.

Informed Consent Statement: Not applicable.

Data Availability Statement: The original contributions presented in this study are included in the article. Further inquiries can be directed to the corresponding author.

Conflicts of Interest: The authors declare no conflicts of interest.

References

1. Tan, S.T.; Tan, S.S.; Tan, C.X. Soy protein, bioactive peptides, and isoflavones: A review of their safety and health benefits. *PharmaNutrition* **2023**, *25*, 100352. [CrossRef]
2. Hu, S.; Liu, C.; Liu, X. The Beneficial Effects of Soybean Proteins and Peptides on Chronic Diseases. *Nutrients* **2023**, *15*, 1811. [CrossRef]
3. Zhang, W.; Huang, L.; Chen, W.; Wang, J.; Wang, S. Influence of ultrasound-assisted ionic liquid pretreatments on the functional properties of soy protein hydrolysates. *Ultrason. Sonochem.* **2021**, *73*, 105546. [CrossRef]
4. Meng, Y.; Wei, Z.; Xue, C. Protein fibrils from different food sources: A review of fibrillation conditions, properties, applications and research trends. *Trends Food Sci. Technol.* **2022**, *121*, 59–75. [CrossRef]
5. Cao, Y.; Mezzenga, R. Food protein amyloid fibrils: Origin, structure, formation, characterization, applications and health implications. *Adv. Colloid Interface Sci.* **2019**, *269*, 334–356. [CrossRef] [PubMed]
6. Li, T.; Wang, L.; Zhang, X.; Geng, H.; Xue, W.; Chen, Z. Assembly behavior, structural characterization and rheological properties of legume proteins based amyloid fibrils. *Food Hydrocoll.* **2021**, *111*, 106396. [CrossRef]

7. Wei, Z.; Chen, Y.; Wijaya, W.; Cheng, Y.; Xiao, J.; Huang, Q. Hydrogels assembled from ovotransferrin fibrils and xanthan gum as dihydromyricetin delivery vehicles. *Food Funct.* **2020**, *11*, 1478–1488. [CrossRef]
8. Chen, C.; Wang, H.; Wang, Q.; Wang, M.; Everett, D.W.; Huang, M.; Zhai, Y.; Li, T.; Fu, Y. Amyloid fibrils for β-carotene delivery—Influence of self-assembled structures on binding and in vitro release behavior. *Food Chem.* **2025**, *464*, 141849. [CrossRef]
9. Ji, F.; Wang, Z.; Bai, X.; Zhao, Y.; Zhong, X.; Luo, S.; Shen, Y.; Jiang, S.; Zheng, Z. Ultrasound-treated soy protein fibrils: A potential vehicle for curcumin with improved water solubility, antioxidant activity and sustained-release property. *Food Hydrocoll.* **2023**, *143*, 108929. [CrossRef]
10. Liu, Z.; Tang, P.; Liu, C.; Tan, Y.; Liu, T.; Cheng, J.; Aadil, R.M.; Liu, X. Promoting ovalbumin amyloid fibrils formation by cold plasma treatment and improving its emulsifying properties. *Food Hydrocoll.* **2025**, *158*, 110531. [CrossRef]
11. Xie, J.; Feng, Y.; Jiang, J. Multiscale analysis based on amorphous and fibrillated aggregates of fava bean proteins: Mechanism of regulation of aggregation morphology and its functional properties. *Food Hydrocoll.* **2026**, *170*, 111652. [CrossRef]
12. Akharume, F.U.; Aluko, R.E.; Adedeji, A.A. Modification of plant proteins for improved functionality: A review. *Compr. Rev. Food Sci. Food Saf.* **2021**, *20*, 198–224. [CrossRef]
13. Yagi, H.; Hasegawa, K.; Yoshimura, Y.; Goto, Y. Acceleration of the depolymerization of amyloid β fibrils by ultrasonication. *Biochim. Biophys. Acta (BBA)-Proteins Proteom.* **2013**, *1834*, 2480–2485. [CrossRef]
14. Afkhami, R.; Varidi, M.J.; Varidi, M.; Hadizadeh, F. Improvement of heat-induced nanofibrils formation of soy protein isolate through NaCl and microwave. *Food Hydrocoll.* **2023**, *139*, 108443. [CrossRef]
15. Cao, Z.; Wang, X.; Zhao, J.; Liang, X.; Zhang, Y.; Jiang, L.; Xu, Z.; Sui, X. Elucidating the modulatory influence of Hofmeister divalent ions on the structural dynamics and rheological properties of soy protein amyloid fibrils. *Food Hydrocoll.* **2024**, *151*, 109871. [CrossRef]
16. Wang, Y.; Shen, Y.; Qi, G.; Li, Y.; Sun, X.S.; Qiu, D.; Li, Y. Formation and physicochemical properties of amyloid fibrils from soy protein. *Int. J. Biol. Macromol.* **2020**, *149*, 609–616. [CrossRef]
17. An, D.; Li, L. Synergistic treatment of pH and ultrasound promotes the formation of insoluble soy protein hydrolysate nanofibrils. *Food Chem.* **2025**, *470*, 142659. [CrossRef]
18. Han, S.; Wang, Y.; Wang, J.; Xue, X.; Guo, P.; Dong, S. Mechanism of zein self-assembly fibrillation induced via plasma-assisted acid-heat treatment: Influence of the fibrillation degree. *Innov. Food Sci. Emerg. Technol.* **2025**, *100*, 103909. [CrossRef]
19. Deng, Y.; Lu, Y.; Jiang, Y.; Yuan, G.; Yang, T.; Gao, B.; Yang, J.; Guo, L.; Fan, F. Effect of cold plasma treatment time on walnut protein isolate: Revealing structural changes and improving functional properties. *Int. J. Biol. Macromol.* **2025**, *311*, 143693. [CrossRef]
20. Xia, W.; Zhang, H.; Chen, J.; Hu, H.; Rasulov, F.; Bi, D.; Huang, X.; Pan, S. Formation of amyloid fibrils from soy protein hydrolysate: Effects of selective proteolysis on β-conglycinin. *Food Res. Int.* **2017**, *100*, 268–276. [CrossRef]
21. Yu, Z.; Li, N.; Liu, Y.; Zhang, B.; Zhang, M.; Wang, X.; Wang, X. Formation, structure and functional characteristics of amyloid fibrils formed based on soy protein isolates. *Int. J. Biol. Macromol.* **2024**, *254*, 127956. [CrossRef]
22. Deng, Y.; Yuan, G.; Yang, T.; Gao, B.; Lu, Y.; Yang, J.; Guo, L.; Ma, Q.; Fan, F. Modifying the Structural and Functional Properties of Walnut Glutenin Through Atmospheric Cold Plasma Treatment: Evaluation of Treatment Times Effects. *Foods* **2025**, *14*, 2289. [CrossRef]
23. Tang, P.; Zhang, L.; Xiong, Y.; Jiang, D.; Liu, X.; Cheng, J.; Aadil, R.M.; Liu, Z. Reduction of antigenicity and emulsibility improvement of ovalbumin by dielectric-barrier discharge plasma treatment induced structure modification. *Innov. Food Sci. Emerg. Technol.* **2024**, *92*, 103602. [CrossRef]
24. Xiang, H.; Sun-Waterhouse, D.; Cui, C.; Wang, W.; Dong, K. Modification of soy protein isolate by glutaminase for nanocomplexation with curcumin. *Food Chem.* **2018**, *268*, 504–512. [CrossRef] [PubMed]
25. Mohammadian, M.; Madadlou, A. Characterization of fibrillated antioxidant whey protein hydrolysate and comparison with fibrillated protein solution. *Food Hydrocoll.* **2016**, *52*, 221–230. [CrossRef]
26. Gu, W.; Liu, X.; Gao, Q.; Gong, S.; Li, J.; Shi, S.Q. Multiple hydrogen bonding enables strong, tough, and recyclable soy protein films. *ACS Sustain. Chem. Eng.* **2020**, *8*, 7680–7689. [CrossRef]
27. Ye, J.; Fan, F.; Xu, X.; Liang, Y. Interactions of black and green tea polyphenols with whole milk. *Food Res. Int.* **2013**, *53*, 449–455. [CrossRef]
28. Qin, X.; Chen, S.; Li, X.; Luo, S.; Zhong, X.; Jiang, S.; Zhao, Y.; Zheng, Z. Gelation properties of transglutaminase-induced soy protein isolate and wheat gluten mixture with ultrahigh pressure pretreatment. *Food Bioprocess Technol.* **2017**, *10*, 866–874. [CrossRef]
29. Oboroceanu, D.; Wang, L.; Brodkorb, A.; Magner, E.; Auty, M.A. Characterization of β-lactoglobulin fibrillar assembly using atomic force microscopy, polyacrylamide gel electrophoresis, and in situ Fourier transform infrared spectroscopy. *J. Agric. Food Chem.* **2010**, *58*, 3667–3673. [CrossRef]
30. Feng, Z.; Li, L.; Zhang, Y.; Li, X.; Liu, C.; Jiang, B.; Xu, J.; Sun, Z. Formation of whey protein isolate nanofibrils by endoproteinase GluC and their emulsifying properties. *Food Hydrocoll.* **2019**, *94*, 71–79. [CrossRef]

31. Tong, X.; Cao, J.; Tian, T.; Lyu, B.; Miao, L.; Lian, Z.; Cui, W.; Liu, S.; Wang, H.; Jiang, L. Changes in structure, rheological property and antioxidant activity of soy protein isolate fibrils by ultrasound pretreatment and EGCG. *Food Hydrocoll.* **2022**, *122*, 107084. [CrossRef]
32. Dong, S.; Xu, H.; Li, B.; Cheng, W.; Zhang, L. Inhibition or improvement for acidic subunits fibril aggregation formation from β-conglycinin, glycinin and basic subunits. *J. Cereal Sci.* **2016**, *70*, 263–269. [CrossRef]
33. Pang, S.; Shao, P.; Sun, Q.; Pu, C.; Tang, W. Relationship between the emulsifying properties and formation time of rice bran protein fibrils. *LWT* **2020**, *122*, 108985. [CrossRef]
34. Huang, L.; Ding, X.; Dai, C.; Ma, H. Changes in the structure and dissociation of soybean protein isolate induced by ultrasound-assisted acid pretreatment. *Food Chem.* **2017**, *232*, 727–732. [CrossRef]
35. Zhang, C.; Fu, Y.; Li, Z.; Li, T.; Shi, Y.; Xie, H.; Li, Y.; Su, H.; Li, Z. Application of whey protein isolate fibrils in encapsulation and protection of β-carotene. *Food Chem.* **2021**, *346*, 128963. [CrossRef]
36. Lavoisier, A.; Vilgis, T.A.; Aguilera, J.M. Effect of cysteine addition and heat treatment on the properties and microstructure of a calcium-induced whey protein cold-set gel. *Curr. Res. Food Sci.* **2019**, *1*, 31–42. [CrossRef]
37. Hu, H.; Li Chan, E.C.; Wan, L.; Tian, M.; Pan, S. The effect of high intensity ultrasonic pre-treatment on the properties of soybean protein isolate gel induced by calcium sulfate. *Food Hydrocoll.* **2013**, *32*, 303–311. [CrossRef]
38. Li, B.; Peng, L.; Cao, Y.; Liu, S.; Zhu, Y.; Dou, J.; Yang, Z.; Zhou, C. Insights into Cold Plasma Treatment on the Cereal and Legume Proteins Modification: Principle, Mechanism, and Application. *Foods* **2024**, *13*, 1522. [CrossRef]
39. Meng, A.; Luan, B.; Zhang, W.; Zheng, Y.; Guo, B.; Zhang, B. Exploring changes in aggregation and gel network morphology of soybean protein isolate induced by pH, NaCl, and temperature in view of interactions. *Int. J. Biol. Macromol.* **2024**, *273*, 132911. [CrossRef]
40. Wei, Z.; Huang, Q. Impact of covalent or non-covalent bound epigallocatechin-3-gallate (EGCG) on assembly, physicochemical characteristics and digestion of ovotransferrin fibrils. *Food Hydrocoll.* **2020**, *98*, 105314. [CrossRef]
41. Hu, Y.; He, C.; Woo, M.W.; Xiong, H.; Hu, J.; Zhao, Q. Formation of fibrils derived from whey protein isolate: Structural characteristics and protease resistance. *Food Funct.* **2019**, *10*, 8106–8115. [CrossRef]
42. Greenwald, J.; Riek, R. Biology of Amyloid: Structure, Function, and Regulation. *Structure* **2010**, *18*, 1244–1260. [CrossRef]
43. Deb, S.; Kumar, Y.; Saxena, D.C. Functional, thermal and structural properties of fractionated protein from waste banana peel. *Food Chem. X* **2022**, *13*, 100205. [CrossRef]
44. Hu, A.; Li, L. Effects of ultrasound pretreatment on functional property, antioxidant activity, and digestibility of soy protein isolate nanofibrils. *Ultrason. Sonochem.* **2022**, *90*, 106193. [CrossRef] [PubMed]
45. Ji, F.; Xu, J.; Ouyang, Y.; Mu, D.; Li, X.; Luo, S.; Shen, Y.; Zheng, Z. Effects of NaCl concentration and temperature on fibrillation, structure, and functional properties of soy protein isolate fibril dispersions. *LWT* **2021**, *149*, 111862. [CrossRef]
46. Kiel, S.; Grinberg, O.; Perkas, N.; Charmet, J.; Kepner, H.; Gedanken, A. Forming nanoparticles of water-soluble ionic molecules and embedding them into polymer and glass substrates. *Beilstein J. Nanotechnol.* **2012**, *3*, 267–276. [CrossRef] [PubMed]
47. Li, B.; Yang, Y.; Ding, Y.; Ge, Y.; Xu, Y.; Xie, Y.; Shi, Y.; Le, G. Dityrosine in food: A review of its occurrence, health effects, detection methods, and mitigation strategies. *Compr. Rev. Food Sci. Food Saf.* **2023**, *22*, 355–379. [CrossRef]
48. Jiang, H.; Pan, J.; Hu, X.; Zhu, M.; Gong, D.; Zhang, G. A combination of alkaline pH-shifting/acidic pH and thermal treatments improves the solubility and emulsification properties of wheat glutenin. *Food Chem.* **2022**, *393*, 133358. [CrossRef]
49. Bolder, S.G.; Sagis, L.M.; Venema, P.; van der Linden, E. Effect of stirring and seeding on whey protein fibril formation. *J. Agric. Food Chem.* **2007**, *55*, 5661–5669. [CrossRef]
50. Dabbour, M.; He, R.; Mintah, B.; Xiang, J.; Ma, H. Changes in functionalities, conformational characteristics and antioxidative capacities of sunflower protein by controlled enzymolysis and ultrasonication action. *Ultrason. Sonochem.* **2019**, *58*, 104625. [CrossRef]
51. Mantovani, R.A.; Furtado, G.d.F.; Netto, F.M.; Cunha, R.L. Assessing the potential of whey protein fibril as emulsifier. *J. Food Eng.* **2018**, *223*, 99–108. [CrossRef]
52. Miao, L.; Zhu, J.; Peng, X.; Feng, J.; Dong, H.; Tong, X.; Wang, H.; Jiang, L. Effects of $CaCl_2$ concentration on fibrils formation and characteristics of soybean protein isolate and β-conglycinin/glycinin. *Food Hydrocoll.* **2023**, *142*, 108769. [CrossRef]
53. Wang, Y.; Yang, Q.; Du, Y.; Chen, H. Evaluation of the impact of stirring on the formation, structural changes and rheological properties of ovalbumin fibrils. *Food Hydrocoll.* **2022**, *128*, 107615. [CrossRef]
54. Li, H.; Wang, X.; Liu, Y.; Zhang, P.; Chen, F.; Zhang, N.; Zhao, B.; Guo, Y. Cysteine Thiol-Based Oxidative Post-Translational Modifications Fine-Tune Protein Functions in Plants. *Agronomy* **2024**, *14*, 2757. [CrossRef]
55. Oboroceanu, D.; Wang, L.; Magner, E.; Auty, M.A.E. Fibrillization of whey proteins improves foaming capacity and foam stability at low protein concentrations. *J. Food Eng.* **2014**, *121*, 102–111. [CrossRef]
56. Rullier, B.; Axelos, M.A.; Langevin, D.; Novales, B. β-Lactoglobulin aggregates in foam films: Effect of the concentration and size of the protein aggregates. *J. Colloid Interface Sci.* **2010**, *343*, 330–337. [CrossRef]

57. Moro, A.; Báez, G.D.; Busti, P.A.; Ballerini, G.A.; Delorenzi, N.J. Effects of heat-treated β-lactoglobulin and its aggregates on foaming properties. *Food Hydrocoll.* **2011**, *25*, 1009–1015. [CrossRef]
58. Liu, G.; Hu, M.; Du, X.; Liao, Y.; Yan, S.; Zhang, S.; Qi, B.; Li, Y. Correlating structure and emulsification of soybean protein isolate: Synergism between low-pH-shifting treatment and ultrasonication improves emulsifying properties. *Colloids Surf. A Physicochem. Eng. Asp.* **2022**, *646*, 128963. [CrossRef]

Disclaimer/Publisher's Note: The statements, opinions and data contained in all publications are solely those of the individual author(s) and contributor(s) and not of MDPI and/or the editor(s). MDPI and/or the editor(s) disclaim responsibility for any injury to people or property resulting from any ideas, methods, instructions or products referred to in the content.

foods

MDPI

Article

Modifying the Structural and Functional Properties of Walnut Glutenin Through Atmospheric Cold Plasma Treatment: Evaluation of Treatment Times Effects

Yanmei Deng [1], Guohui Yuan [1], Tongqin Yang [1], Baoyu Gao [1], Yanling Lu [1], Jiaojiao Yang [1], Lei Guo [1,2], Qian Ma [1,2,*] and Fangyu Fan [1,2,3,*]

1 College of Biological and Food Engineering, Southwest Forestry University, Kunming 650224, China; 14736649465@163.com (Y.D.); 15894518605@163.com (G.Y.); 18385703790@163.com (T.Y.); gaobaoyu66@163.com (B.G.); luyanling20220222@163.com (Y.L.); aw11p6@163.com (J.Y.); guoleigift.student@sina.com (L.G.)

2 Key Laboratory of Forest Disaster Warning and Control of Yunnan Province, Kunming 650224, China

3 Forest Resources Exploitation and Utilization Engineering Research Center for Grand Health of Yunnan Provincial Universities, Kunming 650224, China

* Correspondence: maqian0224@163.com (Q.M.); ffy118@163.com (F.F.)

Abstract: Walnut gluten (WGLU) is a plant-based protein rich in essential amino acids for the human body. Due to its poor water solubility and functional properties, its application in the food industry is limited. For the first time, this study looks into how different durations (0, 30, 60, 90, and 120 s) of atmospheric cold plasma (ACP) treatment affect the structure and functional properties of WGLU. ACP processing destroys the spatial structure of the WGLU and alters its functional properties. The comprehensive performance reached its best after 60 s of ACP treatment, the main manifestations included increased β-sheet content, reduced α-helix content, and unfolding of the tertiary structure, which ultimately improved the stability of emulsification and foam. Meanwhile, the solubility (86.35%), water retention rate (2.15 g/g), oil retention rate (5.31 g/g), emulsification rate (10.59 m^2/g), and foaming rate (24.67%) of WGLU reached their maximum values. However, longer treatment times (90 and 120 s) induce WGLU aggregation, followed by decreased functional properties. In summary, the physicochemical and functional properties of WGLU can be significantly enhanced through ACP treatment, enhancing the bioavailability of gluten and providing an effective strategy for its application in food processing.

Keywords: atmospheric cold plasma; walnut glutenin; physicochemical properties; secondary and tertiary structural; differential scanning calorimetry

1. Introduction

Plant proteins are defined as natural, renewable, and biodegradable functional macromolecules. They commonly replace or supplement animal proteins and plant-based meats, dairy alternatives, protein powders, energy bars, and baked goods [1–3]. Compared to other plant proteins, the walnut protein includes eight essential amino acids, especially rich in arginine content, which makes it a high-quality, nutritionally and biologically valuable source of plant protein [4,5]. In recent years, the modification of walnut proteins to improve their functional properties has become a hot research topic. Zhang et al. [6] showed that ultrasonic treatment improved the water solubility, emulsification, and emulsion stability of walnut isolate protein. Feng et al. [7] found that walnut protein was enzymatically digested by papain, and it had better emulsification and antioxidant properties. These studies provide some basis for studying walnut proteins. However, the specific mechanisms affecting the functional properties of walnut proteins still need to be explored in depth. Walnut gluten (WGLU) constitutes the primary part of walnut protein, and an in-depth exploration of its structure and functional properties can not only elucidate the physicochemical properties and biological activities of walnut protein but also provide an important theoretical basis for the high-value utilization of walnut resources and the development of functional foods. However, WGLU is a water-insoluble protein with poor solubility, emulsification, and foaming properties, and much of it ends up as feed or is discarded, causing a considerable waste of walnut protein resources [8].

Atmospheric cold plasma (ACP) treatment, as a novel non-thermal treatment technology, has been widely used to alter protein structure and function. ACP induces protein covalent bond rupture by ionizing the air to produce high-energy reactants, such as reactive oxygen species (ROS) and reactive nitrogen species (RNS), ultraviolet photons, and so on, which can disrupt the three-dimensional structure of proteins. In addition, these ROS and RNS transfer peptide bonds, thereby generating compelling oxygen radical-induced structural unfolding of proteins, which alters their secondary and tertiary structures, and improves their emulsification, solubility, and foaming properties [9]. Recently, emphasis has been placed on operating at low temperatures, being energy efficient, and preserving compounds sensitive to heat. ACP treatment utilizes air as the reaction gas medium. During the discharge process, the temperature of the electrons generated by the high-energy electric field is considerably higher than that of the ions. Consequently, the electrons experience minimal energy loss when colliding with ions or neutral particles, thereby preventing an increase in the air's temperature [10]. It has been reported to improve sunflower proteins' emulsification properties by ACP treatment [11]. In addition, prolonged ACP treatment time (1–10 min) leads to changes in the secondary and tertiary structure of soybean isolate proteins (SPI), which improves emulsification capacity and foaming ability [12]. However, there are fewer studies on ACP altering the thermal stability and water solubility of glutenin. Meanwhile, few studies have been reported on the interaction mechanism of ACP treatment on the changes in structural and physicochemical properties of WGLU and the time relationship.

Therefore, this research systematically assessed the different ACP treatment times (0, 30, 60, 90, and 120 s) on the structural and functional attributes of WGLU. Firstly, fluorescence spectroscopy, Fourier transform infrared spectroscopy (FTIR), and X-ray diffraction (XRD) were used to investigate the structural effects of different ACP treatment durations on WGLU. Simultaneously, scanning electron microscopy (SEM) was used to observe the microstructural changes of WGLU under different ACP treatment times. On this basis, the influence of ACP treatment at different times on thermal stability, surface hydrophobicity, solubility, water-holding capacity, foaming characteristics, and emulsification characteristics of WGLU was further investigated. The outcomes of this research offer innovative approaches to change the structure and functional properties of WGLU, as well as laying the groundwork for expanding the use of walnut protein resources into new areas.

2. Materials and Methods

2.1. Materials

Walnuts were purchased from Yangbi, Dali, Yunnan Province. Food-grade soybean oil was sourced from Yihai Kerry Food Industry Co., Ltd. (Kunming, China). Other chemicals used were of analytical grade.

2.2. Preparation of WGLU

As shown in Figure 1A, WGLU was prepared by first extracting defatted walnut powder sequentially with deionized water, 1 M NaCl, and 70% ethanol in a 1:10 ratio with magnetic stirring for 1.5 h. After each extraction step, centrifuge at 8000 rpm for 10 min and recover the precipitate. Finally, extraction was carried out with 0.1 M NaOH, and the supernatant was taken. The pH was adjusted to 4.5 (1 mol/L HCl), and the precipitate was left to settle for 12 h. The precipitate was centrifuged at 5000 rpm for 20 min, and the precipitate was collected. A total of 50 mL of distilled water was added to the precipitate and stirred magnetically at 500 rpm for 10 min, the pH was adjusted to 7.0 (0.1 mol/L NaOH), it was dialyzed for 36 h, and it was vacuum freeze-dried (48 h) to obtain WGLU [13].

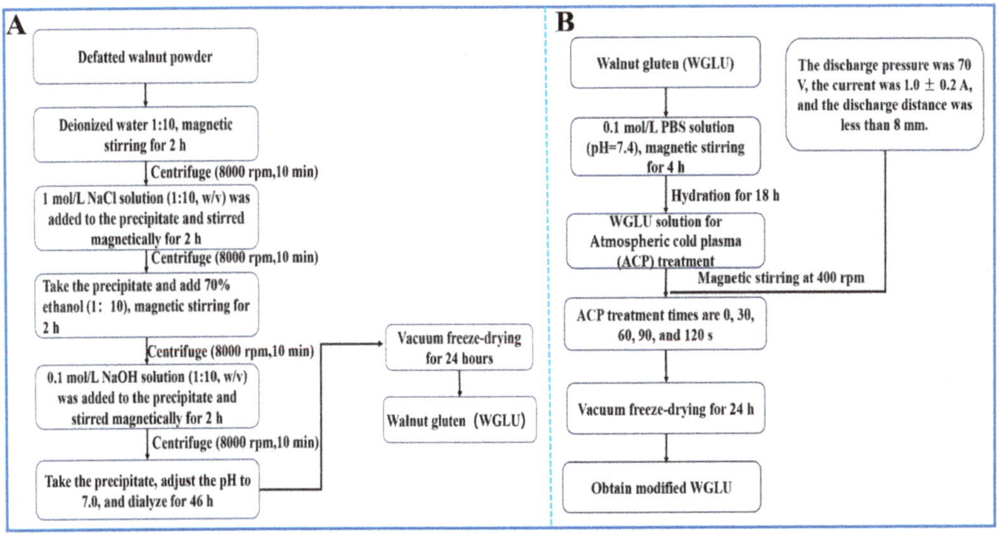

Figure 1. Preparation process of WGLU (**A**) and WGLU treated at different times by ACP (**B**). WGLU, walnut glutenin; ACP, atmospheric cold plasma.

2.3. Cold Plasma Treatment of WGLU

As shown in Figure 1B, the WGLU solution, with a concentration of 0.1 g/mL, was made using a 0.1 mol/L phosphate buffer solution (PBS) at pH 7.4, and magnetically stirred at 500 rpm for 4 h. After being hydrated for 18 h at 4 °C, 200 mL of WGLU solution was subjected to ACP treatment under magnetic stirring at 400 rpm (CTP-2000K, Nanjing Suman Electronics Company, Ltd., Nanjing, China). The treatment times were set at 30, 60, 90, and 120 s, with a duration of 0 s serving as the control condition. Carrier gas is air, temperature is 25 ± 2 °C, relative humidity is 45–5%, discharge pressure is 70 V, current is 1.0 ± 0.2 A, and discharge distance is less than 8 mm. The WGLU solution was vacuum freeze-dried after ACP treatment [14].

2.4. Structural Properties of WGLU

2.4.1. Measurement of Fluorescence Spectroscopy

A WGLU solution with a 0.01 g/mL concentration was made using a 0.1 mol/L PBS solution at pH 7.4. The fluorescence spectrum of the WGLU was obtained using an enzyme marker (EPOCH2, BioTek Instruments, Inc., Winooski, VT, USA). An excitation wavelength of 280 nm was used, with an emission wavelength range between 300 and 450 nm, and a slit width of 2 nm [13].

2.4.2. Measurement of Fourier Transform Infrared (FTIR) Spectroscopy

The functional group of the WGLU was analyzed using an infrared spectrometer (650, Tianjin, China). The WGLU was subjected to 32 scans at a resolution of 4 cm^{-1}, spanning wavenumbers from 4000 cm^{-1} to 400 cm^{-1}. To assess changes in secondary structure content, Gaussian area fitting was conducted using Peakfitv4.12 [5].

2.4.3. Measurement of X-Ray Diffraction (XRD)

The crystalline structure was examined using an X-ray diffractometer (Ultima IV, Rigaku Corporation, Akishima, Japan) with a scanning rate of 5°/min and a range from 5° to 90°. The crystallinity was examined using JADE 6.0 software [14].

2.4.4. Measurement of Scanning Electron Microscope (SEM) Analysis

Using a scanning electron microscope (FE-SEM, Sigma 300, Zeiss, Germany) at 15 kV and 10,000× magnification, the microstructures of the WGLU were observed after gold plating [14].

2.5. *Physicochemical Properties of Walnut Glutenin (WGLU)*

2.5.1. Measurement of Particle Size and Zeta Potential

A WGLU solution with a concentration of 0.01 g/mL was made using a 0.1 mol/L PBS solution at pH 7.4, and the size of particles and zeta potential for WGLU were assessed with a nanoparticle size and zeta potential analyzer (Brookhaven Instruments, New York, NY, USA) [5].

2.5.2. Measurement of Total and Active Sulfhydryl (-SH)

A mixture was prepared by adding 1 mL of 5 mg/mL WGLU solution to 5 mL of Tris-glycine buffer (8 M urea, 0.086 mol/L Tris, 0.09 mol/L glycine, 4 mmol/L EDTA, pH 8.0) and 200 μL of Ellman's reagent. Then, the mixture was reacted for 30 min and centrifuged at 8000 rpm for 10 min. The supernatant was collected and the absorbance at 412 nm was determined. A similar method was used to determine active free thiol groups without adding urea. The total -SH groups and active -SH groups were calculated according to Equation (1) [13].

$$-\text{SH}\ (\mu\text{mol/g}) = \frac{73.53 \times A_{412} \times D}{C} \quad (1)$$

where A_{412} indicates the absorbance at 412 nm, with D as the dilution factor and C as the concentration of the WGLU solution (5 mg/mL).

2.5.3. Measurement of Thermal Stability

WGLU was analyzed by thermal stability using a differential scanning calorimeter (DSC) (3500 Sirius, Netzsch, Germany). The nitrogen flow rate was set at 20 mL/min, the heating rate was set at 10 °C/min, and the heating temperature was set at 20–180 °C [5].

2.5.4. Measurement of Surface Hydrophobicity (H_0)

Prepare a WGLU solution at a concentration of 0.1 to 0.5 mg/mL in 0.1 mol/L PBS at pH 7.4. Then, 20 μL of 8 mM 8-(phenylamino)-1-naphthalene sulfonate (ANS) was added to 4 mL of WGLU solution and incubated in the dark at room temperature for 20 min. An enzyme marker was used to measure fluorescence intensities at the excitation and emission wavelengths of 390 nm and 470 nm. A primary function was fitted with WGLU concentration and fluorescence intensity, with H_0 calculated as the slope of the resulting curve [15].

2.5.5. Measurement of Solubility

WGLU solution (0.01 g/mL) was prepared in 0.1 mol/L PBS (pH 7.4) and centrifuged at 8000 rpm for 10 min to separate the supernatant. The WGLU content in the supernatant was determined by the Coomassie brilliant blue method. WGLU solubility (%) was calculated according to Equation (2) [16].

$$\text{Solubility (\%)} = \frac{\text{Weigh of protein in the supernatant}}{\text{Weigh of total protein}} \times 100 \quad (2)$$

2.5.6. Measurement of Water-Holding Capacity (WHC) and Oil-Holding Capacity (OHC)

In a centrifuge tube, 0.5 g of WGLU was combined with 20 mL of distilled water and allowed to hydrate at 27 °C for 2 h. Subsequently, the mixture was centrifuged at 8000 rpm for 10 min, and after the removal of the supernatant the weight of the centrifuge tube containing the sediment was measured [17]. WHC was calculated according to Equation (3). OHC was measured using a similar method, where soybean oil replaced distilled water. Equation (4) was used to compute the OHC value.

$$\text{WHC (g/g)} = \frac{W_3 - W_1 - W_2}{W_1} \quad (3)$$

$$\text{OHC (g/g)} = \frac{W_3' - W_1 - W_2}{W_1} \quad (4)$$

where W_1 is the weight of the WGLU (5 g), W_2 is the weight of the centrifuge tube (g), and W_3 and $W_3{}'$ indicate the mass of the centrifuge tube and the sediment that remains (g).

2.5.7. Measurement of Emulsifying Properties

A WGLU solution with a concentration of 0.01 g/mL was made using a 0.1 mol/L PBS solution at pH 7.4, and soybean (5 mL) oil was added to the WGLU solution (15 mL) and homogenized at 10,000 rpm for 5 min. A 0.05 mL portion of the emulsion was drawn from the bottom and mixed with a 0.1% sodium dodecyl sulfate solution. The absorbance at a 500-nm wavelength was measured using an ultraviolet spectrophotometer (UV-2600, Shimadzu, Kyoto, Japan). After 10 min, 0.05 mL of the lower emulsion was sampled again, combined with sodium dodecyl sulfate solution, and its absorbance was recorded. The emulsifying properties were calculated according to Equations (5) and (6) [18].

$$\text{EAI (m}^2\text{/g)} = 2 \times 2.303 \frac{A_0 \times N}{C \times \varphi \times 10000} \quad (5)$$

$$\text{ESI (min)} = \frac{A_0}{A_0 - A_{10}} \times 10 \quad (6)$$

where EAI is the emulsifying activity index, N is the WGLU dilution coefficient (100), C is the WGLU concentration (0.02 g/mL), φ is the oil volume fraction (0.25), ESI is the emulsion stability index, and A_0 and A_{10} represent absorbances of the emulsions at 0 and 10 min.

2.5.8. Measurement of Foaming Properties

WGLU solution (0.01 g/mL) was prepared in 0.1 mol/L PBS (pH 7.4), and 10 mL of WGLU solution was taken and dispersed at 8000 rpm for 2 min, then quickly transferred to a measuring cylinder. The Foaming properties were calculated according to Equations (7) and (8) [19].

$$\text{FC (\%)} = \frac{V_0}{V} \times 100 \quad (7)$$

$$\text{FS (\%)} = \frac{V_t}{V_0} \times 100 \quad (8)$$

where FC (%) is the foaming capacity, FS (%) is foam stability, V represents the WGLU solution's volume (mL), and V_0 and V_t are the foam volumes at 0 and 10 min (mL).

2.6. Statistical Analysis

Each experiment was conducted three times, and SPSS version 26.0 was used for data analysis, with significant differences in results indicated at $p < 0.05$, and plotted using Origin 2021.

3. Results

3.1. Structural Characterization

3.1.1. Fluorescence Spectra Analysis

A protein's fluorescence spectral properties can be used to characterize changes in its tertiary structure. In Figure 2A, it is shown that the control group has a maximum emission wavelength (λmax) of 341 nm, and that of WGLU after ACP treatment showed a slight red shift (λmax = 342 nm). The reason is that WGLU is extracted by strong alkaline solutions, which may damage its structure, and the oxidation effect is enhanced after ACP treatment, resulting in tryptophan residues shifting toward a more polar hydrophilic environment [12,20]. Furthermore, compared to the control group, ACP treatment caused a rise in WGLU fluorescence intensity and showed an increase followed by a decrease. These discrepancies may be related to the fact that ACP treatment produces an etching effect that depolymerizes glutamine in WGLU, exposing the chromophores and enhancing fluorescence intensity [21]. It was observed that the WGLU fluorescence intensity was not the highest at 30 s of ACP treatment, which may be due to the short duration of ACP treatment and incomplete exposure of the chromogenic groups within the protein. ACP treatment for 60 s, the spatial structure of WGLU becomes looser, resulting in maximum fluorescence intensity. When the treatment time reaches 90 s and 120 s, the fluorescence intensity of WGLU decreases. This may be related to the intense oxidation of WGLU during prolonged treatment, leading to carbonyl group formation, aggregation of proteins, and concealment of amino acid residues, leading to a reduction in fluorescence [22].

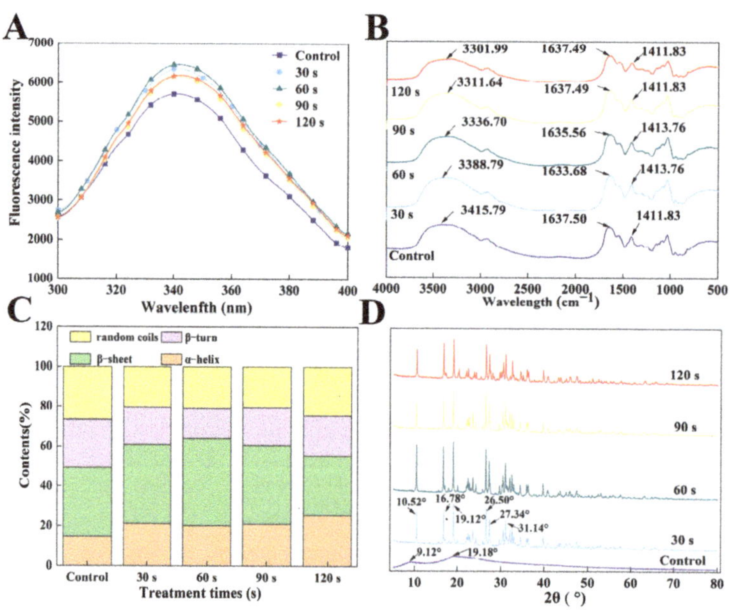

Figure 2. Fluorescence spectra (**A**), Fourier transform infrared spectroscopy (FTIR) spectra (**B**), secondary structure content (**C**), and X-ray diffraction (XRD) spectra (**D**) of WGLU treated with ACP at different times. WGLU, walnut glutenin; ACP, atmospheric cold plasma.

3.1.2. Fourier Transform Infrared Spectroscopy Analysis

The effects of ACP treatment on the chemical composition and conformation of WGLU were analyzed using FTIR spectroscopy. Figure 2B indicates that the control group exhibits a pronounced characteristic peak within the 3415.79 cm^{-1} band, mainly due to the vibrations from N-H and O-H stretching. After ACP treatment, the absorption peak of WGLU in the 3415.79 cm^{-1} band shifted to lower wavelengths compared to the control group, which indicated that ACP treatment disrupted the hydrogen bonding of WGLU. This disruption can improve the hydrophilic property of the protein by inducing polymer chains or specific functional groups to covalently attach to the protein surface [23]. At 1637.50 cm^{-1}, the control group showed a unique absorption peak, corresponding to C-O bond stretching in the protein amide I band, at 1411.83 cm^{-1}, the peak largely caused by N-H bending and C-N stretching within the amide II band [24]. After ACP treatment, the absorption peak of WGLU at 1637.50 cm^{-1} showed a blue shift, with the greatest shift occurring at 30 s and 60 s. The result indicated that ACP treatment induced dissociation of the WGLU structure, increasing the electron cloud density of C-O, which caused the blue shift in the absorption peak [13]. At 90 s and 120 s, the absorption peak shifted blue by only 0.01 cm^{-1}. This may be due to prolonged processing, causing the cross-linking and aggregation of WGLU structures. Additionally, compared with the control group, the absorption peak of WGLU at 1411.83 cm^{-1} showed a red shift at 30 s and 60 s after ACP modification. High-energy ions and particles may collide with proteins, thereby transferring energy to intermolecular bonds, leading to further unfolding of the WGLU structure [18].

According to Figure 2C, the control group's secondary structure was primarily β-turn and β-sheet. The β-sheet content of WGLU initially went up and then declined as ACP treatment time was extended, while the α-helix content first decreased and then increased. It has been reported that an increase in α-helix and β-turn indicates protein cross-linking to form macromolecular aggregates, causing peptide chains to contract and become more tightly ordered [13]. A rise in β-sheets and random coil is seen as a sign of increased flexibility, structural disorder, and unfolding of protein [25]. The secondary structure of proteins is dynamically balanced. α-helices, β-folds, β-turns, and random curls together form the local conformation of the protein backbone. They can be interconverted with each other, thus affecting the protein's structural conformation. The control group had a β-sheet content of 46.62%, an α-helix content of 14.03%, a random coils content of 13.24%, and a β-turn content of 26.11% (Table 1). The secondary structure of the WGLUs was altered after ACP treatment, with the β-sheet content peaking at 60 s of ACP treatment (47.64%), while the α-helix content decreased to 13.81%. Attributed to ACP oxidation of side-chain amino acids destroys the groups responsible for maintaining the ordered spatial structure, resulting in the loss of non-covalent bonds that form the basic secondary structural units [9,26]. After 120 s of ACP treatment, β-sheet and β-turn were transformed into an α-helix, which may cause WGLU re-aggregation and increased protein ordering by prolonged ACP treatment.

Table 1. Secondary structure content of WGLU after ACP treatment at different times.

Treatment Times (s)	β-Sheet (%)	Random Coil Content (%)	α-Helix (%)	β-Turn (%)
Control	46.62	13.24	14.03	26.11
30 s	45.57	13.07	13.98	27.38
60 s	47.64	13.21	13.81	25.34
90 s	46.94	12.98	13.80	26.28
120 s	44.98	13.19	28.28	13.55

3.1.3. X-Ray Diffraction Analysis

In Figure 2D, the control group is shown to have diffraction peaks at 2θ = 9.12° and 19.18°, which are indicative of an amorphous or semi-crystalline structure [27]. After ACP treatment, WGLU exhibited more diffraction peaks at 10.52°, 16.78°, 19.12°, 26.50°, 27.34°, and 31.14°, and these peaks were sharper, indicating an enhanced crystalline structure. This may be because ACP treatment destroys the chemical bonds of WGLU, weakening the regularity of the crystal arrangement and leading to drastic changes in the spatial conformation [28]. It is worth noting that, after ACP treatment, the crystallinity of WGLU increased compared to the control group (38.16%), with values of 78.51%, 91.57%, 79.79%, and 77.26%, respectively. The crystallinity reached the maximum at 60 s, probably due to the oxidative effect produced by ACP, which made the protein structure looser. The crystallinity decreased at 90 s and 120 s. However, due to the prolongation of the ACP treatment time, the disorder of the crystal arrangement of WGLU decreased [29].

3.1.4. Scanning Electron Microscope Analysis

Figure 3 shows that the microstructure of the control group displays a rigid spherical structure. After ACP treatment, the closed structure of WGLU was disrupted, presenting an irregular shape with cracks appearing on the surface. Meanwhile, the size of the particles first reduces and then grows as the processing time extends. This effect is caused by the high-energy particles generated by ACP that interact with the surface of WGLU, reducing the size of large particles and significantly modifying their external appearance [30]. At 60 s of ACP treatment, WGLU was mostly fragmented into small particles with the smallest particle size and uniform distribution. This could be because a longer ACP treatment time generates more active substances that bombard the WGLU surface, resulting in a stronger etching effect and creating cavities within the protein [18]. Active substances can also diffuse from these cavities into WGLU, disrupting intermolecular hydrogen bonds and leading to WGLU structural unfolding [31]. However, at 90 s and 120 s of ACP treatment time, WGLU formed aggregates, mostly lumps. This may result from the prolonged treatment period, which strengthens inter-protein interactions and increases the chance of protein-protein binding. Also, when ACP treatment was prolonged to 120 s, the WGLU surface exhibited numerous small fragments and voids. The cause is the presence of a significant quantity of electrons, ions, and other active substances that attacked the surface of WGLU for a long time, generating many low fragments, which could not be aggregated in time, thus leading to their adhesion to the protein surface [32].

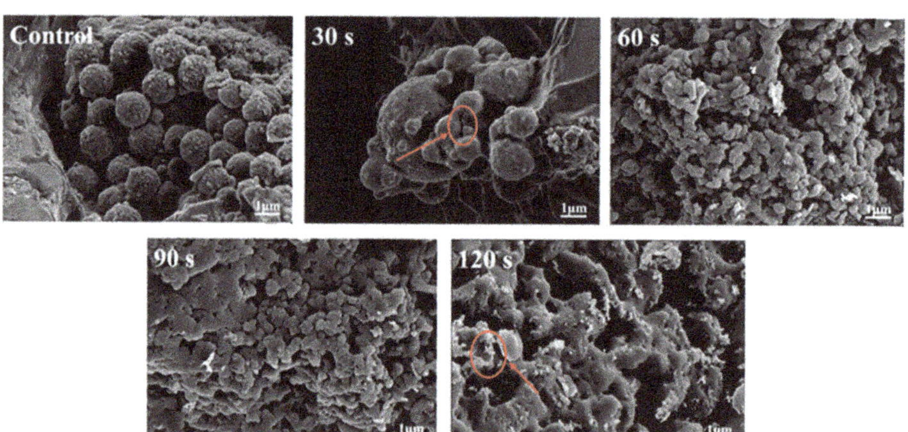

Figure 3. Scanning electron microscopy (SEM) of WGLU after ACP treatment at different times. WGLU, walnut glutenin; ACP, atmospheric cold plasma. The red arrows/circles in the figure indicate significant changes in WalPI after ACP treatment.

3.2. Physical and Chemical Properties

3.2.1. Particle Size and Zeta Potential Analysis

Figure 4A–E shows the particle size distribution of WGLU with different times of ACP treatment, where the protein particle size distribution was gradually shifted to the left toward smaller particle sizes after ACP treatment compared to the control. This change suggests that ACP treatment may have contributed to the depolymerization or dispersion of protein particles, resulting in the overall particle size reduction. In addition, the peak widths of the particle size distributions tended to decrease and then increase with time, and the peak widths were the smallest at 60 s. The decrease in peak width may mean that the ACP treatment resulted in a more uniform distribution of protein particles at the initial stage, while the subsequent increase in peak width may indicate that the protein particles were further depolymerized or partially aggregated with the extension of the treatment time, resulting in a wider range of particle size distribution [33]. As shown in Table 2, the average particle size of the control group was 6201.45 nm, and the average particle size of WGLU decreased after ACP treatment and exhibited a trend of first declining and then rising as treatment time increased. This occurred primarily because electrons were released, ions, and neutral particles during ACP discharge, which eroded the protein surface and disrupted the apparent densities of thin layers, leading to a reduction of the large particles [34]. The smallest average particle size of 341.70 nm for WGLU was recorded at 60 s of ACP treatment. The cause could be the elevated density of charged particles and ions resulting from ACP, which disrupts the molecular forces between proteins and reduces WGLU particle aggregation [18]. WGLU's particle size averaged 543.08 nm and 573.48 nm at 90 s and 120 s of ACP treatment. This increase is due to the large amount of reactive chemicals, such as O_3 and H_2O_2, which can oxidize amino acid side chains, especially cysteine, promoting protein–protein binding. Furthermore, the binding of polar compounds to protein surfaces can increase surface energy and adhesive strength [35]. The surface potential of a protein in solution is represented by its zeta potential. Usually, the physical stability of proteins in an aqueous solution is enhanced by a rise in the absolute zeta potential value [36].

Table 2. Average particle size of WGLU after atmospheric cold plasma treatment at different times.

Treatment Times (s)	Average Particle Size (nm)
Control	6201.45 ± 98.63 [a]
30 s	960.00 ± 84.27 [b]
60 s	341.70 ± 6.95 [c]
90 s	543.08 ± 29.85 [d]
120 s	573.66 ± 4.77 [c]

Different lowercase letters ([a]–[d]) indicate significant differences between various treatment conditions ($p < 0.05$).

Figure 4F shows that the control group's zeta potential was −11.06 mV, and the zeta potentials of WGLU increased after ACP treatment, initially rising and then falling as ACP treatment time increased. This was largely because ACP treatment led to the creation of active particles like N^{2+}, O_3, and H_2O_2, which led to the breakdown of soluble aggregates, revealing additional polar groups on the WGLU surface and enhancing the zeta potential [11]. At 60 s of ACP treatment, zeta potential reached its maximum (−15.33 mV), attributed to the high density of chemical reactants generated by ACP, which increased the oxidation of amino acid residues and converted them into negatively charged byproducts [22]. Furthermore, ACP treatment can lead to the degradation of WGLU aggregates or conformational changes in secondary structures, resulting in the exposure of some previously buried amino acids and an increase in zeta potential. At 90 s and 120 s of ACP treatment, the zeta potential of WGLU was −14.10 and −14.08 mV, which may be due to WGLU aggregation and reduced exposure of polar groups, leading to a decrease in zeta potential.

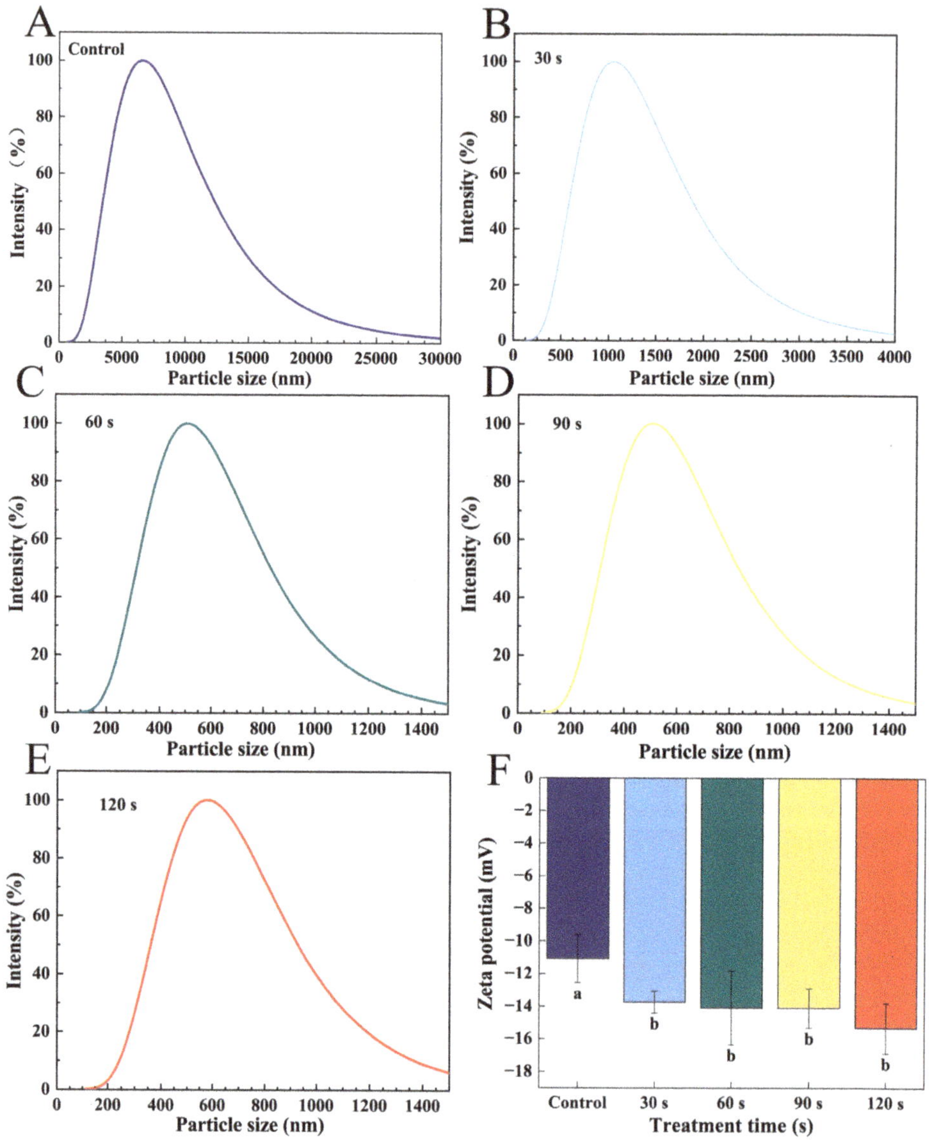

Figure 4. Particle size distribution (**A–E**) and zeta potential (**F**) of WGLU treated with ACP at different times. Different lowercase letters (a,b) indicate significant differences between different treatment conditions ($p < 0.05$). WGLU, walnut glutenin; ACP, atmospheric cold plasma.

3.2.2. Total and Active -SH Analysis

Alterations in free -SH groups can influence the functional characteristics of proteins. In Figure 5A, it is shown that the active -SH group of the control group was 12.55 µmol/g and the total -SH group was 15.77 µmol/g. However, following treatment with ACP, WGLU showed a significant increase ($p < 0.05$) in its active and total -SH groups and displayed an initial rise followed by a decline with longer treatment durations. It may result from the tertiary structure unfolding and the disulfide bonds between amino acid side chains breaking [37]. Meanwhile, active -SH groups were positively correlated with total -SH groups. WGLU's active and total -SH groups peaked at 60 s during ACP treatment, showing an increase of 23.90% and 23.72% compared to the control group. This increase is primarily attributed to the disaggregation of WGLU aggregates and the reduction in particle size, which makes the -SH groups more accessible to react with the active substances generated by ACP [22]. It is noteworthy that prolonged treatment may lead to increased oxidation, prompting the reaggregation of WGLU and an increase in disulfide bonds, thereby decreasing the amount of free -SH groups.

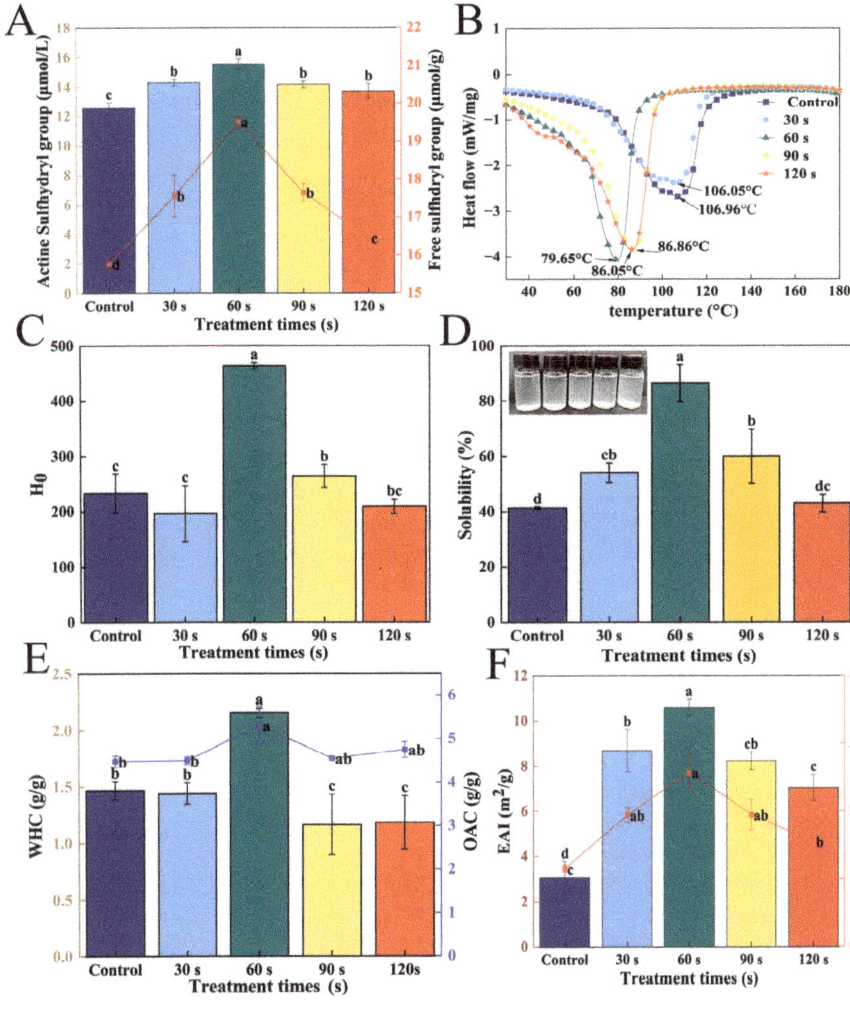

Figure 5. Total and active sulfhydryl (**A**), thermal stability (DSC) (**B**), surface hydrophobicity (H_0) (**C**), solubility (**D**), water-holding capacity (WHC), oil-holding capacity (OHC) (**E**), and emulsification activity index (EAI) and emulsification stability index (ESI) (**F**) of WGLU treated with ACP at different times. Different lowercase letters (a–d) indicate significant differences between various treatment conditions ($p < 0.05$). WGLU, walnut glutenin; ACP, atmospheric cold plasma.

3.2.3. Thermal Stability Analysis

The thermal stability of WGLU was evaluated using DSC, and its thermal stability was related to the hydrogen bond content within the protein structure [38]. As shown in Figure 5B, all WGLU exhibited a single broad endothermic peak, which was primarily attributed to the denaturation of the protein's tertiary or quaternary structure. The peak denaturation temperature (Td) reflects the thermal stability of proteins, with higher Td indicating denser protein structures and more stable configurations [39]. In the control group, Td was 106.96 °C. After ACP treatment, WGLU's Td decreased, exhibiting a trend of initially decreasing followed by an increase as the ACP treatment time extends. This could be a result of ACP's high-energy particles impacting WGLU's surface structure, breaking its hydrogen bonds and weakening intermolecular forces, resulting in the WGLU structure transitioning from a dense state to a loose state [28]. The Td of WGLU was the lowest at 60 s of ACP treatment (79.65 °C). This may be ACP-induced oxidation, which increases electrostatic repulsion and steric hindrance between protein molecules, leading to a more open WGLU structure and reduced particle size, thereby reducing its ability to withstand higher temperatures. The Td of WGLU increased at 90 and 120 s of ACP treatment, which may be due to the prolonged time of ACP treatment, which causes WGLU to cross-link oxidatively and form aggregates.

3.2.4. Surface Hydrophobicity Analysis

Differences in aggregation and folding of protein tertiary structures can be assessed using H_0 [40]. Figure 5C illustrates that the control group's H_0 was measured at 233.43. Following ACP treatment, the H_0 of WGLU first rose and then fell as the duration of treatment grew. The H_0 of WGLU was lower than that of the control group at 30 s of ACP treatment, probably because the energetic particles generated by ACP rapidly broke the C-C and C-H bonds on the surface of the material and caused them to react with reactive oxygen species in the plasma environment or oxygen in the treated air. The introduction of polar functional groups in WGLU led to its rapid oxidation, resulting in a significant decrease in H_0 ($p < 0.05$) [22]. The H_0 of WGLU was all greater than that of the control group with the increase in ACP treatment time, which may be due to the ACP etching effect leading to an increase in the roughness of the protein surface or the dissociation of the protein subunits, which leads to the observed increase in H_0 [41]. The H_0 of WGLU peaked at 60 s of ACP treatment (463.50), an increase of 98.56% compared to the control group. The rise is attributed to the adequate processing time changing the spatial structure of WGLU, which causes the WGLU structure to unfold and exposes hydrophobic amino acids, increasing the polarity of the water environment, and enabling ANS to fully bind with hydrophobic amino acids [42]. Li et al. [43] studied the effects of various modification methods on walnut protein and found that the hydrophobicity of the walnut protein surface decreased after ACP modification. This decrease is likely due to ACP-induced oxidation and cross-linking, which increase the oxygen-containing groups on walnut protein, expose polar groups, and reduce the hydrophobic regions. At 90 s and 120 s of ACP treatment, the H_0 of WGLU were 263.77 and 209.20, exhibiting a notable downward trend ($p < 0.05$). The decrease in hydrophobicity is due to the aggregation of WGLUs by electrostatic attraction, and the hydrophobic sites bound by ANS probes may be covered, resulting in a decrease in hydrophobicity.

3.2.5. Solubility Analysis

Protein solubility can evaluate the interaction between amino acid residues on the protein surface and water, as well as protein-protein interactions [42]. In Figure 5D, it is shown that the control group had a solubility of 41.38%. The solubility of WGLU increased significantly after ACP treatment ($p < 0.05$) and showed an upward trend initially, followed by a downward trend as ACP treatment time progressed. This increase was primarily attributed to the high-energy active particles generated by ACP, which disrupted the spatial structure of WGLU. This disruption caused protein collisions, reduced the particle size, and increased the specific surface area and the number of charged groups, thereby strengthening the interaction between WGLU and water and leading to better solubility [44]. The WGLU solubility reached the maximum at 60 s of ACP treatment (86.35%). It may be that ACP produced more energetic electrons, ions, and a significant amount of free radicals, which increase the surface energy of the hydration layer on the protein surface, speeding up the movement of water and facilitating the binding of WGLU particles to water [45]. Additionally, the unfolding of the WGLU structure, reduction in size, increase in surface potential, and increase in H_0 all contributed to enhanced solubility. The WGLU solubility was significantly reduced ($p < 0.05$) at 90 s and 120 s of ACP treatment. Prolonged treatment likely exposes too many active sites, causing WGLU to aggregate and reducing its interaction with water.

3.2.6. Water and Oil Holding Capacity

The ability of proteins to gel, emulsify, and foam is largely dependent on WHC and OHC, which affect their interactions with water and oil, respectively. Figure 5E shows that the WHC of the control group was 1.47 g/g, while the OHC was 4.49 g/g. After ACP treatment, the WHC and OHC of WGLU initially increased and then decreased, showing a clear trend concerning the treatment time. WGLU exhibited a WHC of 2.15 g/g and an OHC of 5.31 g/g, at 60 s of ACP treatment, which were increased by 46.26% and 18.26% compared to the control group. This is attributed to the unfolding of the tertiary structure of WGLU, which increases hydrophobicity and allows the protein to absorb more water and oil [46]. In addition, this may also be related to the increase in protein zeta potential. The WHC and OHC of WGLU decreased at 90 and 120 s of ACP treatment due to protein aggregation and the reduction of hydrophobic sites available for binding water and oil. The reduction in WHC might result from the decreased solubility of WGLU and the creation of protein aggregates. The decrease in OHC may be due to the reduced hydrophobicity and increased hydrophilicity of the protein surface with prolonged ACP treatment. Since OHC is positively correlated with H_0, the decrease in hydrophobicity likely leads to reduced OHC [47].

3.2.7. Emulsification Activity Index and Emulsion Stability Index Analysis

EAI and ESI can assess the ability of proteins to form oil-water interfaces and stabilize emulsion droplets [48]. As shown in Figure 5F, the control group's EAI was 3.09 m^2/g, and its ESI was 11.54 min. After ACP treatment, both EAI and ESI of WGLU exhibited notable increases ($p < 0.05$) and displayed a trend of increasing before decreasing as the duration of ACP treatment extended. The improvement in emulsification performance is the result of significant increases in WGLU zeta potential, H_0, and OHC. These changes promoted the dispersion and adsorption of protein at the water-oil interface [49]. The WGLU's EAI and ESI achieved their peak values at 60 s of ACP treatment, which were increased by 2.43 and 1.23 times compared to the control group. The reason for this is the increased solubility of WGLU, which facilitates its migration to the water-oil interface. Nevertheless, the EAI and ESI of the WGLU significantly decreased ($p < 0.05$) at 90 s and 120 s of ACP treatment. This suggests that the prolonged ACP treatment time resulted in the formation of covalent cross-links within or between proteins. Consequently, the interaction between the protein and small oil droplets is hindered, leading to a reduction in the emulsifying capacity of WGLU [50].

3.2.8. Foaming and Foam Stability Analysis

Proteins are amphiphilic and can stabilize air–water (foam) interfaces by forming a thin, flexible, and stable interfacial film that reduces surface tension, facilitating foam formation and stabilization [43]. Figure 6A is the foam visual diagram of WGLU, which can help more intuitively understand the foam characteristics between different samples. The foam in the control group had nearly dissipated after 10 min, and the stability of WGLU was poor. Figure 6B,C illustrates that the control group exhibited an FC of 12% and an FS of 16.74%. In addition, both the FC and FS of WGLU increased after ACP treatment and displayed a tendency to rise initially and then decrease with prolonged ACP treatment time. This occurs because the protein side chains unfold, resulting in a more flexible structure and exposing more hydrophobic sites, which facilitates adsorption at the air–water interface, resulting in more foam [18]. The increased FC is associated with enhanced H_0, as the exposure of hydrophobic residues strengthens protein–air interactions, thereby promoting foam formation [51]. ACP treatment has been shown to cause oxidation in the side chains of amino acids, thus enhancing the flexibility of proteins and improving their adsorption rate at the air–water interface [52]. At 60 s of ACP treatment, the FC of WGLU was 24.67%, likely as a result of the WGLU secondary structure unraveling, which enhances hydration and protein–water interactions, thereby maximizing FC. The FS of WGLU reached a maximum (56.67%) at 60 s, which was attributed to the increase in H_0, enhanced structural integrity of the interfacial membrane, and the formation of a protein network that acts as a mechanical barrier to bubble breakage and aggregation, thus improving the foam stability of the WGLU [2]. The FC and FS of WGLU decreased at 90 s and 120 s of ACP treatment, likely due to the formation of aggregates, a reduction in hydrophobicity, and an increase in particle size, which hindered the protein's ability to adsorb efficiently at the air–water interface. Extended treatment resulted in reduced solubility of WGLU, which further hindered its adsorption and negatively affected foam formation and stability.

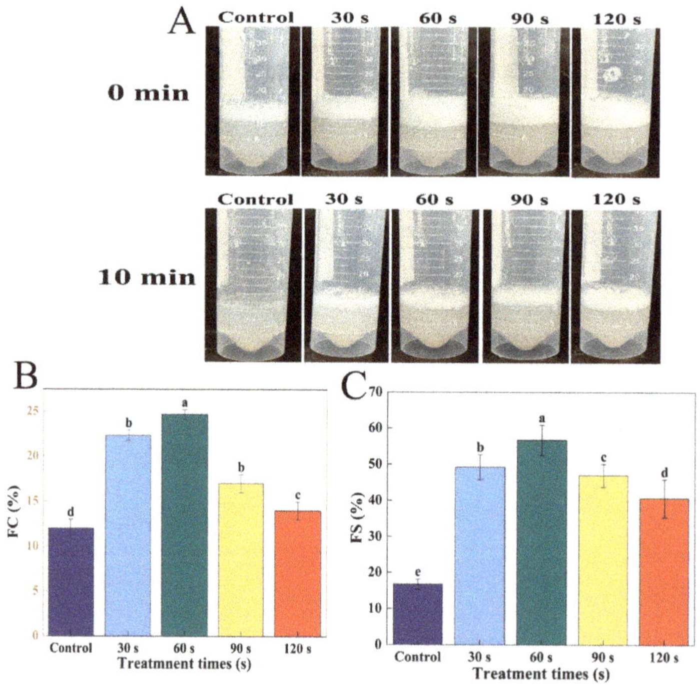

Figure 6. Foaming visualization diagram (**A**) and foaming capacity (FC) (**B**) and foam stability (FS) (**C**) of WGLU treated with ACP at different times. Different lowercase letters (a–e) indicate significant differences between different treatment conditions ($p < 0.05$). WGLU, walnut glutenin; ACP, atmospheric cold plasma.

4. Conclusions

The ACP treatment effectively disrupted the spatial structure of WGLU, disrupting its secondary structure and unfolding its tertiary structure, thereby significantly enhancing surface hydrophobicity. Secondly, the duration of ACP treatment is closely linked to the structural alterations and functional characteristics of WGLU. At 60 s into ACP treatment, WGLU reached its smallest particle size and greatest zeta potential, enhancing its physicochemical properties, notably solubility and emulsifying activity. However, prolonged treatment resulted in the formation of WGLU aggregates, and, in particular, their WHC and FS were significantly reduced. Overall, ACP treatment for the appropriate time, especially 60 s, represents an effective strategy to improve the structural and functional characteristics of WGLU. This study can provide valuable insights for the future development of plant-based food. However, further studies on the long-term stability and oxidation resistance of ACP-treated WGLUs are needed. Follow-up studies could use molecular dynamics to investigate the effect of ACP treatment on the structure of WGLU.

Author Contributions: Y.D.: writing—review and editing, writing—original draft, visualization, software, methodology, investigation, formal analysis. G.Y.: data curation, conceptualization, validation, software, methodology. T.Y.: formal analysis, data curation, methodology. B.G.: writing—review and editing, data curation. Y.L.: investigation, formal analysis. J.Y.: validation, methodology. L.G.: investigation, formal analysis. Q.M. and F.F.: writing—review and editing, validation, resources, supervision, project administration, methodology, funding acquisition, data curation, conceptualization. All authors have read and agreed to the published version of the manuscript.

Funding: The authors thank the Yunnan Fundamental Research Projects (grant NO. 202401AS070012), Ten thousand people plan of Yunnan Province, young top-notch personnel (YNWR-QNBJ-2018-046), Yunnan Provincial Talent Reserve Project for Young and Middle-aged Academic and Technical Leaders (202405AC350032).

Institutional Review Board Statement: Not applicable.

Informed Consent Statement: Not applicable.

Data Availability Statement: The original contributions presented in this study are included in the article. Further inquiries can be directed to the corresponding authors.

Conflicts of Interest: The authors declare no conflicts of interest.

References

1. Lei, Y.; Gao, S.; Xiang, X.; Li, X.; Yu, X.; Li, S. Physicochemical, structural and adhesion properties of walnut protein isolate-xanthan gum composite adhesives using walnut protein modified by ethanol. *Int. J. Biol. Macromol.* **2021**, *192*, 644–653. [CrossRef]
2. Alabi, O.O.; Annor, G.A.; Amonsou, E.O. Effect of cold plasma-activated water on the physicochemical and functional properties of Bambara groundnut globulin. *Food Struct.* **2023**, *36*, 32. [CrossRef]
3. Zheng, J.; Zhu, T.; Yang, G.; Zhao, L.; Li, F.; Park, Y.-M.; Tabung, F.K.; Steck, S.E.; Li, X.; Wang, H. The Isocaloric Substitution of Plant-Based and Animal-Based Protein in Relation to Aging-Related Health Outcomes: A Systematic Review. *Nutrients* **2022**, *14*, 272. [CrossRef] [PubMed]
4. Li, X.; Guo, M.; Chi, J.; Ma, J. Bioactive Peptides from Walnut Residue Protein. *Molecules* **2020**, *25*, 1285. [CrossRef]
5. Lu, Y.; Jiang, Y.; Liu, J.; Yang, X.; Zhao, Y.; Fan, F. Preparation and Properties of Walnut Protein Isolate–Whey Protein Isolate Nanoparticles Stabilizing High Internal Phase Pickering Emulsions. *Foods* **2024**, *13*, 2389. [CrossRef] [PubMed]
6. Zhang, M.; Zhu, Z.; Pan, F.; Zhou, Q.; Zhao, L.; Zhao, L. Enhancing walnut protein isolate functionality with ultrasound treatment: An integrated experimental and molecular dynamics simulation study. *Food Hydrocoll.* **2025**, *163*, 111135. [CrossRef]
7. Feng, L.; Wu, Y.; Han, Y.; Yao, X.; Li, Q.; Liu, M.; Cao, Y. Structural characteristics, functional properties and nutritional value of walnut protein by limited enzymatic hydrolysis. *LWT* **2024**, *197*, 115923. [CrossRef]
8. Li, Y.; Zhang, C.; Liang, Y.; Wang, L.; Xiong, W. Solubility and conformational characterization of rice glutelin after high temperature treatment. *Int. J. Biol. Macromol.* **2022**, *223*, 1720–1726. [CrossRef]
9. Chen, Y.; Chen, Y.; Jiang, L.; Wang, J.; Zhang, W. Investigating binding mechanism between coconut globulin and tannic acid mediated by atmospheric cold plasma: Protein structure and stability. *Food Chem.* **2025**, *464*, 141670. [CrossRef]

10. Zou, J.-J.; Liu, C.-J.; Eliasson, B. Modification of starch by glow discharge plasma. *Carbohydr. Polym.* **2004**, *55*, 23–26. [CrossRef]
11. Wang, P.; Wang, Y.; Du, J.; Han, C.; Yu, D. Effect of cold plasma treatment of sunflower seed protein modification on its structural and functional properties and its mechanism. *Food Hydrocoll.* **2024**, *155*, 110175. [CrossRef]
12. Zhang, Q.; Cheng, Z.; Zhang, J.; Nasiru, M.M.; Wang, Y.; Fu, L. Atmospheric cold plasma treatment of soybean protein isolate: Insights into the structural, physicochemical, and allergenic characteristics. *J. Food Sci.* **2020**, *86*, 68–77. [CrossRef] [PubMed]
13. Wang, Y.; Tao, L.; Wang, Z.; Wang, Y.; Lin, X.; Dai, J.; Shi, C.; Dai, T.; Sheng, J.; Tian, Y. Effect of succinylation-assisted glycosylation on the structural characteristics, emulsifying, and gel properties of walnut glutenin. *Food Chem.* **2024**, *446*, 138856. [CrossRef]
14. Deng, Y.; Lu, Y.; Jiang, Y.; Yuan, G.; Yang, T.; Gao, B.; Yang, J.; Guo, L.; Fan, F. Effect of cold plasma treatment time on walnut protein isolate: Revealing structural changes and improving functional properties. *Int. J. Biol. Macromol.* **2025**, *311*, 143693. [CrossRef]
15. Meng, Y.; Zhao, X.; Jiang, Y.; Ban, Q.; Wang, X. Effect of Maillard reaction conditions on the gelation and thermal stability of whey protein isolate/d-tagatose conjugates. *Food Chem.* **2023**, *405*, 134928. [CrossRef]
16. Hu, G.; Zhang, J.; Wang, Q.; Ma, M.; Ma, L.; Li, S. Succinylation Modified Ovalbumin: Structural, Interfacial, and Functional Properties. *Foods* **2022**, *11*, 2724. [CrossRef]
17. Jiang, Y.; Lu, Y.; Liu, J.; Zhao, Y.; Fan, F. Characterization of bamboo shoot cellulose nanofibers modified by TEMPO oxidation and ball milling method and its application in W/O emulsion. *LWT* **2024**, *205*, 116563. [CrossRef]
18. Sharafodin, H.; Soltanizadeh, N. Potential application of DBD Plasma Technique for modifying structural and physicochemical properties of Soy Protein Isolate. *Food Hydrocoll.* **2022**, *122*, 107077. [CrossRef]
19. Wang, Y.; Wang, S.; Li, R.; Wang, Y.; Xiang, Q.; Li, K.; Bai, Y. Effects of combined treatment with ultrasound and pH shifting on foaming properties of chickpea protein isolate. *Food Hydrocoll.* **2022**, *124*, 107351. [CrossRef]
20. Zhang, S.; Huang, W.; Feizollahi, E.; Roopesh, M.S.; Chen, L. Improvement of pea protein gelation at reduced temperature by atmospheric cold plasma and the gelling mechanism study. *Innov. Food Sci. Emerg. Technol.* **2021**, *67*, 102567. [CrossRef]
21. Zhang, Q.-T.; Tu, Z.-C.; Xiao, H.; Wang, H.; Huang, X.-Q.; Liu, G.-X.; Liu, C.-M.; Shi, Y.; Fan, L.-L.; Lin, D.-R. Influence of ultrasonic treatment on the structure and emulsifying properties of peanut protein isolate. *Food Bioprod. Process.* **2014**, *92*, 30–37. [CrossRef]
22. Mahdavian Mehr, H.; Koocheki, A. Effect of atmospheric cold plasma on structure, interfacial and emulsifying properties of Grass pea (*Lathyrus sativus* L.) protein isolate. *Food Hydrocoll.* **2020**, *106*, 105899. [CrossRef]
23. Desmet, T.; Morent, R.; De Geyter, N.; Leys, C.; Schacht, E.; Dubruel, P. Nonthermal plasma technology as a versatile strategy for polymeric biomaterials surface modification: A review. *Biomacromolecules* **2009**, *10*, 2351–2378. [CrossRef] [PubMed]
24. Ma, G.; Tang, C.; Sun, X.; Zhang, J. The interaction mechanism of β-casein with oligomeric proanthocyanidins and its effect on proanthocyanidin bioaccessibility. *Food Hydrocoll.* **2021**, *113*, 106485. [CrossRef]
25. Han, Y.-X.; Cheng, J.-H.; Sun, D.-W. Changes in activity, structure and morphology of horseradish peroxidase induced by cold plasma. *Food Chem.* **2019**, *301*, 125240. [CrossRef]
26. Dong, S.; Wang, J.-m.; Cheng, L.-m.; Lu, Y.-l.; Li, S.-h.; Chen, Y. Behavior of Zein in Aqueous Ethanol under Atmospheric Pressure Cold Plasma Treatment. *J. Agric. Food Chem.* **2017**, *65*, 7352–7360. [CrossRef]
27. Yu, J.-J.; Zhang, Z.-Y.; Lin, X.-N.; Ji, Y.-Q.; Zhang, R.-R.; Ji, H.; Chen, Y. Changes in the structure and hydration properties of high-temperature peanut protein induced by cold plasma oxidation. *Int. J. Biol. Macromol.* **2023**, *253*, 127500. [CrossRef]
28. Dong, S.; Gao, A.; Zhao, Y.; Li, Y.-t.; Chen, Y. Characterization of physicochemical and structural properties of atmospheric cold plasma (ACP) modified zein. *Food Bioprod. Process.* **2017**, *106*, 65–74. [CrossRef]
29. Yang, T.; Zhou, D.-D.; Wu, C.-E.; Li, T.-T.; Fan, G.-J.; Li, X.-J.; Cong, K.-P.; Yan, Z.-C.; Cheng, X. Structural characterization of modified soy protein isolate composite coatings and its application on fresh-cut cantaloupe (*Cucumis melo* cv. Xiaomi). *Food Biosci.* **2024**, *60*, 104361. [CrossRef]
30. De Geyter, N.; Morent, R.; Desmet, T.; Trentesaux, M.; Gengembre, L.; Dubruel, P.; Leys, C.; Payen, E. Plasma modification of polylactic acid in a medium pressure DBD. *Surf. Coat. Technol.* **2010**, *204*, 3272–3279. [CrossRef]
31. Barnett, G.V.; Qi, W.; Amin, S.; Neil Lewis, E.; Roberts, C.J. Aggregate structure, morphology and the effect of aggregation mechanisms on viscosity at elevated protein concentrations. *Biophys. Chem.* **2015**, *207*, 21–29. [CrossRef] [PubMed]
32. Yang, R.; Liu, Y.; Meng, D.; Wang, D.; Blanchard, C.L.; Zhou, Z. Effect of atmospheric cold plasma on structure, activity, and reversible assembly of the phytoferritin. *Food Chem.* **2018**, *264*, 41–48. [CrossRef]
33. Zhang, Y.; Li, Y.; Guo, J.; Feng, Y.; Xie, Q.; Guo, M.; Yin, J.; Liu, G. Effect of two-stage low-temperature tempering process assisted by electrostatic field application on physicochemical and structural properties of myofibrillar protein in frozen longissimus dorsi of tan mutton. *Food Chem.* **2024**, *456*, 140001. [CrossRef]
34. Shelar, A.; Singh, A.V.; Dietrich, P.; Maharjan, R.S.; Thissen, A.; Didwal, P.N.; Shinde, M.; Laux, P.; Luch, A.; Mathe, V.; et al. Emerging cold plasma treatment and machine learning prospects for seed priming: A step towards sustainable food production. *RSC Adv.* **2022**, *12*, 10467–10488. [CrossRef]
35. Ren, C.S.; Wang, K.; Nie, Q.Y.; Wang, D.Z.; Guo, S.H. Surface modification of PE film by DBD plasma in air. *Appl. Surf. Sci.* **2008**, *255*, 3421–3425. [CrossRef]

36. Li, Y.; Cheng, Y.; Zhang, Z.; Wang, Y.; Mintah, B.K.; Dabbour, M.; Jiang, H.; He, R.; Ma, H. Modification of rapeseed protein by ultrasound-assisted pH shift treatment: Ultrasonic mode and frequency screening, changes in protein solubility and structural characteristics. *Ultrason. Sonochem.* **2020**, *69*, 105240. [CrossRef] [PubMed]
37. Yin, Y.; Yang, X.; Li, L. Noncovalent interaction between proanthocyanidins and soy protein isolate fibers: Structure, functionality and interaction mechanism. *Food Hydrocoll.* **2025**, *160*, 110663. [CrossRef]
38. Privalov, P.L.; Khechinashvili, N.N. A thermodynamic approach to the problem of stabilization of globular protein structure: A calorimetric study. *J. Mol. Biol.* **1974**, *86*, 665–684. [CrossRef]
39. Zhang, Y.; Wu, C.; Shen, X.; McClements, D.J.; Liu, X.; Liu, F. Effects of combined hot alkaline and pH-shift treatments on structure and functionality of legume protein-EGCG conjugates: Soybean-, pea-, and chickpea protein-EGCG systems. *Food Hydrocoll.* **2025**, *158*, 110424. [CrossRef]
40. Bao, Q.; Yan, J.; Ma, S. Effect of heat treatment on conformation and aggregation properties of wheat bran dietary fiber-gluten protein. *Int. J. Biol. Macromol.* **2023**, *253*, 127164. [CrossRef]
41. Chang, C.; Tu, S.; Ghosh, S.; Nickerson, M. Effect of pH on the inter-relationships between the physicochemical, interfacial and emulsifying properties for pea, soy, lentil and canola protein isolates. *Food Res. Int.* **2015**, *77*, 360–367. [CrossRef]
42. Li, C.; Li, W.; Zhang, X.; Wang, G.; Liu, X.; Wang, Y.; Sun, L. The changed structures of *Cyperus esculentus* protein decide its modified physicochemical characters: Effects of ball-milling, high pressure homogenization and cold plasma treatments on structural and functional properties of the protein. *Food Chem.* **2024**, *430*, 137042. [CrossRef] [PubMed]
43. Li, S.; Liu, Z.; Hei, X.; Wu, C.; Ma, X.; Hu, H.; Jiao, B.; Zhu, J.; Adhikari, B.; Wang, Q.; et al. Effect of Physical Modifications on Physicochemical and Functional Properties of Walnut Protein. *Foods* **2023**, *12*, 3709. [CrossRef]
44. Oliete, B.; Potin, F.; Cases, E.; Saurel, R. Modulation of the emulsifying properties of pea globulin soluble aggregates by dynamic high-pressure fluidization. *Innov. Food Sci. Emerg. Technol.* **2018**, *47*, 292–300. [CrossRef]
45. Misra, N.N.; Kaur, S.; Tiwari, B.K.; Kaur, A.; Singh, N.; Cullen, P.J. Atmospheric pressure cold plasma (ACP) treatment of wheat flour. *Food Hydrocoll.* **2015**, *44*, 115–121. [CrossRef]
46. Pankaj, S.K.; Bueno-Ferrer, C.; Misra, N.N.; Bourke, P.; Cullen, P.J. Zein film: Effects of dielectric barrier discharge atmospheric cold plasma. *J. Appl. Polym. Sci.* **2014**, *131*, 40803. [CrossRef]
47. Julakanti, S.; Charles, A.P.R.; Zhao, J.; Bullock, F.; Syed, R.; Myles, Y.; Wu, Y. Hempseed protein (*Cannabis sativa* L.): Influence of extraction pH and ball milling on physicochemical and functional properties. *Food Hydrocoll.* **2023**, *143*, 108835. [CrossRef]
48. Nawaz, M.A.; Buckow, R.; Jegasothy, H.; Stockmann, R. Enzymatic hydrolysis improves the stability of UHT treated faba bean protein emulsions. *Food Bioprod. Process.* **2022**, *132*, 200–210. [CrossRef]
49. Zhang, R.-Y.; Wang, Y.; Jiang, Y.; Min, E.-H.; Rao, S.-Q. Effects of dual succinylation and ultrasonication modification on the structural and functional properties of ovalbumin. *Food Res. Int.* **2023**, *165*, 112511. [CrossRef]
50. Yu, J.; Sun, B.; Zhang, S.; Liu, X.; Xie, P. The Effect of Different Induction Methods on the Structure and Physicochemical Properties of Glycosylated Soybean Isolate Gels. *Foods* **2022**, *11*, 3595. [CrossRef]
51. Zhu, Y.; Fu, S.; Wu, C.; Qi, B.; Teng, F.; Wang, Z.; Li, Y.; Jiang, L. The investigation of protein flexibility of various soybean cultivars in relation to physicochemical and conformational properties. *Food Hydrocoll.* **2020**, *103*, 105709. [CrossRef]
52. Chizoba Ekezie, F.-G.; Cheng, J.-H.; Sun, D.-W. Effects of Mild Oxidative and Structural Modifications Induced by Argon Plasma on Physicochemical Properties of Actomyosin from King Prawn (*Litopenaeus vannamei*). *J. Agric. Food. Chem.* **2018**, *66*, 13285–13294. [CrossRef] [PubMed]

Disclaimer/Publisher's Note: The statements, opinions and data contained in all publications are solely those of the individual author(s) and contributor(s) and not of MDPI and/or the editor(s). MDPI and/or the editor(s) disclaim responsibility for any injury to people or property resulting from any ideas, methods, instructions or products referred to in the content.

foods

MDPI

Article

Exploration of Pea Protein Isolate–Sodium Alginate Complexes as a Novel Strategy to Substitute Sugar in Plant Cream: Synergistic Interactions Between the Two at the Interface

Jingru Sun [1], Xiyuan Yang [1], Jingjing Diao [1,2], Yichang Wang [3] and Changyuan Wang [1,2,*]

[1] College of Food, Heilongjiang Bayi Agricultural University, Xinfeng Road 5, Daqing 163319, China; s15545723965@163.com (J.S.); 13555517359@163.com (X.Y.); diaojing62@163.com (J.D.)
[2] National Coarse Cereals Engineering Research Center, Daqing 163319, China
[3] College of Food Science, Northeast Agricultural University, Harbin 150030, China; 13351999556@163.com
[*] Correspondence: byndwcy@163.com

Abstract: This study aims to explore the feasibility of using pea protein isolate (PPI)/sodium alginate (SA) complex as a sugar substitute to develop low sugar plant fat cream. Firstly, this study analyzed the influence of SA on the structure and physicochemical properties of PPI and evaluated the types of interaction forces between PPI and SA. The addition of SA effectively induces the unfolding and structural rearrangement of PPI, causing structural changes and subunit dissociation of PPI, resulting in the exposure of internal-SH groups. In addition, the addition of SA increased the content of β-folding in PPI, making the structure of PPI more flexible and reducing interfacial tension. The ITC results indicate that the binding between PPI and SA exhibits characteristics of rapid binding and slow dissociation, which is spontaneous and accompanied by heat release. Next, the effect of PPI/SA ratio on the whipping performance and quality of low sugar plant fat creams was studied by using PPI/SA complex instead of 20% sugar in the cream. When using a PPI/SA complex with a mass ratio of 1:0.3 instead of sugar, the stirring performance, texture, and stability of plant fat cream reach their optimum. Finally, the relevant analysis results indicate that the flexibility and interface characteristics of PPI are key factors affecting the quality of cream. This study can provide theoretical support for finding suitable sugar substitute products and developing low sugar plant fat cream.

Keywords: pea protein isolate; sodium alginate; whipped cream; interface characteristics; synergistic interactions

1. Introduction

Plant fat cream is a product with complex oil–water emulsion structure formed by mixing, homogenizing, cooling, and stirring plant fat and sugar as raw materials, adding protein, emulsifier, hydrocolloid, water, and other components [1]. Plant fat cream has the characteristics of low cost, good stability, and easy processing. It is very well known in food processing fields such as baking and catering. It is widely used in the production of various processed foods, such as ice cream, cakes, coffee, etc. Traditional fat cream contains 25–30 wt% of sugar compounds. Its high sugar content and high glycemic index will increase the risk of many health problems, such as obesity or type II diabetes [2]. Simply reducing the sugar compounds in fresh cream is the most direct and simple way to reduce health risks. However, simply reducing the amount of sugar will have adverse effects on the quality and acceptability of its fresh cream due to the decrease in soluble solids content

https://doi.org/10.3390/foods14060991

and viscosity [3]. Cream will exhibit extremely low hardness and stability. At this point, many formulas have been developed to replace sugar in the production of healthier and higher quality low sugar plant cream [4].

Protein–polysaccharide complexes can serve as excellent carbohydrates substitutes, as their sweet taste characteristics and low impact on blood sugar levels can simultaneously meet consumer taste expectations and better control blood sugar levels [5]. In addition, the protein–polysaccharide complex, due to its high stability and solubility, can be uniformly dispersed in plant cream, ensuring the consistency and stability of the product. Therefore, protein–polysaccharide complexes have enormous potential for sugar substitution in plant cream. The recent study proved that whey protein [5] and phycocyanin [6] can be used as sugar substitutes in low sugar creams. However, compared with sugar substitutes such as whey protein, PPI has been widely used in food and other fields due to its high nutritional value and low allergenicity [7]. However, the fresh cream prepared from PPI will flocculate or coalesce when the emulsion is stored, mainly due to the collapse of partially coalesced fat spheres and foam structure formed by phase separation. This may be because the O/W interface in the cream structure is easy to break, which promotes the fat spheres to gather together, leading to the instability of emulsion [1]. Therefore, comprehensive control of interface performance is crucial for preparing higher quality fresh cream. Sodium alginate (SA) is a linear anionic copolymer widely used as a food emulsifier. In a mixed system, SA may undergo electrostatic interactions with proteins, resulting in the formation of new structures and textures [8]. Complex structures undergo significant structural unfolding at the oil–water interface [9]. Ultimately, a viscous and viscoelastic interfacial film may be generated to impede the coalescence of oil droplets [10]. Therefore, SA is widely used in protein emulsion food to regulate its structure, function, and stability. In this research, we used the PPI/SA binary complex as a sugar substitute in the cream formulation.

At present, many polysaccharides such as xanthan gum, pectin, locust bean gum, carboxymethyl cellulose, and kappa carrageenan have been used to couple with proteins to stabilize cream [11–13]. The addition of these polysaccharides not only further enhances the functional properties of the cream but also improves its sensory attributes. But most research focuses mainly on the evaluation of cream applications. There have been no reports yet on the impact of protein polysaccharide complexes on the properties of fresh cream from the perspective of protein structural differences, interactions between components, and the correlation between O/W interface characteristics. Therefore, it is of great significance to explore the interactions, structural differences, and interfacial behavior mechanisms among the multi-components of PPI/SA, as well as the structure–activity relationship of the macroscopic properties of plant cream. This study chose PPI and SA as research models, established the relationship between their interactions, complex properties, and interfacial behavior, and further investigated the effect of protein–polysaccharide complexes on the quality of plant cream. This provides theoretical support for finding suitable sugar substitute products to develop low sugar plant fat cream and provides new ideas for the production and design of low sugar plant cream, which hold significant practical implications.

2. Materials and Methods

2.1. Materials

Pea protein (with a protein content of 81.2%) was purchased from Shandong Yuwang Ecological Food Industry Co., Ltd. (Shandong, China). Sodium alginate (purity > 98%), sodium stearate lactate (purity 99%), xanthan gum (purity 99%), and D-sorbitol (purity 99%) were purchased from Beijing Yuanye Biotechnology Co., Ltd. (Beijing, China). We purchased hydrogenated palm oil, white sugar, and sucrose syrup from a supermarket in Harbin.

2.2. Preparation of PPI/SA Complexes

Based on the method proposed by Wang et al. [14], we prepared the PPI/SA complexes. Appropriate amounts of PPI and SA powders were dissolved in phosphate-buffered solution (PBS, 0.01 M, pH 7.0) and stirred for 2 h to prepare PPI and SA solutions, with concentrations of 5% (w/v) and 2% (w/v) in the PPI and SA solutions, respectively. PPI and SA solutions were mixed in a certain proportion at room temperature to prepare PPI/SA complexes with different mass ratios (1:0, 1:0.1, 1:0.2, 1:0.3, 1:0.4, and 1:0.5). The concentration of PPI in the complex solution is fixed at 1% (w/v). The pH of all samples was adjusted to 7 using a solution of 0.1 M HCl/NaOH. Table 1 shows the specific preparation formula of PPI/SA complexes.

Table 1. The Specific preparation formula of PPI/SA complexes.

PPI/SA Mass Ratios	5% PPI/ml	2% SA/mL	PBS/mL
1:0	20	0	80
1:0.1	20	5	75
1:0.2	20	10	70
1:0.3	20	15	65
1:0.4	20	20	60
1:0.5	20	25	55

2.3. Characterization of PPI/SA Complexes

2.3.1. Fluorescence Spectroscopy

All samples were diluted to 0.1 mg/mL with PBS and detected by a fluorescence spectrophotometer (F-4500, Hitachi, Japan) at wavelength. The measured parameters included an excitation wavelength of 280 nm and an emission wavelength range of 300–400 nm.

2.3.2. Zeta Potential

We measured the zeta potential of the sample solution using the method by Zhu et al. [8]. Firstly, dilute all samples to 1 mg/mL. Subsequently, we used a 1 mL glass syringe to aspirate the diluted sample solution and slowly injected it into the potential cup until the solution volume reached 2/3 of the potential cup scale. Each sample is scanned 3 times, and the average value is recorded.

2.3.3. Measurement of Flexibility

Protein samples were diluted to 1 mg/mL using phosphate-buffered solution (PBS). Subsequently, 250 μL of trypsin (1 mg/mL) solution was added to 4 mL of the sample solution, and the solution was placed in a water bath at 38 °C for 10 min. Enzymatic hydrolysis was then stopped with 4 mL of trichloroacetic acid (50 mg/mL). Then, the whole solution was spun in a centrifuge at 2500× g for 30 min. The amount of digested protein in the clear liquid on top was found out using the BCA method.

2.3.4. Determination of SH Content

Measure the content of free thiols and total thiols according to the measurement method of Hu et al. [15]. Measure its absorbance at 412 nm. Calculate the content of thiol and disulfide bonds according to the following formulas:

$$SH(\mu mol/g) = 73.53 \times D \times A_{412}$$

$$SS(\mu\text{mol/g}) = \frac{SH_1 - SH_2}{2}$$

Among them, D represents the dilution factor, and A_{412} represents the absorbance measured at a wavelength of 412 nm.

2.3.5. Circular Dichroism (CD) Spectra

The CD spectra were obtained using a CD spectropolarimeter (JASCO J-810, RI, USA). The wavelength range was set to 190–250 nm for each solution.

2.3.6. Isothermal Titration Calorimetry (ITC)

The ITC experiment was conducted at room temperature using an isothermal titrator (MicroCal Inc., Northampton, UK). All samples should be degassed before the experiment begins [16]. The injection volume was 2 μL, with a total of 25 times and an interval of 350 s. The stirring speed was 250 rpm. A series of thermodynamic parameters were analyzed using NanoAnalyze V01.02.01.01 software.

2.3.7. Interfacial Tension

The interfacial tension analyzer (Sigma 700, Biolin, Sweden) was used to measure interfacial tension. Corn oil was used instead of the fat blend and purified before measuring. If the interfacial tension remained at 29.0 ± 0.5 mN/m for 10,800 s, the corn oil was considered acceptable [17]. The complex solutions of different mass ratios were added into a micro needle sampler, then the needle was inserted into an optical glass colorimetric dish, and 25 μL of the sample was pushed onto the needle tip to form a complete droplet. The CCD camera system captures droplet images at 25 °C for 10,800 s, and the software automatically calculates the interfacial tension value.

2.3.8. Interfacial Protein Concentrations

Interfacial protein concentrations (Γ) were ascertained using the previously described method of Yan et al. [18]. A 15 mL sample was transferred into a 50 mL centrifuge tube and subjected to centrifugation at 4 °C and 8000× g for 60 min. Subsequently, the aqueous phase and the protein precipitate were gathered. The Kjeldahl method was employed to analyze the protein content. Based on the results, the interfacial protein concentrations could be calculated using the following equation:

$$\Gamma\left(\text{mg/m}^2\right) = \frac{V_c \times (C_i - C_{eq})}{SSA \times V_{oil}}$$

where C_i represents the initial protein concentration (mg/mL), C_{eq} represents the concentration of non-adsorbed proteins (mg/mL), V_c and V_{oil} represents the volumes of the aqueous and oil phases (mL), respectively, and SSA represents the surface area of the droplets (m^2/g).

2.4. Preparation and Characterization of Whipped Creams

2.4.1. Preparation of Whipped Creams

PPI/SA complex solutions with different mass ratios were created according to the method in 2.2, and then xanthan gum, D-sorbitol, white sugar, and sucrose syrup were added and completely dissolved at 75 °C to obtain the aqueous phase. Hydrogenated palm oil and sodium stearyl lactate (emulsifier) were melted and mixed evenly at a temperature of 75 °C to obtain the oil phase. At 60 °C., the water phase was slowly added to the oil phase while stirring. After mixing the two phases, they were heated at 75 °C for 30 min to ensure thorough and even mixing. The mixture was homogenized at a speed of 8000 rpm for 2 min, then frozen at −18 °C for 12 h, thawed to 4 °C, and stirred until beaten to obtain plant cream. The PPI/SA mass ratio in the prepared plant fat cream is 1:0, 1:0.1, 1:0.2, 1:0.3, 1:0.4, and 1:0.5, with whole sugar plant fat cream as the control. Table 2 shows the formula for plant fat cream.

Table 2. The formula for plant fat cream.

Component	Content
D-sorbitol	0.95%
White sugar	7.51%
Sucrose syrup	15.02%
Xanthan gum	0.81%
Hydrogenated palm oil	25%
Sodium stearoyl lactylate	1.08%
PPI/SA complexes	Add according to experimental requirements
Water	Supplement to 100%

2.4.2. Determination of Hardness of the Whipped Creams

The hardness of the whipped cream was assessed using texture analyzer (Stable Micro System Co., London, UK). The probe was immersed into the sample at a speed of 3 mm/s to a depth of 15 mm, and the hardness of the whipped cream was determined by the required force (N) [19].

2.4.3. Determination of Overrun of the Whipped Creams

Whipped cream was filled into a 25 mL aluminum dish with dimensions of 40 mm in diameter and 22 mm in depth until reaching full capacity. The weight of the samples was recorded, and the foam overrun was determined using the provided equation. Foam overrun was defined as the ratio of gas weight to liquid weight, expressed as a percentage.

$$\textbf{Overrun}(\%) = \frac{m_1 - m_2}{m_2} \quad (1)$$

$\mathbf{m_1}$: the mass of unwhipped cream emulsion (g)
$\mathbf{m_2}$: the mass of whipped cream sample (g)

2.4.4. Measurement of Stirring Time and Beating Rate

A proper amount of cream emulsion was taken, and a cream mixer was used to fully stir the cream so that the cream can stand upright as the end of the mixing. The time from the beginning to the end was recorded as the stirring time. The beating rate is the ratio of the mass of the same volume of accumulated water to the mass of the same volume of cream product.

2.4.5. Separation Index of Cream Slurry

According to the method of Van Aken et al. [20] the mass of plant fat cream and the proportion of water droplets in the total mass were calculated to determine the separation coefficient of the cream slurry.

2.4.6. Observation of Stability of Mounting and Stability of Mounting Cutting Surface

The cream was placed in a laminating bag and squeezed into a flower shape through a nozzle to observe the change in flower shape at 0 h and 6 h. After 6 h, the cream was cut to observe the cross-section and photographed for recording.

2.4.7. Apparent Viscosity of the Whipped Creams

The dynamic shear rheometer (DV3THA, Brookfield Corp, USA) was utilized to measure the apparent viscosity of emulsions at 25 °C. The shear rate was systematically increased from 0.1 to 100 s^{-1} over a period of 100 s.

2.4.8. Frequency Sweeps of the Whipped Creams

Using a dynamic shear rheometer (DV3THA, Brookfield Corp, USA), the rheological properties of cream were measured at 25 °C. Frequency sweep testing (1.0–10 Hz) was performed at 0.1% strain to obtain the elastic modulus (G') and viscous modulus (G'').

2.4.9. Confocal Laser Scanning Microscopy (CLSM)

The cream microstructure was analyzed utilizing CLSM (Zeiss 780, Leica Corp., Germany). After staining, take 5 μL of the sample and place it on a glass slide for measurement and observation.

2.5. Statistical Analysis

Each sample was measured three times, and the results were presented in the form of $X \pm SD$. SPSS 19.0 software was used for data analysis, and ANOVA and Duncan tests were used for significant difference comparison. Using software such as Origin 2022b, Paekfit 4.12, Adobe Illustrator 2020 for data processing and chart production.

3. Results

3.1. Fluorescence Spectroscopy

Usually, the fluorescence spectrum of proteins is due to the presence of aromatic amino acids, whose positions largely depend on the degree of protein folding and can be used as a sensitive monitor for protein conformational changes [21]. In addition, the change in protein tertiary structure will affect the migration rate of protein to the oil–water interface and the ability to form an interfacial film, thus, affecting the quality of cream. Figure 1 shows the fluorescence spectra of PPI and PPI/SA complexes. The results indicate that a gradual decrease in peak fluorescence emission intensity of PPI at 333 nm is observed as the concentration of SA increases. This indicates that the interaction between sodium alginate and PPI occurs near the aromatic amino acid region, and the aromatic amino acid groups in the protein structure are buried by sodium alginate molecules. In addition, the addition of SA can cause a change in the spatial conformation of PPI [22]. It is generally believed that folded proteins exhibit higher fluorescence intensity (FI), and partial or complete unfolding of proteins can lead to a decrease in FI. The experimental results indicate that the addition of SA leads to the unfolding of PPI and alters the spatial configuration of amino acids within the protein and exposes previously shielded hydrophobic amino acids to a hydrophilic polar environment [23]. This unfolded structural feature will promote the reduction in interfacial tension, accelerate the migration rate of PPI from the water phase to the oil drop surface, and, thus, improve the ability of protein to form interfacial film.

3.2. Zeta Potential

The Zeta potential serves as a crucial indicator for assessing system stability. The Zeta potential values higher than ±30 mV suggest that the electrostatic repulsions between molecules are sufficiently strong [24]. Figure 2A shows the Zeta potential of PPI/SA complexes with different mass ratios. Compared to PPI, when SA is present in the solution system, the absolute value of the Zeta potential increases. As the concentration of SA increases, the absolute value of the Zeta potential of PPI exhibited an initial increase and subsequent decline. And the absolute value of the Zeta potential reaches the maximum value when the PPI/SA mass ratio is 1:0.3. Firstly, the increase in Zeta potential may be due to the exposure of hydrophilic polar groups. The functional groups on SA can undergo hydrogen bonding or hydrophobic interactions with proteins, leading to structural changes in proteins and promoting the exposure of hydrophilic groups. Secondly, as SA is a natural anionic polysaccharide, it can attach to the surface of PPI and enhance its negative charge. When the mass ratio of PPI/SA exceeds about 1:0.3, the positive charge on the PPI is saturated with SA, and any excess SA may promote the aggregation of the complex through a depletion mechanism. This may lead to excessive SA increasing the coverage of the protein surface and exerting a shielding effect, thereby reducing the absolute value of the negative Zeta potential.

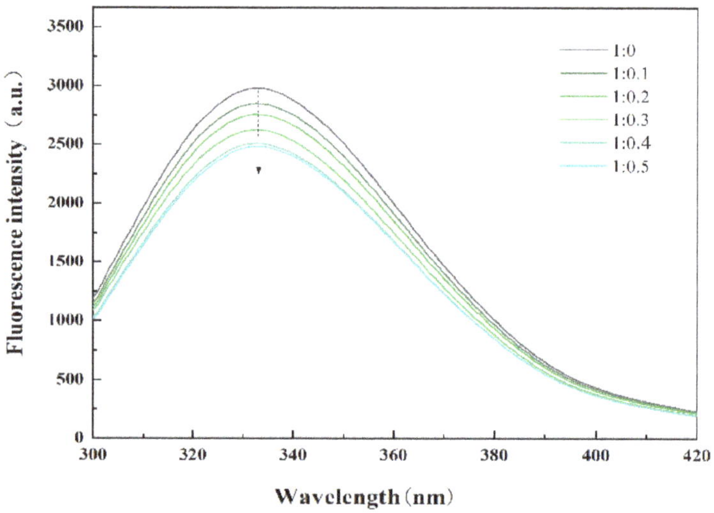

Figure 1. Fluorescence spectrum of PPI/SA complexes with different mass ratios.

3.3. *Protein Flexibility*

The flexibility of proteins influences the adsorption and rearrangement kinetics of proteins at the interface, consequently impacting the stability of protein-based creams at the interface [25]. Figure 2B shows the molecular flexibility of PPI/SA composites. As concentration of SA increases, the flexibility of proteins first increases and then decreases compared with natural PPI ($p < 0.05$). When the mass ratio of PPI/SA complex is 1:0.4, the highest molecular flexibility occurs. The increased flexibility of the complexes may be due to the interaction between PPI and SA, which leads to a certain degree of change in the distribution of amino acid residues in the main peptide chain of PPI. This promotes the dissociation of protein subunits and changes in secondary structure. The dissociation of subunits will gradually expand the conformation of the protein, exposing the non-polar residues previously buried in the protein, and the protein will transition from an ordered structure to a disordered structure [26]. These lead to an increase in the flexibility of the PPI structure. PPI with higher flexibility can quickly adsorb and rearrange at the oil–water interface, thereby reducing interfacial tension and enhancing emulsifying performance. When the mass ratio of PPI/SA complex is 1:0.5, the molecular flexibility of PPI decreases. This may be due to the excessive reaction between SA and PPI, which generates strong electrostatic repulsion and steric hindrance, inhibiting the expansion and extension of PPI molecules, thereby reducing their flexibility [27]. This phenomenon will result in a decrease in the diffusion rate of PPI towards the oil–water interface, hindering the adsorption of proteins at said interface.

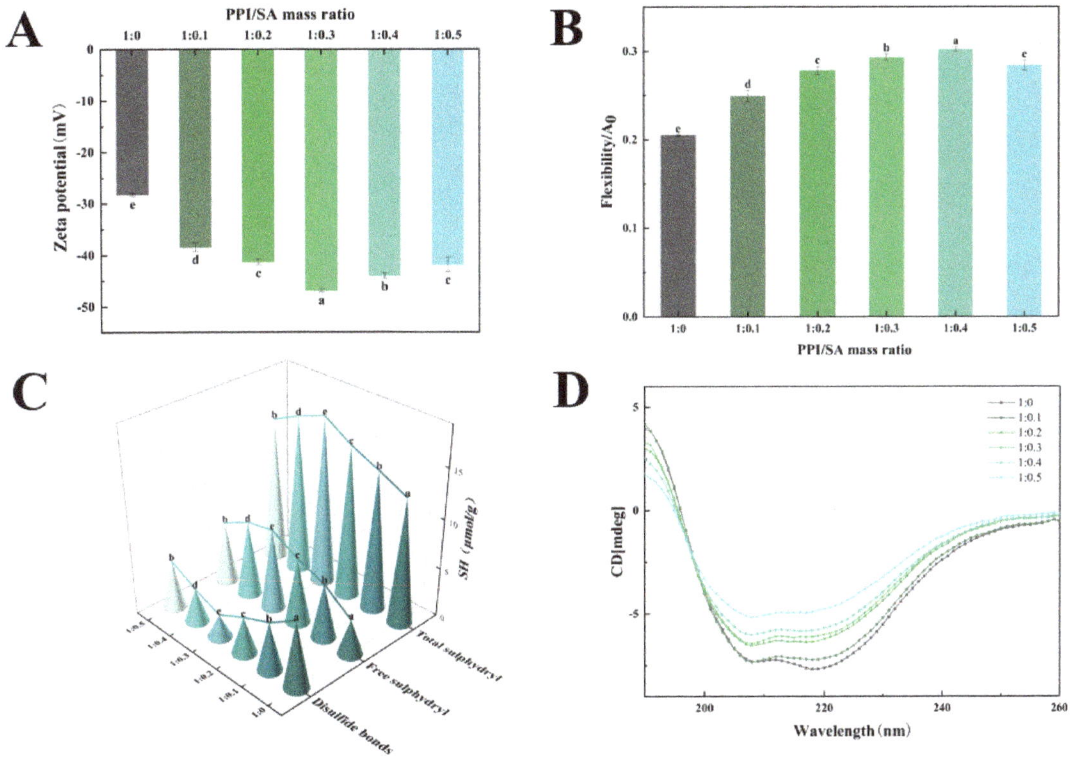

Figure 2. Physicochemical and structural characterization of PPI/SA complexes: zeta potential (**A**), protein flexibility (**B**), sulfhydryl and disulfide bond content (**C**), and CD spectra (**D**). Note: Different letters indicate significant differences ($p < 0.05$) between sample mean values, the same applies below.

3.4. Free Sulfhydryl Content

Sulfhydryl and disulfide bonds are important chemical bonds that maintain the spatial structure of proteins and endow them with certain functional properties, and their content is dynamically changing [28]. Figure 2C shows the sulfhydryl groups and disulfide bond contents of PPI/SA complexes with different mass ratios. The results indicate that as the concentration of SA increased, the total sulfhydryl groups content and free sulfhydryl groups content first increased and then decreased, while the disulfide bond content showed the opposite change. When the mass ratio of PPI/SA complex is 1:0.3, the total sulfhydryl groups and free sulfhydryl groups content reaches its peak, and the disulfide bond content reaches its lowest value, which is 27.84 μmol/g, 0.55 μmol/g and 6.74 μmol/g, respectively. The reason for this phenomenon may be due to the interaction between PPI and SA, which leads to changes in protein structure and subunit dissociation, promoting the exposure of -SH groups and the breaking of disulfide bonds in proteins [29]. However, when the mass ratio of PPI/SA complexes exceeds 1:0.3, the content of disulfide bonds increases, while the total sulfhydryl groups content and free sulfhydryl groups content decrease. This may be due to the formation of larger particle size aggregates in the presence of high concentrations of SA, which weakens the interaction between SA and PPI. This is not conducive to the complete unfolding of protein and the exposure of free sulfhydryl groups and leads to the transformation of free thiol groups into disulfide bonds [30].

3.5. Secondary Structure

The changes in the secondary structure of proteins can affect their interfacial properties to a certain extent, so CD spectroscopy is used to analyze changes in the secondary structure of PPI/SA [31]. Figure 2D shows the CD spectroscopy results of PPI/SA complexes. As the SA concentration gradually increases, the maximum negative average residual ellipticity of PPI/SA complexes also significantly increases. Meanwhile, within the wavelength range of 213 to 230 nm, the negative ellipticity of the shoulder peak also increases, indicating a change in the secondary structural conformation. Table 3 shows the secondary structure content of proteins. As the SA concentration gradually increases, the β-sheet content increased from 13.87% to 25.02%, while α-helix content decreased from 24.83% to 17.72%. At the same time, the content of random coil has decreased. The α-helix of protein is maintained by intra-chain hydrogen bonds. The decrease in the content of α-helix indicated that the interaction between PPI and SA led to the break of intra-chain hydrogen bonds and the formation of interchain hydrogen bonds [32]. This is also the reason for the increase in β-sheet content. This transformation indicates that during the unfolding process of PPI, the deeply masked intramolecular hydrogen bonds within the protein may be disrupted. A lower α-helix content and increased flexibility of a more disordered structure may accelerate the protein rearrangement at the oil–water interface, thus, improving the stability and functionality of the emulsion [33].

Table 3. The secondary structure content of PPI/SA complexes with different mass ratios.

PPI/SA Mass Ratios	α-Helix (%)	β-Sheet (%)	β-Turn (%)	Random Coil (%)
1:0	24.83 ± 0.06 ^{a}	13.87 ± 0.15 ^{e}	14.64 ± 0.05 ^{c}	44.05 ± 0.15 ^{c}
1:0.1	23.67 ± 0.07 ^{b}	15.61 ± 1.07 ^{d}	15.42 ± 0.23 ^{bc}	43.92 ± 1.76 ^{bc}
1:0.2	21.03 ± 0.53 ^{c}	17.94 ± 0.27 ^{c}	17.09 ± 0.81 ^{ab}	40.10 ± 0.35 ^{b}
1:0.3	20.36 ± 0.71 ^{c}	18.93 ± 0.80 ^{bc}	18.29 ± 1.76 ^{a}	40.97 ± 0.95 ^{b}
1:0.4	20.79 ± 1.14 ^{c}	19.92 ± 0.34 ^{b}	16.66 ± 0.44 ^{a}	39.60 ± 0.37 ^{a}
1:0.5	17.72 ± 0.21 ^{d}	25.02 ± 0.56 ^{a}	17.91 ± 0.89 ^{a}	38.19 ± 0.53 ^{a}

Note: Comparisons were carried out between values of the same column, values with different letters indicate a significant difference at $p \leq 0.05$.

3.6. ITC

Isothermal titration calorimetry (ITC) is a method of qualitatively characterizing protein/ligand interactions through thermodynamics. The main advantage of ITC is that it can sensitively measure the thermal changes caused by intermolecular interactions [34]. Figure 3A shows the thermal spectrum of the heat flow rate and time when titrating 40 mg/mL SA to PPI. Obviously, each peak corresponds to the thermal changes associated with the calorimeter unit when injecting SA solution equally into the PPI solution. The heat flow curve shows a negative heat flow change, indicating that the entire process is a combination of exothermic processes. This means that the binding between SA and PPI is beneficial for the reduction in system energy [34]. In addition, the heat flow will gradually decrease with the increase in SA titration amount and eventually remain stable, indicating that the sites on the surface of PPI gradually decrease with the binding of SA and eventually reach saturation. The experiment obtained thermodynamic binding parameters (Figure 3B) by fitting the titration curve using a unit point independent binding model, such as binding stoichiometry (N), binding constant (Ka), enthalpy change (ΔH), and entropy change (ΔS) [35]. The binding ration of the reaction was calculated to be 1.56 (± 12.8), indicating that each PPI molecule can bind approximately 1.56 SA monomer molecules, further indicating the formation of a complex between PPI and SA. The interaction between two molecules with a binding constant, Ka, greater than 104 M^{-1} is considered to have a high affinity and can undergo strong interactions. The Ka value measured in this experiment is $2.72 \pm 0.32 \times 103$ M^{-1}, indicating a non-specific interaction and relatively weak binding between PPI and SA. The values of TΔS, ΔH, and ΔG are 3.28, −0.218, and −3.5 kcal/mol, respectively, indicating that PPI/SA binding is a mainly entropy driven exothermic process. Therefore, current results indicate that SA mainly binds to the hydrophobic region of PPI through non-specific interactions, which is consistent with results of fluorescence spectroscopy [36]. Similar behaviors also occur in the interaction between seaweed lipid biosurfactants and BSA [37].

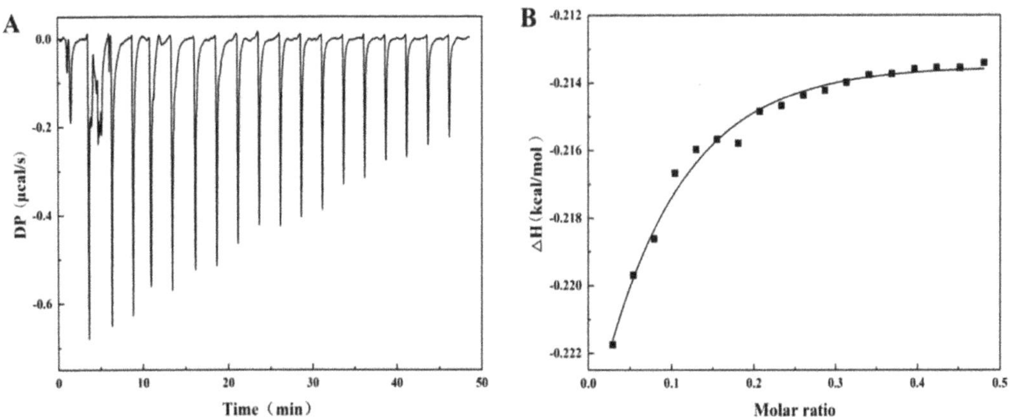

Figure 3. The curve of heat flow vs. time (**A**) and corresponding binding isotherms (**B**) during the titration of PPI (0.29 mM) with SA (0.02 mM) in phosphate-buffer solution (10 mM, pH 7.0).

3.7. Adsorption Kinetics of Protein/Emulsifier Mixture

Whipped cream is a typical O/W cloudy system before whipped, so the interfacial tension of solution is crucial to the stability of cream [38]. In order to investigate the effect of the adsorption kinetics of proteins and polysaccharides at the oil–water interface on the stability of aerated cream, we measured the interfacial tension changes in PPI/SA complexes at different mass ratios. As shown in Figure 4A, the characteristic of interfacial tension change was a rapid decrease at first, followed by a slowdown in the rate of reduction over time and eventually reaching equilibrium. This is related to the different stages of PPI adsorption (diffusion, penetration, unfolding, and rearrangement) [39]. Driven by external energy, PPI can quickly adsorb to the oil-water interface, decreasing thermodynamic unfavorable contact between different types of molecules at the interface and reconfiguring themselves to reduce interfacial tension and forming a protective interfacial film through intermolecular interactions [40]. Throughout the entire time frame, the interfacial tension of PPI and SA composite is lower than that of PPI alone. These findings indicate that SA may coexist with proteins at the oil–water interface in a cooperative manner to promote a decrease in interfacial tension. The presence of sodium alginate can greatly affect interfacial tension through a competitive adsorption with proteins and/or polysaccharide–protein complexation effects. Moreover, polysaccharides can access the protein stabilized interface via small gaps in the entangled protein layer, leading to decreased interfacial tension [41]. However, when the mass ratio of PPI to SA exceeds 1:0.3, the interfacial tension of the complex increases. This may be due to the excessive accumulation of polysaccharides on the hydrophobic side chains of proteins after their binding, which limits the adsorption and rearrangement of proteins at the oil–water interface. This has a negative impact on the diffusion of PPI/SA complex from bulk phase to interface. Secondly, it may be that the protein on the oil–water interface film will be replaced by too much SA, which will lead to instability of emulsion and cream [42].

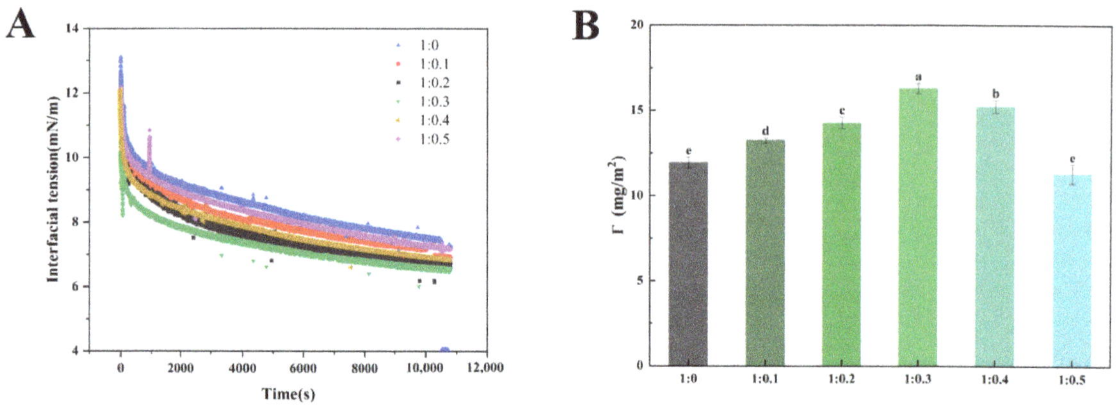

Figure 4. Interfacial tension (**A**) and protein concentration (**B**) at different mass ratios of PPI/SA (**B**). The letters indicate significant differences ($p < 0.05$).

3.8. Interfacial Protein Concentration

Interfacial protein concentration refers to the protein content per unit area, which is the reflection of protein adsorption at the O/W interface in the emulsion system [43]. Figure 4B shows that the interface protein concentration varies among different samples. As the concentration of SA increases, the interface protein concentration shows a trend of first increasing and then decreasing. When the mass ratio of PPI/SA complex is 1:0.3, the interface protein concentration reaches its peak. The underlying cause of the above—mentioned phenomenon lies in the fact that the hydrophobic groups of the protein possess the ability to break free from the poor solvent. As a result, they tend to spontaneously adsorb at the oil—water interface. The significant increase in interface protein concentration may be partially attributed to the increase in spontaneous protein adsorption. In addition, when PPI and SA coexist, the interfacial synergistic/competitive adsorption behavior usually occurs on the interface membrane. When the amount of SA added is low, PPI dominates at the interface, and SA can adsorb on the surface and gaps of proteins. This leads to an increase in thermodynamic incompatibility: the adsorbed PPI is concentrated by polysaccharides, thereby accelerating the diffusion and adsorption of PPI at the interface. Generally, a higher quantity of protein present at the oil—water interface was more conducive to the formation of a thicker interfacial protein layer that could enwrap the oil droplets to achieve the emulsion stability [44]. As the concentration of SA gradually increases, excessive SA disrupts the protein–protein interactions at the oil–water interface. This leads to PPI being replaced and desorbed by SA from the surface of fat globules to the continuous phase [45]. The interface film with polysaccharides or aggregates as the main body was slowly formed, leading to a decrease in the concentration of interface proteins. The decrease in interfacial protein concentration will reduce the strength of lotion interfacial film, which promotes the formation of partial coalescence structure of fat, leading to the decline of cream quality [46]. The above research results confirm the importance of different SA addition amounts in enhancing the affinity of PPI to the oil–water interface, thereby helping to improve the stability of cream.

3.9. Texture Analysis of Cream

Hardness refers to the maximum pressure applied to cream to initiate deformation and plays a crucial role in determining the processing characteristics and edible quality of cream products [47]. It can be utilized to assess the capacity of plant-based cream to retain foam structure and stability after whipped, thereby influencing the ultimate acceptance of the cream. The hardness of cream with different mass ratios of PPI/SA complexes replacing sugars and whole sugar cream is shown in Figure 5A. As the concentration of SA increases, the hardness value of cream first increases and then decreases. Among them, the hardness value of the 1:0.3 sample is the highest (303.40 $\pm$ 0.88). This might be due to the fact that SA makes PPI stretch and reveals some hydrophobic groups. Consequently, its binding ability to fat globules is enhanced. This will promote the movement of fat globules towards the surface of bubbles and further aggregate to create a dense network structure, thereby improving the anti-deformation ability of the cream [48]. On the other hand, PPI and SA will be adsorbed on the surface of the fat sphere to form a hard oil–water interface film. The PPI/SA complex helps to form a thicker facial mask interface, thus, improving the hardness of whipped cream. From the results, it can be seen that the overall improvement of SA on the hardness of fresh cream indicates that SA is effective in changing the quality of fresh cream [49]. When the proportion of SA continues to increase, the hardness value of cream significantly decreases ($p < 0.05$), which may be due to the excessive aggregation of large protein aggregates induced by excessive SA during the stirring process, rather than partial aggregation. The strength of the oil–water interfacial film structure formed by excessive coalescence is weaker than that of the oil–water interfacial film structure formed by partial coalescence, so the network structure formed is looser and has lower hardness.

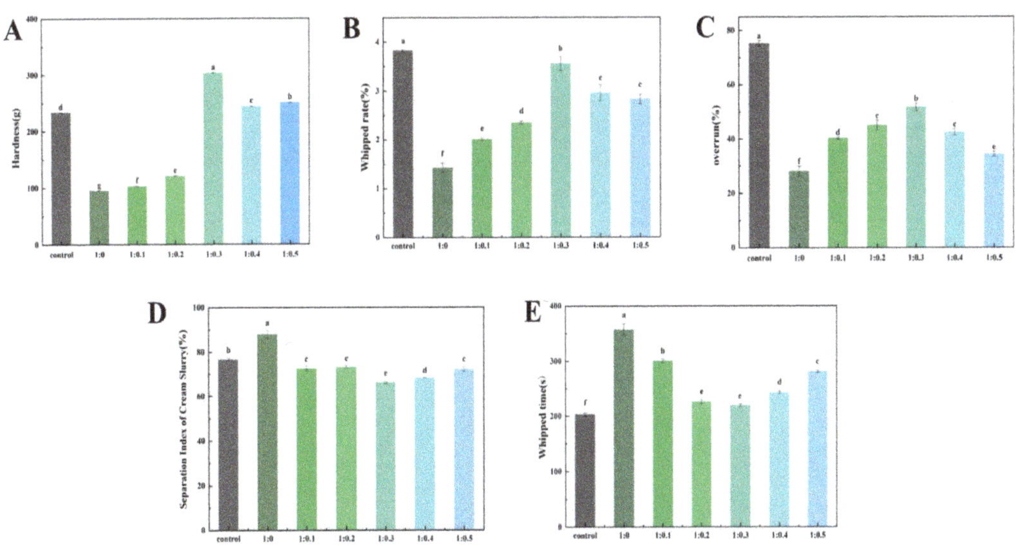

Figure 5. The effect of PPI/SA complex mass ratio on the hardness (**A**), whipped rate (**B**), overrun (**C**), separation index (**D**), and whipped time of whipped cream (**E**). The letters indicate significant differences ($p < 0.05$).

3.10. Analysis of Cream Whipped Rate

The whipped rate refers to the ability of cream to wrap air during whipping, and its size is an important performance index reflecting the quality of whipped cream. The cream with higher whipped rate can obtain good foam hardness, luster, fineness, and instant melting in the mouth [50]. The effect of the PPI/SA complex mass ratio on the whipped rate of whipped plant fat cream is shown in Figure 5B. As the addition of SA increases, the whipped rate first increases and then decreases. When the PPI/SA is 1:0.3, the whipped rate is closest to the control sample. This may be due to the fact that as the concentration of SA increases, the continuous network formed in the continuous phase can effectively suppress the collision frequency of fat globules, resulting in a gradual increase in whipped rate. When the amount of SA added is too high, the beating rate decreases. The main reason is that the increase in SA addition leads to an increase in the apparent viscosity of the system, making it difficult to quickly fill air during the stirring process and resulting in an increase in bubble rupture rate. Secondly, when the amount of SA added is too high, the concentration of liquid phase protein decreases, and the volume of air filled in the first stage of stirring decreases. The lower part of the aggregation rate is not conducive to wrapping bubbles and forming fat network structures [51].

3.11. Analysis of Cream Overrun

Overrun rate is an indicator that provides information on the gas content or gas percentage in fresh cream. It is an important parameter of foam food, because it affects its texture, taste, stability, and overall quality. Cream with a higher overrun rate typically has a lighter and more fluffy texture [52]. As shown in Figure 5C, as the concentration of SA increases, the overrun rate of the sample first increases and then decreases. In the whipped cream system, the partial coalescence of fat globules in the oil phase contributes to the formation of a network structure of fat crystals. Meanwhile, a high oil phase mass fraction promotes the formation of a network structure of fat crystals, thereby accelerating the partial coalescence of fat globules [53]. In contrast, the overrun rate of protein samples is only 28.15%. This is because the coalescence speed of fat globules is very slow, and they cannot quickly form a mesh structure to wrap around enough air. After adding SA, the overrun rate significantly increased, which may be because SA can interact with PPI to form aggregates. The cream network structure is formed through aggregation, and the aggregated fat ball can have a stronger skeleton, which makes the foam more concentrated. However, as the concentration of SA increases, the degree of protein aggregation becomes too high, resulting in uneven protein distribution and a lack of continuous three-dimensional network structure [54]. In addition, its rapidly forming network structure resulted in the encapsulation of less air. Nooshkam et al. [5] confirmed that the ternary compound bioactive foam can partially replace the fat and sugar in ice cream, while maintaining a high overrun. These results show that an appropriate amount of SA can rapidly stabilize on the surface of AW to promote the formation of foam. The synergistic effect of PPI and SA promotes the aggregation/coalescence of fat globules, encapsulating sufficient air.

3.12. Analysis of Separation Index of Cream Slurry

The slurry separation index of cream is a parameter that describes the degree of slurry separation in liquid mixtures (cream) [20]. The degree of separation can affect the quality, taste, and appearance of cream. For cream, a lower slurry separation coefficient is usually ideal. Because cream with a lower slurry separation coefficient can maintain a uniform texture and is less prone to significant phase separation and structural changes. The effect of the PPI/SA complex mass ratio on the separation index of whipped plant fat cream slurry is shown in Figure 5D. Cream prepared from pure protein samples quickly leaks out of the sieve, resulting in a high separation index of the slurry. As the amount of SA increases, the slurry separation index of the sample shows a trend of first decreasing and then increasing. This may be due to the fact that with the addition of SA, the cream will soon enter the stable agglomeration stage after being stirred. At this time, the hardness and viscosity of the cream will increase, and the bubbles will start to form stable foam. The whipped cream at this stage can be considered as a system with very small and tightly arranged bubbles, and a complex is adsorbed on the surface of the bubbles, in a stable state. The results show that SA plays a key role in stabilizing bubbles, and PPI/SA complex can stabilize the microstructure of liquid foam in low sugar system. When the amount of SA added is too high, the separation coefficient of the slurry shows an upward trend. At this time, stirring will cause excessive agglomeration of fat globules, resulting in a rapid decrease in foaming rate during stirring [55]. The agglomerates also begin to form larger aggregates, leading to an increase in the separation coefficient of the slurry. This leads to the inability of plant fat creams to form a continuous network and effectively capture water, resulting in an increase in the slurry separation coefficient [56].

3.13. Analysis of Cream Whipped Time

The whipping time refers to the time required to whip the cream into a normal shape. A shorter whipped time indicates that the whipped cream system can be filled by air faster, thereby maintaining the stability of the system [57]. Therefore, whipped time is one of the important indicators for estimating the whipping performance of cream. The effect of the PPI/SA complex mass ratio on the whipped time of plant fat cream is shown in Figure 5E. The PPI sample requires about 6 min or even longer whipping time to reach the whipped endpoint. This is because the PPI structure is loose and cannot form a strong boundary film, and the system viscosity is small, which leads to easy leakage of bubbles and requires a long whipping time. Especially when only protein is present in the alternative system, the large fat globules formed by excessive aggregation will destroy the bubble structure, resulting in oil separation and defoaming phenomenon. As the concentration of SA increases, the whipping time of the sample first decreases and then increases, with the shortest whipping time occurring at 1:0.3. This is because increasing the proportion of SA appropriately can ensure sufficient gas entry and promote partial aggregation of fat globules, thereby quickly forming a three-dimensional network structure that wraps around the captured bubbles [58]. Secondly, the hydrophilic/hydrophobic structure of SA makes it easier to adsorb at the oil–water interface and reduces the surface tension of the liquid phase. An appropriate amount of SA can provide PPI/SA complexes with more flexible interface behavior, faster adsorption rate and higher rearrangement efficiency. This leads to an improved resistance to crystal puncture at the interface. Finally, SA is an effective surfactant that can replace proteins at the interface, leading to partial crystallization of lipids. When the amount of SA is low, proteins dominate the interfacial properties. However, due to its high interfacial activity, SA can replace proteins and reduce steric hindrance between droplets. This is beneficial for the flow of liquid fat, accelerating fat crystallization and shortening whipped time. As the SA content continues to increase, the whipped time begins to increase. This is mainly because excessive SA increases the viscosity of the system, making it difficult for bubbles to enter the cream system [59]. This makes it difficult for cream to wrap bubbles, making it easier for fat balls to come into contact with each other and aggregate, thereby increasing whipping time. In addition, when the concentration of SA is high enough to replace the main interface properties of protein, the viscoelasticity of the interface film will increase instead [60]. This will prevent the aggregation of fat to some extent, so the whipping time will be extended. Camacho et al. [61] also found that increasing the viscosity of cream will promote the rapid development of elastic properties of whipped products. That is to say, after adding less air, foam will be quickly separated from the mixer, reaching the air limit.

3.14. Observation and Analysis of the Appearance and Section of Mounted Flowers

Fresh cream is typically consumed within a brief time, making short-term storage stability a crucial characteristic of the product. The storage stability of cream produced from PPI/SA complexes with varying proportions can be assessed through observing alterations in their rose-shaped morphology and extent of collapse over different storage durations [62]. Figure 6A displays the visual appearance of cream at varying storage times (0 h, 6 h), while Figure 6B illustrates the roughness of the cream cross-section. It can be observed that the control sample has sharper edges, stronger peaks, clearer textures, and no surface bubbles. On the contrary, PPI samples cannot exhibit a rose-like structure due to their rapid collapse, resulting in a rough surface and the generation of many bubbles. After adding SA, the cream stabilized by PPI/SA complexes initially exhibited a rose-like appearance, with clear textures and firm peaks. With an increase in the storage time of cream, the storage stability of all cream samples decreases. This is due to the consumption of bubbles (drainage, coalescence, and disproportionation) leading to varying degrees of collapse. Compared to other samples, when the mass ratio of PPI/SA is 1:0.3, the degree of cream collapse is lower. It can be clearly seen from the figure that the peak height and overall appearance of the shape of the sample stored for 6 h have not changed much compared to the shape at 0 h. This indicates that a PPI/SA complex mass ratio of 1:0.3 is more conducive to improving the storage stability of cream in the cream system [63]. After adding SA, the appearance of the cream cross-section also showed a similar trend. The pore distribution inside the cream is more uniform, and the cross-section is smoother. Among them, when the mass ratio of PPI/SA is 1:0.3, the sample has the most uniform pore distribution and smoothest cross-section. The main reason for this phenomenon is that when an appropriate amount of SA acts as a bridge between PPI and protein aggregation, a three-dimensional network may be formed. By forming these three-dimensional networks, fat balls wrapped in bubbles are inhibited from freely flowing and aggregating, resulting in uniform pores distribution. At the same time, it improves the stability of foam by providing a solid framework [64]. When the proportion of SA continues to increase (PPI/SA is 1:0.5), the cream becomes less stable, and its smoothness decreases as well, with pores that are larger and unevenly distributed. It may be due to the PPI aggregates that form at high SA concentrations that hinder network continuity. This resulted in uneven holes on the cross-section of the cream.

3.15. Analysis of Cream Whipped Time

The apparent viscosity of cream is also an important indicator for evaluating its whipped performance, texture quality, and acceptability. According to general theory, an increase in the apparent viscosity of cream leads to a decrease in the migration of fat globules. Reducing the chance of collisions between fat globules helps maintain the long-term storage stability of whipped cream [65].

The effect of PPI/SA complexes mass ratio on the apparent viscosity of whipped plant cream is shown in Figure 6C. All samples exhibit a trend of shear thinning, where the apparent viscosity gradually decreases with an increasing shear rate. This is because plant cream is a pseudo plastic fluid, where shear action disrupts or recombines the internal network structure of plant cream, resulting in a decrease in apparent viscosity. As the amount of SA increases, the apparent viscosity of cream appears to increase and then decreases. When PPI/SA is 1:0.3, the apparent viscosity of cream is the highest. This may be due to the interaction between SA and PPI, which facilitates the binding between droplets and inhibits the flow of cream. When PPI and SA are combined in the cream system, the droplets will be closer together, resulting in increased flow resistance between the droplets. This leads to an increase in the apparent viscosity of the cream. The reason may be that the addition of SA can promote the unfolding of protein structure, thereby enhancing the ability to adsorb at the oil–water interface and stabilizing small droplets through tight arrangement [66]. This changes the enhanced droplet–droplet interaction per unit volume, thereby reducing droplet fluidity. This will make the cream more resistant to damage under the same shear force, leading to higher apparent viscosity. However, when the amount of SA is too high, the apparent viscosity decreases. This may be because excessive SA may cause the droplet structure to become disorganized, thereby increasing voids and reducing the degree of arrangement tightness. Therefore, when excessive SA is added, the degree of polymerization between droplet structures decreases, weakening the formed three-dimensional network structure and resulting in lower apparent viscosity [67].

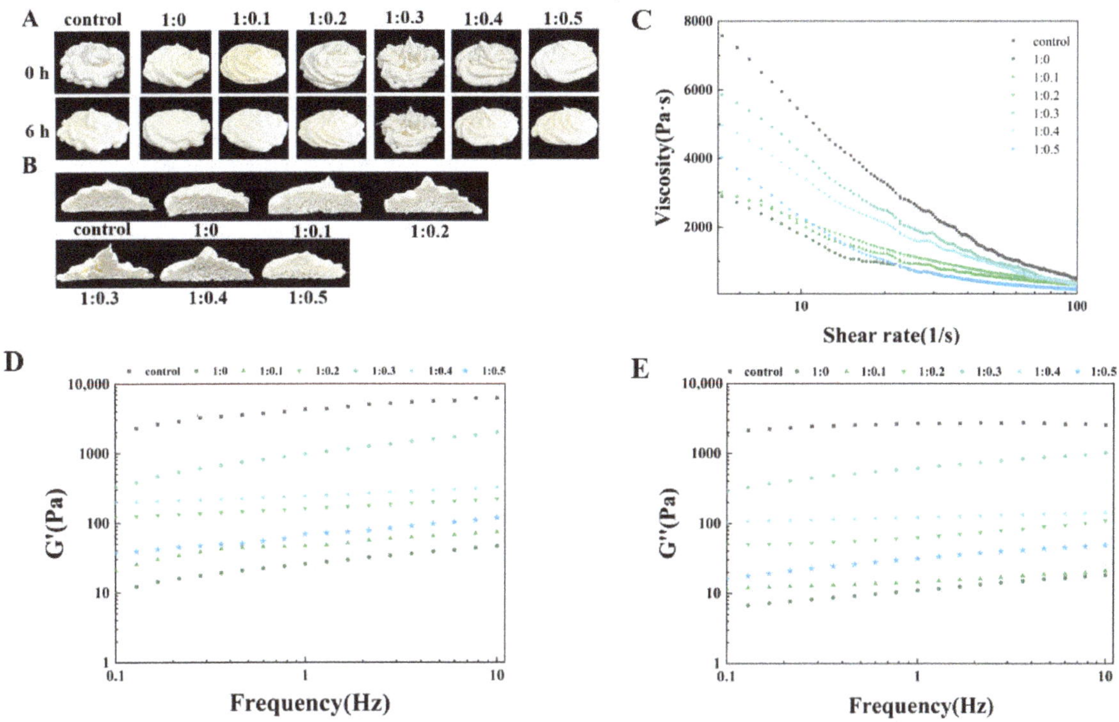

Figure 6. The effect of PPI/SA complex mass ratio on the decorating stability (**A**,**B**), apparent viscosity (**C**), and elastic/viscous modulus (G′ and G″) (**D**,**E**) of whipped cream.

3.16. Analysis of Frequency Scanning of Cream

Frequency scanning can be further used to investigate the effect of PPI/SA complex mass ratio on the internal structural stability of whipped plant cream. The elastic modulus (G′) reflects the ability of cream to maintain its shape and resist deformation during storage. The viscosity modulus (G″) reflects the viscosity of cream. The elastic modulus and viscosity modulus of cream are influenced by the internal network structure and intermolecular interactions of cream [68]. The effect of the PPI/SA composite mass ratio on the viscosity/elastic modulus of whipped plant cream is shown in Figure 6D,E. In the whole frequency range (0–10 Hz), all creams exhibit a slight increase in their elastic modulus and viscous modulus as shear frequency increases. Furthermore, it is observed that the elastic modulus of all cream samples surpasses the corresponding viscosity modulus. The G′ and G″ of the control cream sample are higher than those of the pure protein sample. This indicates that the internal network structure strength of the control sample is stronger than that of the pure protein sample. As SA increases, the G′ and G″ of cream first increase and then decrease, reaching their maximum value when PPI/SA is 1:0.3. It can be inferred that at appropriate concentrations, SA acts as a bridge to further stabilize the three-dimensional network structure formed by the aggregation of PPI. Therefore, a double stable strong network structure is formed inside the cream, which improves the viscosity and elasticity of the cream and enhances the stability of the internal structure of the cream [68]. However, when the amount of SA is too high, the formed macromolecular aggregates may weaken the interaction between SA and PPI, reducing the viscoelasticity of cream. Reference [30] also reported similar behaviors of foam food.

3.17. CLSM

Both protein distribution and fat globule aggregation influence stability of whipped cream [54]. Figure 7 displays the microstructure of whipped cream as observed through CLSM, illustrating the red fluorescence of proteins and the green fluorescence of fats. Furthermore, black holes are air bubbles surrounded by protein and fat globules. At the same level of magnification, it was noted that the control sample exhibited numerous densely packed small air bubbles in close proximity to one another. The observed phenomenon suggests an increased level of network formation among air bubbles, thereby enhancing the structural integrity of the foam. When PPI/SA is 1:0, the air bubbles in the cream were wrapped by protein in liquid form, exhibiting irregular aggregation and flocculation. The observed deformation of the bubbles suggests a deficiency in foam stability [69]. When SA was incorporated, the micrographs showed air bubbles with a more homogenous and spherical form. Meanwhile, the fat globule aggregates, and proteins were uniformly arranged, which led to the construction of a network. According to the findings presented above, it is evident that the inclusion of SA enhances the protein aggregation and reduces excessive fat accumulation within the foam structure. This is beneficial for forming a stable bubble network [70]. Among them, when PPI/SA is 1:0.3, the surface of the bubbles covered by the complexes is completer and denser, forming a clear protein interface layer. One possible explanation is that the presence of SA may lead to conformational alterations and molecular unfolding of PPI, thereby exposing hydrophobic and hydrophilic residues. These exposed hydrophilic and hydrophobic residues can cross the air–liquid interface, which promotes the adsorption, expansion, and reorganization of the PPI/SA complex on the interface, thus, establishing a compact interface layer and improving the stability of foam. However, no such variability was observed in sample with PPI/SA of 1:0.3. This could be attributed to SA enhancing the viscosity of the foam liquid film's surface, thereby impeding the liquid's flow within the film. Specifically, the small and compact bubbles offer a significant level of resistance to deformation and serve to maintain the structural integrity of the foam. In addition, the distribution of free complexes in cream is relatively uniform. Han et al. [71] have confirmed that the number of complexes that can be adsorbed onto the interface is limited, which means that some complexes particles will be free as non-adsorbed particles in the cream system. These non-adsorbed complexes may tend to produce three-dimensional network structures, which also help stabilize foam in cream. However, when the addition of SA is too high, the image shows agglomeration and an increase in particle size, indicating the formation of irregularly shaped large aggregates. Extensive protein adhesion and excessive aggregation of fat globules were observed within the foam structure. It is evident that the quantity of total protein and fat globules in whipped cream remains consistent. A high concentration of SA treatment leads to the re-aggregation of PPI, forming large protein aggregates. When a large amount of protein and fat clusters gather in a relatively small area, it is likely that there will be a shortage of bubbles, stabilizing substances in other areas [72].

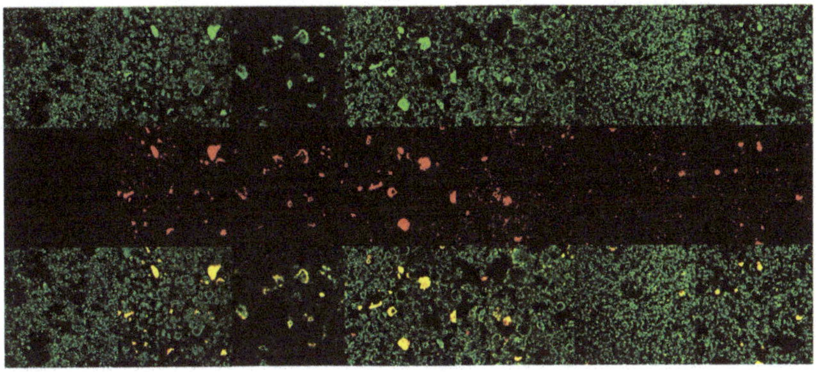

Figure 7. The CLSM images of whipped cream.

3.18. Correlation Analysis

The relevant analysis figure depicts Pearson's correlation coefficient (r), ranging from −1 to 1, and uses color codes to enhance understanding of the degree of correlation, as shown in Figure 8 [73]. It can be seen from the figure that the structural change in protein (flexibility, potential, and free sulfhydryl group) is negatively related to the interfacial tension of the complexes and is positively related to the interfacial protein concentration of the emulsion, the viscosity, and the storage modulus of the cream. The results indicate that after adding SA, PPI has stronger flexibility and higher potential, which is more conducive to obtaining complexes with better interface performance and stability. The augmentation of electrostatic repulsion is advantageous for promoting a more homogeneous and compact adsorption of PPI at the interface, whereas the reduction in interfacial tension will enhance the rate of PPI adsorption at the interface [65]. Meanwhile, the interfacial tension of composite materials is positively correlated with the hardness of cream, beating rate, and exceeding the limit, while negatively correlated with separation coefficient and stirring time. This indicates that the better the interfacial performance of PPI/SA complexes, the better the stirring performance, texture, and stability of cream prepared as a sugar substitute. During the interaction between PPI and SA, proteins undergo unfolding and structural rearrangement, causing structural changes and subunit dissociation, resulting in exposure of internal -SH groups and disulfide bond cleavage. When the SA content is moderate, SA forms a more complex and robust molecular network in the cream system, which is less susceptible to shear damage and ensures the safe entry of trapped bubbles. This enhances the stability of cream foam, which will promote cream to form a stronger network structure and better wrap the air [74]. In addition, the fine crystals of SA easily pierce the interface film between fat spheres during collision, forming crystal bridges with adjacent crystals and quickly forming a coalescence network of fat parts. This enhances the texture characteristics of cream. When the content of SA is high, excessive polymerization of cream can lead to an incomplete or loose network structure, making it difficult for bubbles to maintain stability [75]. Therefore, the different ratios of PPI and SA have a certain impact on the internal structure of the cream, which in turn affects the quality and performance of the cream.

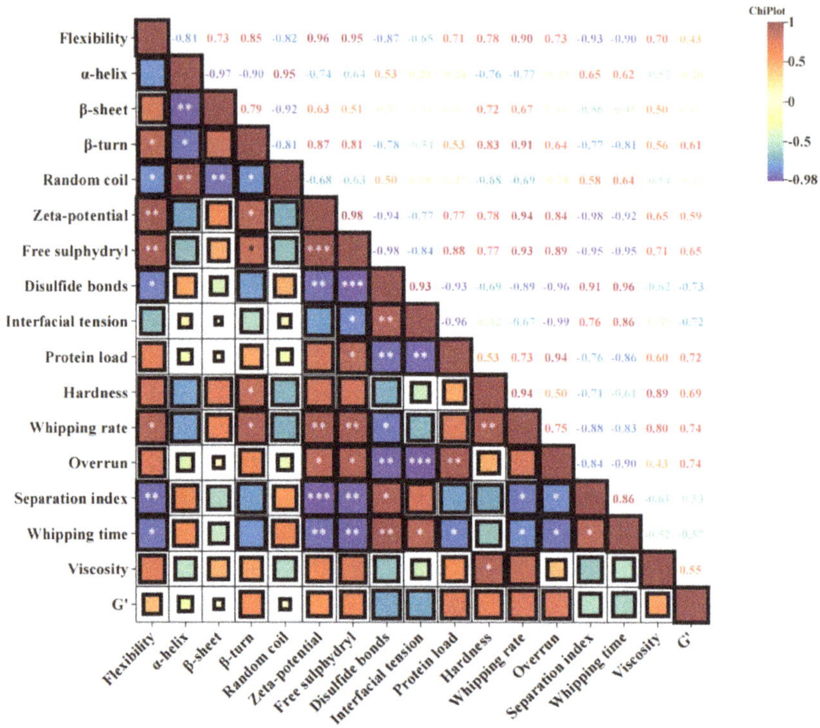

Figure 8. Correlation map between various indicators of PPI/SA complexes and creams. Note: * represents $p < 0.05$, ** represents $p < 0.01$, *** represents $p < 0.001$.

4. Conclusions

This study demonstrated the feasibility of using PPI/SA complexes instead of sugar in low sugar creams by studying the structure and interface properties of PPI/SA complexes with different mass ratios, as well as the influence of PPI/SA complexes on cream properties. Systematically explored the effects and regulatory mechanisms of PPI/SA complexes with different ratios in replacing sugars in plant cream. SA alters the physical and chemical properties of protein molecules, such as flexibility and intermolecular interactions, by affecting the electrostatic repulsion, steric hindrance, and structure of PPI. When the ratio of PPI/SA is 1:0.3, the mixing performance and quality of plant cream are best, including drainage rate, beating rate, mixing time, and slurry separation index. The bridging effect of SA enhances the interaction between PPI and SA, endowing the interface layer with higher rigidity and flexibility. The strong interface characteristics enable it to quickly and accurately respond to external deformation, effectively prevent interface cracking due to stress, and ensure the long-term stability of the cream system at the macro level. As the concentration of SA further increases, the degree of fat partial aggregation decreases, and the formed partial aggregation network structure is incomplete, leading to a decrease in the overall performance of the cream system. This study elucidates the correlation between the structure, interaction, interface properties, and basic properties of PPI/SA complexes. The conclusions of this study provide valuable theoretical support and reference for the research, production and application of functional properties of emulsion food.

Author Contributions: J.S.: Conceptualization, Software, Writing—original draft. X.Y.: Data curation. J.D.: Methodology. Y.W.: Investigation. C.W.: Investigation, Supervision, Funding acquisition, Project administration. All authors have read and agreed to the published version of the manuscript.

Funding: Heilongjiang Provincial Key R&D Program Projects (2022ZX02B18) funded this research.

Institutional Review Board Statement: Not applicable.

Informed Consent Statement: Not applicable.

Data Availability Statement: The data presented in this study are available in [Exploration of pea protein isolate-sodium alginate complexes as a novel strategy to substitute sugar in plant cream: Synergistic interactions between the two at the interface].

Conflicts of Interest: The authors declare no conflicts of interest.

References

1. Zeng, Y.; Zeng, D.; Liu, T.; Cai, Y.; Li, Y.; Zhao, M.; Zhao, Q. Effects of glucose and corn syrup on the physical characteristics and whipping properties of vegetable-fat based whipped creams. *Foods* **2022**, *11*, 1195. [CrossRef] [PubMed]
2. DiNicolantonio, J.J.; Lucan, S.C.; O'Keefe, J.H. The evidence for saturated fat and for sugar related to coronary heart disease. *Prog. Cardiovasc. Dis.* **2016**, *58*, 464–472. [CrossRef] [PubMed]
3. López-Martínez, M.I.; Moreno-Fernández, S.; Miguel, M. Development of functional ice cream with egg white hydrolysates. *Int. J. Gastron. Food Sci.* **2021**, *25*, 100334. [CrossRef]
4. Aaltonen, T.; Kytö, E.; Ylisjunttila-Huusko, S.; Outinen, M. Effect of the milk-based ash-protein ratio on the quality and acceptance of chocolate with a reduced sugar content. *Int. Dairy J.* **2020**, *105*, 104663. [CrossRef]
5. Nooshkam, M.; Varidi, M.; Alkobeisi, F. Licorice extract/whey protein isolate/sodium alginate ternary complex-based bioactive food foams as a novel strategy to substitute fat and sugar in ice cream. *Food Hydrocoll.* **2023**, *135*, 108206. [CrossRef]
6. da Silva Faresin, L.; Devos, R.J.B.; Reinehr, C.O.; Colla, L.M. Development of ice cream with reduction of sugar and fat by the addition of inulin, Spirulina platensis or phycocyanin. *Int. J. Gastron. Food Sci.* **2022**, *27*, 100445. [CrossRef]
7. Yan, X.; Yan, J.; Shi, X.; Song, Y.; McClements, D.J.; Ma, C.; Liu, X.; Chen, S.; Xu, D.; Liu, F. High internal phase double emulsions stabilized by modified pea protein-alginate complexes: Application for co-encapsulation of riboflavin and β-carotene. *Int. J. Biol. Macromol.* **2024**, *270*, 132313. [CrossRef]
8. Zhu, Q.; Qiu, Y.; Zhang, L.; Lu, W.; Pan, Y.; Liu, X.; Li, Z.; Yang, H. Encapsulation of lycopene in Pickering emulsion stabilized by complexes of whey protein isolate fibrils and sodium alginate: Physicochemical property, structural characterization and in vitro digestion property. *Food Res. Int.* **2024**, *191*, 114675. [CrossRef]
9. Gao, Z.; Huang, Y.; Hu, B.; Zhang, K.; Xu, X.; Fang, Y.; Nishinari, K.; Phillips, G.O.; Yang, J. Interfacial and emulsifying properties of the electrostatic complex of β-lactoglobulin fibril and gum Arabic (Acacia Seyal). *Colloids Surf. A Physicochem. Eng. Asp.* **2019**, *562*, 1–7. [CrossRef]
10. Dong, Y.; Wei, Z.; Xue, C. Effect of interaction between ovotransferrin fibrils and pectin on properties of oleogel-based Pickering emulsions. *Food Hydrocoll.* **2023**, *140*, 108620. [CrossRef]
11. Sun, C.; Liu, R.; Wu, T.; Liang, B.; Shi, C.; Cong, X.; Hou, T.; Zhang, M. Combined superfine grinding and heat-shearing treatment for the Microparticulation of whey proteins. *Food Bioprocess Technol.* **2015**, *9*, 378–386. [CrossRef]
12. Rezvani, F.; Abbasi, H.; Nourani, M. Effects of protein-polysaccharide interactions on the physical and textural characteristics of low-fat whipped cream. *J. Food Process. Preserv.* **2020**, *44*, 14743. [CrossRef]
13. Ghribi, A.M.; Zouari, M.; Attia, H.; Besbes, S. Study of protein/κ-carrageenan mixture's effect on low-fat whipping cream formulation. *LWT Food Sci. Tehcnol.* **2021**, *147*, 111647. [CrossRef]
14. Wang, S.; Yang, J.; Shao, G.; Qu, D.; Zhao, H.; Yang, L.; Zhu, L.; He, Y.; Liu, H.; Zhu, D. Soy protein isolated-soy hull polysaccharides stabilized O/W emulsion: Effect of polysaccharides concentration on the storage stability and interfacial rheological properties. *Food Hydrocoll.* **2020**, *101*, 105490. [CrossRef]
15. Hu, H.; Cheung, I.W.; Pan, S.; Li-Chan, E.C. Effect of high intensity ultrasound on physicochemical and functional properties of aggregated soybean β-conglycinin and glycinin. *Food Hydrocoll.* **2015**, *45*, 102–110. [CrossRef]
16. Wang, X.; Chen, J.; Geng, Q.; He, X.; Tang, Y.; Deng, L.; Li, T.; Liu, C.; Dai, T. Non-covalent interaction of the plant protein composites-proanthocyanidins and the utilization in oil in water emulsions. *LWT* **2024**, *198*, 116062. [CrossRef]
17. Zeng, D.; Cai, Y.; Liu, T.; Huang, L.; Wang, J.; Zhao, M.; Zhu, S.; Zhao, Q. Effect of hydrophobic sucrose esters with different fat acid composition and esterification degree on whipped cream properties. *Food Hydrocoll.* **2024**, *146*, 109183. [CrossRef]
18. Yan, G.; Li, Y.; Wang, H.; Cui, S.; Li, Y.; Zhang, L.; Yan, J. Multiscale approach to the characterization of the interfacial properties of micellar casein and whey protein blends and their effects on recombined dairy creams. *Food Res. Int.* **2024**, *188*, 114453. [CrossRef]

19. Cao, Z.; Liu, Z.; Zhang, H.; Wang, J.; Ren, S. Protein particles ameliorate the mechanical properties of highly polyunsaturated oil-based whipped cream: A possible mode of action. *Food Hydrocoll.* **2020**, *99*, 105350. [CrossRef]
20. Van Aken, G.A. Aeration of emulsions by whipping. *Colloids Surf. A Physicochem. Eng. Asp.* **2001**, *190*, 333–354. [CrossRef]
21. Agyare, K.K.; Xiong, Y.L.; Addo, K. Influence of salt and pH on the solubility and structural characteristics of transglutaminase-treated wheat gluten hydrolysate. *Food Chem.* **2008**, *107*, 1131–1137. [CrossRef]
22. Wan, Y.; Liu, J.; Guo, S. Effects of succinylation on the structure and thermal aggregation of soy protein isolate. *Food Chem.* **2018**, *245*, 542–550. [CrossRef] [PubMed]
23. Li, Y.; Xu, J.; Sun, F.; Guo, Y.; Wang, D.; Cheng, T.; Xu, M.; Wang, Z.; Guo, Z. Spectroscopy combined with spatiotemporal multiscale strategy to study the adsorption mechanism of soybean protein isolate with meat flavor compounds (furan): Differences in position and quantity of the methyl. *Food Chem.* **2024**, *451*, 139415. [CrossRef] [PubMed]
24. Li, Y.; Cheng, Y.; Zhang, Z.; Wang, Y.; Mintah, B.K.; Dabbour, M.; Jiang, H.; He, R.; Ma, H. Modification of rapeseed protein by ultrasound-assisted pH shift treatment: Ultrasonic mode and frequency screening, changes in protein solubility and structural characteristics. *Ultrason. Sonochemistry* **2020**, *69*, 105240. [CrossRef]
25. Yuan, M.; Li, B.; Sun, F.; Cheng, T.; Liu, J.; Wang, D.; Guo, Z.; Wang, Z. Characterization of HIPEs stabilized by soy protein isolate with excellent rheological properties for 3D printing: Bridging the correlation among structure, function and application. *Food Hydrocoll.* **2024**, *153*, 110053. [CrossRef]
26. Li, R.; Cui, Q.; Wang, G.; Liu, J.; Chen, S.; Wang, X.; Wang, X.; Jiang, L. Relationship between surface functional properties and flexibility of soy protein isolate-glucose conjugates. *Food Hydrocoll.* **2019**, *95*, 349–357. [CrossRef]
27. Basak, S.; Singhal, R.S. Succinylation of food proteins-a concise review. *LWT* **2022**, *154*, 112866. [CrossRef]
28. Ai, M.; Xiao, N.; Jiang, A. Molecular structural modification of duck egg white protein conjugates with monosaccharides for improving emulsifying capacity. *Food Hydrocoll.* **2021**, *111*, 106271. [CrossRef]
29. Liu, M.; Shan, S.; Gao, X.; Shi, Y.; Lu, W. The effect of sweet tea polysaccharide on the physicochemical and structural properties of whey protein isolate gels. *Int. J. Biol. Macromol.* **2023**, *240*, 124344. [CrossRef]
30. Jiang, L.; Ren, Y.; Xiao, Y.; Liu, S.; Zhang, J.; Yu, Q.; Chen, Y.; Xie, J. Effects of Mesona chinensis polysaccharide on the thermostability, gelling properties, and molecular forces of whey protein isolate gels. *Carbohydr. Polym.* **2020**, *242*, 116424. [CrossRef]
31. Yi, J.; Chen, X.; Wen, Z.; Fan, Y. Improving the functionality of pea protein with laccase-catalyzed crosslinking mediated by chlorogenic acid. *Food Chem.* **2024**, *433*, 137344. [CrossRef] [PubMed]
32. Zhi, Z.; Yan, L.; Li, H.; Dewettinck, K.; Van der Meeren, P.; Liu, R.; Van Bockstaele, F. A combined approach for modifying pea protein isolate to greatly improve its solubility and emulsifying stability. *Food Chem.* **2022**, *380*, 131832. [CrossRef] [PubMed]
33. Xia, B.; Shen, Y.; Zhao, R.; Deng, J.; Wang, C. Interactions with soy lecithin regulate the emulsification capacity of pea protein: Effects of soy lecithin concentration. *Food Hydrocoll.* **2024**, *155*, 110168. [CrossRef]
34. Huo, C.; Liu, G.; Xu, M.; Li, X.; Zong, W.; Liu, R. Characterizing the binding interactions of sodium benzoate with lysozyme at the molecular level using multi-spectroscopy, ITC and modeling methods. *Spectrochim. Acta Part A Mol. Biomol. Spectrosc.* **2021**, *263*, 120213. [CrossRef]
35. Archut, A.; Rolin, C.; Drusch, S.; Kastner, H. Interaction of sugar beet pectin and pea protein: Impact of neutral sugar side chains and acetyl groups. *Food Hydrocoll.* **2023**, *138*, 108454. [CrossRef]
36. Jabarkhyl, S.; Barigou, M.; Badve, M.; Zhu, S. Rheological properties of wet foams generated from viscous pseudoplastic fluids. *Innov. Food Sci. Emerg. Technol.* **2020**, *64*, 102304. [CrossRef]
37. Zaragoza, A.; Teruel, J.A.; Aranda, F.J.; Marqués, A.; Espuny, M.J.; Manresa, Á.; Ortiz, A. Interaction of a Rhodococcus sp. trehalose lipid biosurfactant with model proteins: Thermodynamic and structural changes. *Langmuir* **2012**, *28*, 1381–1390. [CrossRef]
38. Cai, Z.; Wei, Y.; Shi, A.; Zhong, J.; Rao, P.; Wang, Q.; Zhang, H. Correlation between interfacial layer properties and physical stability of food emulsions: Current trends, challenges, strategies, and further perspectives. *Adv. Colloid Interface Sci.* **2023**, *313*, 102863. [CrossRef]
39. Tao, Y.; Cai, J.; Wang, P.; Chen, J.; Zhou, L.; Yang, Z.; Xu, X. Exploring the relationship between the interfacial properties and emulsion properties of ultrasound-assisted cross-linked myofibrillar protein. *Food Hydrocoll.* **2024**, *146*, 109287. [CrossRef]
40. Lin, J.; Yang, J.; Kong, J.; Shen, M.; Yu, Q.; Chen, Y.; Xie, J. Effect of cellulose nanofibrils on stability and digestive properties of legume protein-based emulsions. *Food Hydrocoll.* **2024**, *151*, 109779. [CrossRef]
41. Lin, L.; Xiong, Y.L. Competitive adsorption and dilatational rheology of pork myofibrillar and sarcoplasmic proteins at the O/W emulsion interface. *Food Hydrocoll.* **2021**, *118*, 106816. [CrossRef]
42. Sun, F.; Pan, C.; Liu, Y.; Yang, N. Carboxymethyl tamarind seed polysaccharide/chitosan complexes through electrostatic interaction stabilize high internal phase emulsions: Roles of the mass ratio and oil-water interfacial activity. *LWT* **2024**, *195*, 115833. [CrossRef]

43. Zhang, X.; Qi, B.; Xie, F.; Hu, M.; Sun, Y.; Han, L.; Li, L.; Zhang, S.; Li, Y. Emulsion stability and dilatational rheological properties of soy/whey protein isolate complexes at the oil-water interface: Influence of pH. *Food Hydrocoll.* **2021**, *113*, 106391. [CrossRef]
44. Li, W.; Wang, Y.; Zhao, H.; He, Z.; Zeng, M.; Qin, F.; Chen, J. Improvement of emulsifying properties of soy protein through selective hydrolysis: Interfacial shear rheology of adsorption layer. *Food Hydrocoll.* **2016**, *60*, 453–460. [CrossRef]
45. Cheng, J.; Cui, J.; Ma, Y.; Yan, T.; Wang, L.; Li, H.; Li, X. Effects of soy-to-milk protein ratio and sucrose fatty acid ester addition on the stability of ice cream emulsions. *Food Hydrocoll.* **2016**, *60*, 425–436. [CrossRef]
46. Li, M.; Li, Y.; Wang, R.; Wang, Y.; Li, Y.; Zhang, L. Effects of triglycerol monostearate on physical properties of recombined dairy cream. *Int. Dairy J.* **2020**, *103*, 104622. [CrossRef]
47. Funami, T.; Nakauma, M. Instrumental food texture evaluation in relation to human perception. *Food Hydrocoll.* **2022**, *124*, 107253. [CrossRef]
48. Han, Y.; Zhu, L.; Zhang, H.; Wu, G. Fabrication and characterization of non-fat whipped cream analogue: Effects of type and concentration of polysaccharide. *Int. J. Biol. Macromol.* **2024**, *276*, 133819. [CrossRef]
49. Liu, X.; Sala, G.; Scholten, E. Impact of soy protein dispersibility on the structural and sensory properties of fat-free ice cream. *Food Hydrocoll.* **2024**, *147*, 109340. [CrossRef]
50. Wang, Z.; Liang, G.; Chen, W.; Qie, X.; Fu, L.; Li, X.; He, Z.; Zeng, M.; Goff, H.D.; Chen, J. Effects of soy proteins and hydrolysates on fat globule coalescence and whipping properties of recombined low-fat whipped cream. *Food Biophys.* **2022**, *17*, 324–334. [CrossRef]
51. Long, Z.; Zhao, M.; Sun-Waterhouse, D.; Lin, Q.; Zhao, Q. Effects of sterilization conditions and milk protein composition on the rheological and whipping properties of whipping cream. *Food Hydrocoll.* **2016**, *52*, 11–18. [CrossRef]
52. Feizi, R.; Goh, K.K.; Mutukumira, A.N. Effect of chia seed mucilage as stabiliser in ice cream. *Int. Dairy J.* **2021**, *120*, 105087. [CrossRef]
53. Warren, M.M.; Hartel, R.W. Effects of emulsifier, overrun and dasher speed on ice cream microstructure and melting properties. *J. Food Sci.* **2018**, *83*, 639–647. [CrossRef] [PubMed]
54. Narala, V.R.; Orlovs, I.; Jugbarde, M.A.; Masin, M. Inulin as a fat replacer in pea protein vegan ice cream and its influence on textural properties and sensory attributes. *Appl. Food Res.* **2022**, *2*, 100066. [CrossRef]
55. Zhang, J.; Cheng, T.; Sun, M.; Li, Y.; Zhang, G.; Hu, Z.; Wang, D.; Guo, Z.; Wang, Z. Application of soy protein isolate-naringenin complexes as fat replacers in low-fat cream: Based on protein conformational changes, aggregation states and interfacial adsorption behavior. *Int. J. Biol. Macromol.* **2024**, *274*, 133315. [CrossRef]
56. Ihara, K.; Hirota, M.; Akitsu, T.; Urakawa, K.; Abe, T.; Sumi, M.; Okawa, T.; Fujii, T. Effects of emulsifying components in the continuous phase of cream on the stability of fat globules and the physical properties of whipped cream. *J. Dairy Sci.* **2015**, *98*, 2875–2883. [CrossRef]
57. Xie, P.; Huang, M.; Liu, J.; Elbarbary, A.; Jin, Q.; Wang, X.; Jin, J. Effects of spans on whipping capabilities of aerated emulsions: Reinforcement of fat crystal-membrane interactions. *J. Food Eng.* **2024**, *372*, 112008. [CrossRef]
58. Zhang, J.; Zhu, R.; Meng, Z. Whipped cream stabilized by faba bean protein isolate microgel particles substituted for sodium caseinate: Whipping performance and foam stabilization analysis. *Food Hydrocoll.* **2024**, *157*, 110388. [CrossRef]
59. Sajedi, M.; Nasirpour, A.; Keramat, J.; Desobry, S. Effect of modified whey protein concentrate on physical properties and stability of whipped cream. *Food Hydrocoll.* **2014**, *36*, 93–101. [CrossRef]
60. Wang, Y.; Hartel, R.W.; Zhang, L. The stability of aerated emulsions: Effects of emulsifier synergy on partial coalescence and crystallization of milk fat. *J. Food Eng.* **2021**, *291*, 110257. [CrossRef]
61. Camacho, M.; Martínez-Navarrete, N.; Chiralt, A. Influence of locust bean gum/λ-carrageenan mixtures on whipping and mechanical properties and stability of dairy creams. *Food Res. Int.* **1998**, *31*, 653–658. [CrossRef]
62. Liu, Z.; Cao, Z.; Zhao, M.; Zhang, H.; Wang, J.; Sun, B. Synergistic influence of protein particles and low-molecular-weight emulsifiers on the stability of a milk fat-based whippable oil-in-water emulsion. *Food Hydrocoll.* **2022**, *127*, 107520. [CrossRef]
63. Wu, S.; Zhang, Z.; Liu, C.; Ma, T. Effect of pH-shifting and sonication-assisted treatment on properties and stability of vegetable oil-based whipped cream stabilized by kidney bean protein aggregates. *Food Hydrocoll.* **2023**, *141*, 108736. [CrossRef]
64. Grossi, M.; Rao, J.; Chen, B. Tuning the properties of plant-based whipped cream through zein nanoparticles-sodium stearate complex. *Food Hydrocoll.* **2024**, *155*, 110219. [CrossRef]
65. Cui, H.; Tang, C.; Wu, S.; McClements, D.J.; Liu, S.; Li, B.; Li, Y. Fabrication of chitosan-cinnamaldehyde-glycerol monolaurate bigels with dual gelling effects and application as cream analogs. *Food Chem.* **2022**, *384*, 132589. [CrossRef]
66. Jiang, L.; Wang, Q.; Rao, Z.; Lei, X.; Zhao, J.; Lei, L.; Ming, J. Formulation and characterization of bigels utilizing whey protein and polysaccharides: Potential applications as cream analogues. *Food Hydrocoll.* **2024**, *152*, 109884. [CrossRef]
67. Hei, X.; Liu, Z.; Li, S.; Wu, C.; Jiao, B.; Hu, H.; Ma, X.; Zhu, J.; Adhikari, B.; Wang, Q. Freeze-thaw stability of Pickering emulsion stabilized by modified soy protein particles and its application in plant-based ice cream. *Int. J. Biol. Macromol.* **2024**, *257*, 128183. [CrossRef]

68. Kurt, A.; Cengiz, A.; Kahyaoglu, T. The effect of gum tragacanth on the rheological properties of salep based ice cream mix. *Carbohydr. Polym.* **2016**, *143*, 116–123. [CrossRef]
69. Chang, Y.; Hartel, R. Stability of air cells in ice cream during hardening and storage. *J. Food Eng.* **2002**, *55*, 59–70. [CrossRef]
70. Wang, J.; Liu, Z.; Zheng, Y.; Hong, Q.; Wang, Q.; Xu, X. Synergistic effects of microcrystalline cellulose and xanthan gum on the stability of milk fat-based UHT whipping cream. *LWT* **2023**, *184*, 114966. [CrossRef]
71. Han, Y.; Zhu, L.; Zhang, H.; Liu, T.; Wu, G. Understanding the foam stability mechanisms of complex formed by soy protein isolate and different charged polysaccharides: Air/water interfacial behavior and rheological characteristics. *Int. J. Biol. Macromol.* **2024**, *268*, 131583. [CrossRef] [PubMed]
72. Zheng, Y.; Li, Z.; Zhang, C.; Zheng, B.; Tian, Y. Effects of microwave-vacuum pre-treatment with different power levels on the structural and emulsifying properties of lotus seed protein isolates. *Food Chem.* **2020**, *311*, 125932. [CrossRef] [PubMed]
73. Lu, F.; Chi, Y.; Chi, Y. High-temperature glycosylation of saccharides to modify molecular conformation of egg white protein and its effect on the stability of high internal phase emulsions. *Food Res. Int.* **2024**, *176*, 113825. [CrossRef] [PubMed]
74. Han, Y.; Zhu, L.; Zhang, H.; Liu, T.; Wu, G. Characteristic of the interaction mechanism between soy protein isolate and functional polysaccharide with different charge characteristics and exploration of the foaming properties. *Food Hydrocoll.* **2024**, *150*, 109615. [CrossRef]
75. Martínez, K.D.; Farías, M.E.; Pilosof, A.M. Effects of soy protein hydrolysis and polysaccharides addition on foaming properties studied by cluster analysis. *Food Hydrocoll.* **2011**, *25*, 1667–1676. [CrossRef]

Disclaimer/Publisher's Note: The statements, opinions and data contained in all publications are solely those of the individual author(s) and contributor(s) and not of MDPI and/or the editor(s). MDPI and/or the editor(s) disclaim responsibility for any injury to people or property resulting from any ideas, methods, instructions or products referred to in the content.

foods

MDPI

Article

Synergistic Effects of Ultrasound and pH-Shifting on the Solubility and Emulsification Properties of Peanut Protein

Zhuoran Jiao, Zhiqiang Feng, Siqi Zhao, Yuwei Wang, Miao Feng, Qian Chen, Baohua Kong and Haotian Liu *

College of Food Science, Northeast Agricultural University, Harbin 150030, China; jzr0100@163.com (Z.J.); ffcyfzq@163.com (Z.F.); zhaosiqi0121@163.com (S.Z.); 13682101156@163.com (Y.W.); 17866928220@163.com (M.F.); chenqianego7@126.com (Q.C.); kongbh63@hotmail.com (B.K.)

* Correspondence: liuht920@neau.edu.cn; Tel.: +86-451-55191794

Abstract: Peanut protein is a byproduct of peanut oil extraction with limited applications within the food sector due to its low solubility and emulsifying properties. This study investigated the influences and mechanisms of high-intensity ultrasound (HIU, 200~600 W) and pH-shifting (pH 12), either individually or jointly, on the structure, solubility, and emulsifying properties of PP. Results indicated that the solubility of PP significantly increased after the combined treatment, particularly when the HIU power was 300 W ($p < 0.05$). Accordingly, emulsions prepared from it exhibited highest storage stability. Structural analysis indicated that the increased PP solubility (9.95% to 54.37%, $p < 0.05$) is mainly attributed to the structural changes that occur during protein unfolding, resulting in the uncovering of hydrophobic groups (7181.43 to 14,083.00, $p < 0.05$) and the reduction of α-helices (24.43% to 18.17%, $p < 0.05$). Moreover, confocal laser scanning microscopy of the emulsions revealed that the combination-treated PP resulted in smaller protein particle sizes (50.09 μm to 15.68 μm, $p < 0.05$), tighter adsorption on the oil–water interface, and a denser and more stable interfacial film compared to the native and the individual treatment, thereby enhancing the stability of the system. A rheological analysis confirmed that the combined treatment improved the interfacial properties of the protein, which was advantageous for emulsion stability. In conclusion, HIU combined with pH_{12}-shifting can appreciably improve the solubility and emulsifying properties of PP to broaden its application prospects.

Keywords: peanut protein; solubility; emulsification; pH-shifting; high-intensity ultrasound

1. Introduction

Peanuts (*Arachis hypogaea* L.) are rich in oil and protein, and as a widely cultivated crop worldwide, they are the third primary source of plant protein in the human diet [1]. Peanut proteins, as a byproduct of the peanut oil extraction process, constitute approximately 47% to 55% of defatted peanut meal, are rich in essential amino acids, low in anti-nutritional factors, with an effective utilization rate of more than 90%, and have a nutritional value that is roughly equivalent to that of animal source proteins [2,3]. However, PP's high molecular weight and limited number of ionizable groups leads to its low solubility in aqueous media and directly affects its emulsifying capabilities, thereby significantly restricting its practical applications in food production [4].

The limited functionality of PP can be ascribed to its highly organized quaternary structure, intermolecular hydrophobic interactions, and disulfide cross-links (S-S), hindering its water dissolution and effective localization at oil–water interfaces of emulsions. Various physical and chemical methods for protein modification have been developed to

 https://doi.org/10.3390/foods14050853

address this, including high pressure, microwave treatment, glycosylation, phosphorylation, and enzymatic hydrolysis [5,6]. However, single-method treatments have limited efficacy, causing a shift toward combined methodologies for optimizing protein structure and enhancing its functional properties. For example, M. Wang et al. [7] modified peanut protein isolate by enzymatic treatment combined with glycosylation, which altered the protein's spatial structure and enhanced its flexibility, thereby improving its emulsifying properties. Chen et al. [8] modified PPI through extrusion cooking pretreatment and papain-induced proteolysis, resulting in a notable enhancement in the hydrolysis extent and protein solubility of the hydrolysate, along with a marked improvement in emulsification. However, these methods are frequently cumbersome to operate and possess various degrees of limitations. For instance, the degree of the glycosylation reaction is not easy to control. It also requires strict high-temperature conditions, which promote the production of carcinogenic byproducts and lead to nutrient degradation [9]. Enzymatic hydrolysis has the disadvantages of high enzyme cost, long hydrolysis time, and low yield [10]. Therefore, better methods are needed to improve the solubility and emulsification of PP.

Ultrasonic waves carry mechanical energy through the local vibration of particles at frequencies exceeding 20 kHz. They propagate through the material, especially liquid media, and reflect off walls and internal reflectors, such as volumetric voids and inclusions [11]. High-intensity ultrasound (HIU) is a nonthermal, physically-based treatment technology known for its simplicity, green nature, and lack of pollution. It has gained widespread adoption for modifying food proteins. Acoustic cavitation, induced by HIU, is a periodic phenomenon involving rapid bubble expansion and violent collapse, which generates substantial mechanical forces, including shear forces, shock waves, and microturbulence. These mechanical forces lead to the disintegration of protein particles and aggregates, exposing the hydrophobic and sulfhydryl groups (-SH), thus improving the functional attributes of the proteins, including their aqueous solubility, emulsification, and foaming properties [12,13]. However, HIU can only impact large-scale aggregation of rigid proteins by altering intermolecular forces, exhibiting minimal effect on intramolecular forces and, consequently, failing to modify the protein's inherent rigid structure and, thus, has limited effect on improving protein function [14].

pH-shifting, recognized as a straightforward, mild, and effective chemical modification technique, has been extensively documented to substantially enhance the functional characteristics of proteins. By altering the pH of the medium, the protein is exposed to extreme acid-base environments, leading to its unfolding. Subsequently, the pH is readjusted to neutral, allowing the protein to refold. This process may lead to the formation of a flexible structure termed a "molten globule" [15]. Yang et al. [16] explored the impact of pH-shifting on egg yolk protein (EYP) and observed that alkaline pH-shifting increased the number of hydrophobic groups exposed in EYP, modifying its tertiary structure. Furthermore, the pH-shifting treatment decreased the size of EYP particles in the solution and improved the protein emulsification properties. Liu et al. [17] demonstrated that pH-shifting causes proteins to unfold, exhibit greater flexibility, expose more hydrophobic groups and amino acids, and thus increase protein complexation with curcumin.

Although HIU and pH-shifting have shown potential in protein modification, their combined effects on PP remain underexplored. Specifically, the synergistic mechanisms of HIU and pH-shifting in modifying the structural and functional properties of PP are not well understood. This knowledge gap limits the development of efficient and scalable modification strategies for industrial applications. Against this background, we propose the hypothesis that the application of HIU during pH-induced protein unfolding disrupts the flexible molecular chains through the cavitation effect. Specifically, after pH-shifting treatment, the protein structure undergoes unfolding, resulting in a more flexible conformation. When HIU is applied, the mechanical forces and cavitation effects act more effectively on disrupting the intermolecular linkages between the protein chains. The subsequent readjustment of the pH to neutral facilitates the flexible refolding of protein units, resulting in the formation of protein particles with adaptable conformations. This process is anticipated to markedly enhance the solubility of PP, thereby improving its functionality within emulsion systems.

Building on the aforementioned background, this study investigates the influences of HIU and pH-shifting, either separately or jointly, on the solubility and emulsification properties of PP. Furthermore, investigations into structural alterations were conducted to reveal the mechanisms underpinning the enhancements in solubility and emulsifying properties. The aim is focused on substantiating our hypothesis and broadening the application of PP in the field of foodstuffs through these theoretical insights.

2. Materials and Methods

2.1. Materials

The PP powder (protein ~55%) was bought from Jiangsu Yuanshengtong Bioengineering Co., Ltd., Nanjing, China and stored at 4 °C. Peanut oil was purchased from Shandong Luhua Group Co., Ltd., Yantai, China. All other chemicals and reagents used in this study were of analytical grade.

2.2. Ultrasonic Treatment and pH-Shifting

The PP powder (50 mg/mL) was dispersed in deionized water at 4 °C with constant agitation for an extended period to ensure thorough hydration and uniformity. Subsequently, according to the result of the preliminary experiments, the PP suspension underwent pH_{12}-shifting, as detailed by Zhang et al. [14]. In brief, the pH of the PP suspension was raised to 12 using 2 M NaOH while maintaining continuous stirring, and the suspension was maintained at this pH for 2 h to ensure reaction homogeneity. During pH adjustment, the PP suspensions were HIU-treated using an ultrasonic cell pulverizer (Scientz-IID, Ningbo Scientz Biotechnology Co., Ltd., Ningbo, China) with a 6 mm diameter probe at 20 kHz and various power levels (200, 300, 400, 500, and 600 W) for 5 min (on-time 2 s and off-time 3 s pulse duration). Throughout the HIU process, the sample was consistently maintained in a chilled water bath to ensure its temperature remained below 20 °C, with real-time monitoring conducted using an integrated temperature probe. Following HIU treatment, the PP solution was stirred for 1 h at 25 °C and then pH-adjusted to 7.0 using 1 M HCl. PU_2, PU_3, PU_4, PU_5, and PU_6 represent samples subjected to pH-shifting and HIU at various powers of 200, 300, 400, 500, and 600 W, respectively. To assess the joint impact of the two treatments, we prepared a series of samples treated by pH-shifting and HIU (at various powers) alone, under the same conditions described above, respectively labeled as P, U_2, U_3, U_4, U_5, and U_6. "C" represents the control group of the untreated native protein. All samples were kept at 4 °C for subsequent analysis.

2.3. Solubility Determination

Solubility was assessed according to the modified method of Ma et al. [18]. After diluting the sample (50 mg/mL) to 10 mg/mL using deionized water, it was centrifuged at 10,000× *g* for 15 min. The Biuret method was used to measure the protein concentration in the supernatant. Protein solubility was determined using the following formula:

$$\text{Solubility}\ (\%) = \frac{\text{Supernatant protein concentration (mg/mL)}}{\text{Total protein concentration (mg/mL)}} \times 100\% \quad (1)$$

2.4. Particle Size and Zeta Potential Determination

A Zetasizer Nano-ZS 90 (Malvern Instruments, Worcestershire, UK) was used to assess the particle size, polydispersity index (PDI), and zeta potential of the samples. Before measurement, the samples (50 mg/mL) were diluted to 0.1 mg/mL using deionized water and centrifuged at 10,000× g for 15 min to minimize multiple scattering effects [11]. All experiments were performed at 25 °C.

2.5. Microscopic Morphological Characterization

The micromorphology and distribution of PP under various treatments were observed using confocal laser scanning microscopy (CLSM; Leica TSC SP8, Leica Microsystems, Heidelberg, Germany). The procedure involved diluting the samples (50 mg/mL) to 10 mg/mL with deionized water, incorporating 20 μL of a protein staining solution (1.0 mg/mL Nile blue solution) into 1.0 mL of a protein suspension, vortexing for 2 min, and then allowing the reaction to proceed in the dark for 30 min. Afterward, 1.5 mm thick microscope slides were coated with 10 μL of the stained samples and covered with coverslips. The distribution of PP was then examined under a 40× objective lens using a 633 nm He-Ne laser as the light source [19].

2.6. Sodium Dodecyl Sulfate-Polyacrylamide Gel Electrophoresis (SDS-PAGE) Analysis

Referring to the approach of Ji et al. [20] with modifications, sodium dodecyl sulfate-polyacrylamide gel electrophoresis (SDS-PAGE) was conducted on PP samples subjected to various treatments. An equal volume of loading buffer (0.05 M Tris-HCl, 10% glycerol, 2% SDS, 0.02% bromophenol blue; pH 6.8) with or without 12.5 mM β-mercaptoethanol was added to the protein samples (2.0 mg/mL), followed by boiling for 3 min. SDS-PAGE was performed using 10% separating and 5% concentrating gels at a constant voltage of 120 V. The gels were stained with Coomassie brilliant blue R-250 for 10 min and decolorized using a decolorizing solution (5% methanol and 7.5% acetic acid).

2.7. Structural Characterization of PP

2.7.1. Fourier Transform Infrared (FTIR)

Fourier transform infrared (FTIR) spectroscopy was used to determine the secondary structure of the protein. The protein sample suspension was freeze-dried for 12 h to yield a lyophilized powder. The FTIR spectra of the powder samples were recorded using a Nicolet iS50 (Thermo Fisher Scientific, Shanghai, China) over a wavelength range of 4000–400 cm^{-1} [21]. Spectra analysis was performed using OMNIC (Analytical Software, Waltham, MA, USA) and PeakFit software (Analytical Software, San Jose, CA, USA) to obtain the percentage of secondary structure for the samples after different treatments.

2.7.2. Intrinsic Fluorescence Spectroscopy

An RF6000-fluorescence spectrophotometer (Hitachi Co., Tokyo, Japan) was used to measure the fluorescence intensity of the samples (0.1 mg/mL, diluted from 50 mg/mL with deionized water) after different treatments, with an excitation wavelength of 280 nm and emission wavelength between 250 and 400 nm [22]. The slit widths for both excitation and emission were set at 10 nm, and the scanning speed was set to 6000 nm/min.

2.7.3. Surface Hydrophobicity (H_0)

For evaluating the surface hydrophobicity (H_0), an established protocol [11] was used with minor modifications. Specifically, the samples (50 mg/mL) were diluted with deionized water to achieve concentrations between 0.2 and 1.0 mg/mL. Into 4 mL of each diluted sample, 20 μL of ANS solution (pH 7.8, 8 mM) was introduced, followed by a 30 min reaction period in the dark. The fluorescence intensity was recorded at an excitation wavelength of 380 nm and an emission wavelength of 480 nm; the slit width was 5 nm and the scanning speed was 6000 nm/min. H_0 was quantified by determining the initial slope of the fluorescence intensity versus protein concentration (mg/mL) plot, as determined by linear regression analysis.

2.7.4. Sulfhydryl Groups (-SH) and Disulfide Bonds (S-S)

The -SH and S-S contents in the samples were assessed using a method adapted from Yang et al. [23]. To determine the free -SH content, 1 mL of protein sample (1 mg/mL) was mixed with 4 mL of Buffer B (0.086 M Tris, 0.09 M Gly, and 4 mM Na_2EDTA). To assess the total -SH content, 1 mL of protein sample (1 mg/mL) was mixed with 4 mL of Buffer B+ (Buffer B supplemented with 0.5% SDS and 6 M urea). Subsequently, 50 µL of Ellman's reagent (4 mg/mL DTNB dissolved in Buffer B) was introduced into each sample, thoroughly mixed, and allowed to react in the dark for 30 min. After centrifugation of the samples at 10,000× *g* for 10 min, the absorbance of the supernatant was measured at 412 nm. The -SH content of the sample was obtained from Equation (2):

$$-\text{SH}\ (\mu\text{mol/g}) = \frac{73.53 \times A \times D}{C} \quad (2)$$

where A represents the absorbance at 412 nm, D denotes the dilution factor, and C indicates the protein concentration (mg/mL).

The S-S content of the samples was obtained from Equation (3):

$$\text{S-S}\ (\mu\text{mol/g}) = \frac{\text{Total} - \text{SH} - \text{Free} - \text{SH}}{2} \quad (3)$$

2.8. Emulsifying Properties Characterization

2.8.1. Preparation of PP Emulsions

Emulsions were prepared following a modified version of the method reported by Zhang et al. [14]. Ten milliliters of peanut oil were mixed with 90 mL of PP suspensions that had undergone various treatments (C, U_3, P, PU_2, PU_3, and PU_4). The mixture was homogenized at 12,000 rpm for 2 min using a T18 basic high-speed homogenizer (IKA Works GmbH & Co., Staufen, Germany) to form a crude emulsion and then passed twice through a high-pressure homogenizer (SPCH-10; Shanghai Donghua High Pressure Homogenizer Factory, Shanghai, China) at 70 MPa. The resultant emulsions were stored at 4 °C for 7 days to evaluate their storage stability, and the appearance of the emulsions was documented during this time.

2.8.2. Particle Size and Zeta Potential of PP Emulsions

The particle size and droplet size distribution of both freshly prepared and 7-day-stored emulsions were assessed using a SYNC laser particle sizer (Microtrac Inc., York, PA, USA). Zeta potential measurements of the emulsions were undertaken using a Zetasizer Nano-ZS 90 (Malvern Instruments). Prior to measurement, the samples were diluted 100-fold using deionized water.

2.8.3. Characterization of the Microstructure of PP Emulsions

CLSM (Leica TSC SP8, Leica Microsystems, Heidelberg, Germany) was implemented to assess the micromorphology and distribution patterns of emulsions stabilized by PP with different treatments. To this end, 20 µL of Nile blue (0.1% *w*/*v* dissolved in water) and 25 µL of Nile red (0.1% *w*/*v* dissolved in alcohol) were blended into 1 mL of the emulsion and vortexed for 2 min, then allowed to react in the dark for 30 min. Next, 10 µL of the stained sample was applied to 1.5 mm thick microscope slides and covered with a coverslip.

The micromorphological distribution of the emulsions was observed under a 40× objective lens using 488 and 633 nm excitation wavelengths as light sources [24,25].

2.8.4. Rheological Properties of PP Emulsions

The rheological properties of the emulsions were measured using a modular rotational rheometer (HAAKE MARS60, Thermo Fisher Scientific, Shanghai, China). Apparent viscosity measurement was taken at shear rates from 0.01 to 100 s^{-1}. Dynamic viscoelasticity was evaluated in the linear viscoelastic region from 0.1 to 10 Hz to record the storage modulus (G') and loss modulus (G'') [26].

2.9. Statistical Analysis

All experiments were repeated three times. All data are expressed as mean ± standard deviation (SD). SPSS Statistics 27 software (Analytical Software, Armonk, NY, USA) was used for statistical analysis. Analysis of variance (ANOVA) was used to test the significance of the data. Tukey's test was used for multiple comparisons with the significance level set at $p < 0.05$. All graphs were plotted using Origin 2023b software (Analytical Software, Northampton, MA, USA).

3. Results and Discussion

3.1. Solubility

Solubility represents a critical characteristic of proteins, with insufficient solubility potentially impacting other functional attributes, including emulsification, foamability, and gelation [27]. As illustrated in Figure 1A, the solubility of native PP is low (9.95%) because the protein's highly ordered, rigid three-dimensional (3D) structure promotes aggregation into insoluble particles with a hydrophobic core, resulting in poor functional properties [28]. During the pH-shifting treatment, PP experiences partial unfolding and assumes a flexible molten globule conformation under highly alkaline conditions, leading to a marked increase in solubility ($p < 0.05$). The mechanical effect of acoustic cavitation during HIU treatment enhances the solubility of PP (Figure 1A), although this enhancement is less significant compared to pH-shifting ($p < 0.05$). Previous studies have extensively reported that the intense mechanical forces generated by HIU can decrease the size of protein particles and increase the specific surface area, thereby enhancing the hydration capacity [29,30]. However, for highly ordered globular proteins, despite the reduction in particle size, aggregation remains inevitable due to the formation of a hydrophobic core, which is the driving force limiting the solubility of PP post-HIU treatment. These results are consistent with the findings of our previous research [14].

The significantly higher solubility of PU_3, reaching a maximum of 54.37% ($p < 0.05$) compared to other treatments can be attributed to the synergistic effects of pH-shifting and HIU, which collectively enhance protein solubility through multiple mechanisms (Figure 1A). Under alkaline conditions (pH 12), strong electrostatic repulsion between protein molecules weakens intermolecular interactions, causing the protein to swell and adopt more flexible secondary and tertiary structures [31]. This expanded conformational state exposes more hydrophilic groups to the aqueous environment, thereby enhancing solubility. Furthermore, HIU contributes to this process through transient cavitation effects, including robust mechanical shear forces and localized high temperatures and pressures. These forces disrupt the flexible polymer structure, preventing protein refolding and altering hydrogen bonds, hydrophobic interactions, electrostatic forces, and other non-covalent bonds. As a result, the protein undergoes conformational changes that increase its spatial expansion at the protein–water interface, as reported by Ma et al. [18], ultimately enhancing solubility. Specifically, in PU_3, the combination of optimal HIU power (300 W) and pH-shifting creates a unique environment where the protein structure is sufficiently expanded without causing reaggregation. However, the solubility enhancement is HIU power-dependent (Figure 1A), with solubility decreasing when the power exceeds 300 W ($p < 0.05$). This may be attributed to protein reaggregation induced by the excessive HIU power, which counteracts the desired effects of pH-shifting and leads to a decrease in solubility [32]. The combined treatment (PU_3) thus achieves an optimal balance between structural flexibility and mechanical disruption, maximizing solubility while avoiding overprocessing-induced reaggregation.

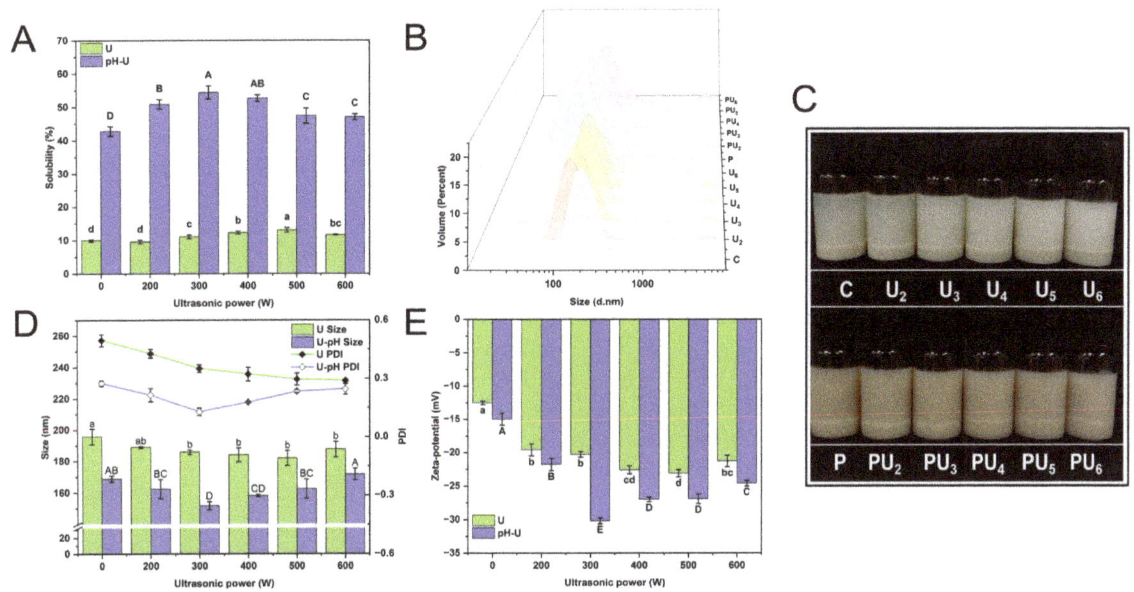

Figure 1. (**A**) solubility, (**B**) droplet distribution, (**C**) appearance, (**D**) mean particle size and polydispersity index (PDI), and (**E**) zeta potential of native (C) and treated PP with high-intensity ultrasound (HIU, 200–600 W) and pH-shifting (pH 12), individually (U_2, U_3, U_4, U_5, U_6, and P) or in combination (PU_2, PU_3, PU_4, PU_5, and PU_6). U and U-pH represent HIU treatment alone and the combined treatment of HIU and pH-shifting, respectively. Different upper (A–E) and lower (a–d) case letters denote marked differences in the solubility, size, and zeta potential of PP with HIU individual and pH-shifting and HIU combined treatment, respectively ($p < 0.05$).

Figure 1C illustrates the visual appearance of PP suspensions subjected to various treatments. Native PP (C) suspensions exhibited significant sedimentation at the bottom of the bottle, indicating pronounced phase separation and thus demonstrating low solubility and poor stability. In contrast, the PP suspensions treated with HIU alone showed no noticeable changes in appearance, suggesting that individual HIU treatment had minimal effects. Upon applying the synergistic effect of pH_{12}-shifting and HIU at various intensities, the suspension's color intensified compared to native PP (C). This indicated a more homogeneous distribution of particles within the solution, consistent with the solubility results.

3.2. Particle Size and Zeta Potential

Particle size and zeta potential are essential for deciphering the intrinsic correlation between protein–protein interactions and their structural attributes. Figure 1B,D display the particle distribution, size, and PDI of PP following various treatment protocols. Among the various treatments, native PP has a larger particle size due to the presence of a higher number of insoluble aggregates. When HIU treatment was applied alone, the particle size decreased gradually, from 195.77 ± 4.90 to 181.93 ± 4.76 nm, as the ultrasonic power increased ($p < 0.05$). This is attributed to ultrasonic cavitation. Cavitation-induced shear stress and turbulence augment the frequency and intensity of PP collisions, potentially disrupting the intermolecular interactions essential for maintaining the structure of protein aggregates in aqueous solution [33,34]. Despite the particle-shearing stresses associated with acoustic cavitation, subsequent hydrophobic aggregation remains inevitable, resulting in limited particle size reduction. When treated by a combination of pH-shifting and HIU, the particle size of PP was markedly diminished and exhibited a more homogenous distribution compared to PP treated with either approach individually (Figure 1B,D). PU_3 exhibited the most diminutive particle size and the most homogeneous distribution ($p < 0.05$), which was in concordance with the solubility outcomes (Figure 1A).

Research has demonstrated that the application of ultrasound in conjunction with acid treatment can markedly diminish the particle size of pine kernel protein, consequently improving its solubility [35]. Similarly, Yang et al. [36] demonstrated that their coprocessing method effectively reduced the particle size of perilla protein from 1218.30 to 71.88 nm, ascribing this reduction to the significant mechanical forces exerted by HIU (900 W for 20 min) and the electrostatic repulsion elicited by pH-shifting. Hence, the discernible decrease in particle size can be ascribed to the subsequent factors: firstly, in highly alkaline conditions, the protein expands and exposes reactive groups, which facilitates the formation of a flexible structure. Then, the vigorous physical forces exerted by HIU disrupt various non-covalent bonds, such as hydrogen bonds, electrostatic interactions, and hydrophobic forces, leading to the disintegration of the flexible structure into smaller protein particles [19]. Additionally, the pH of the alkaline environment deviates considerably from the protein's isoelectric point, at which PP exhibits strong electrostatic repulsion. Concurrently, the HIU treatment enhances the collision frequency among unstable protein aggregates. These superposition effects promote the uniform distribution of protein particles in water [2,35,37].

Regardless of whether HIU is used alone or in combination, however, when the ultrasound power is too large, a significant increase in particle size is observed, particularly for PU_6, which exhibits a prominent peak in the large particle size distribution. During HIU treatment, the decomposition of water generates highly reactive free radicals, which can cause protein oxidation to a certain extent, leading to intermolecular cross-linking and the formation of aggregates, potentially explaining the large particle distribution peak [38]. Accordingly, it is hypothesized that an excessively high HIU power might instigate uncontrollable cavitation, generate highly reactive radical species, provoke intense and excessive collisions of particles, and induce internal excessive heating, which could ultimately lead to the reaggregation of previously dissociated aggregates. The findings align with those reported by Yang et al. [39], which indicated that ultrasound treatment of quinoa protein at 800 W for 20 min led to an increase in particle size.

The zeta potential indicates the surface charge of proteins. Although electrostatic repulsion is influenced by multiple factors, a greater absolute value of the zeta potential typically leads to increased repulsion, thereby enhancing the stability of the protein system and reducing its propensity for aggregation [40]. As shown in Figure 1E, native PP displays a relatively low net negative charge ($p < 0.05$) attributed to the deprotonation of carboxyl groups. After treatment, the overall negative zeta potential value exhibited a significant increase ($p < 0.05$), regardless of whether PP was treated by HIU alone, pH-shifting alone, or a combination of these two treatments. HIU can induce physical shearing that disrupts protein structure, resulting in the exposure of polar residues [41]. In contrast, pH-shifting induces protein refolding, exposing more negative charges and intensifying electrostatic repulsion on PP surfaces. The resulting looser and more flexible molecular structure enhances the sensitivity of PP to subsequent HIU treatment, allowing the mechanical forces and cavitation effects to more effectively disrupt intermolecular interactions and further modify the protein structure [42]. Hence, as the polar groups on the surface ionize, the molecular surface accumulates charges, resulting in an increased net negative charge. Systems with a substantial net negative charge generally display diminished aggregation and augmented stability, attributable to the robust electrostatic repulsion force between protein particles, as reported by Hussain Badar et al. [43]. This is in agreement with the data presented in Figure 1A,B. Similarly, when the ultrasonic power was too large, the increased intensity and subsequent thermal effects led to the reaggregation of the dispersed system, resulting in the reburial of polar groups previously exposed on the protein's molecular surface.

3.3. Confocal Laser Scanning Electron Microscopy (CLSM)

The microstructure and distribution of PP were imaged using CLSM, and the results are illustrated in Figure 2A. Native PP possesses a dense, irregular protein structure, with large particles and a heterogeneous particle size distribution in the protein suspension. The structure contains numerous cavities, which may be remnants of lipid globules from the peanut cells [44]. Upon HIU treatment, the structure of the PP unfolds and becomes looser, with mechanical shear forces fragmenting the protein and causing a slight reduction in particle size. In comparison, pH-shifting has a more direct effect, as it leads to a marked reduction in protein particle size and results in a system that is more uniform and better dispersed. Subsequent HIU generates mechanical shear forces and turbulence, which further destroy the interactions, such as hydrogen bonding and hydrophobic forces, within the protein, leading to particle morphology changes in PP [45]. Consistent with the preceding solubility findings (Figure 1), PU_3 exhibited smaller and more uniformly distributed particles compared to other treatments. However, when the HIU power in the combined treatment exceeded 400 W, PP exhibited reaggregation and increased particle size, aligning with the findings of Sun et al. [46], who proposed that extended ultrasonication (3.17 W/cm^3, 50 min) might induce the reassembly and stacking of protein particles.

Figure 2B illustrates the changes in PP in the supernatant following various treatments. Consistent with the solubility results (Figure 1A), native PP has a low content of soluble protein and an uneven size distribution. The HIU treatment alone slightly increased the content of soluble protein. However, after the combination of pH-shifting and HIU treatment, a system with a higher content of soluble protein and uniform size dispersion was obtained. Similarly, Dabbour et al. [47] have reported that the synergistic application of pH-shifting and ultrasound treatment markedly augments the solubility of cottonseed protein. They attributed this to the exposure of polar groups following alkaline pH changes, leading to the formation of a molten globule structure, which increases protein molecular flexibility. Subsequent ultrasound introduces cavitation-induced mechanical effects, potentially altering the 3D structure of globular proteins, reducing their particle size, and ultimately improving interactions with water molecules. Moreover, the increase in local pressure and temperature during ultrasound treatment leads to further protein unfolding, promoting the exposure of more hydrophilic amino acids, thereby converting initially insoluble aggregates into more soluble forms [48].

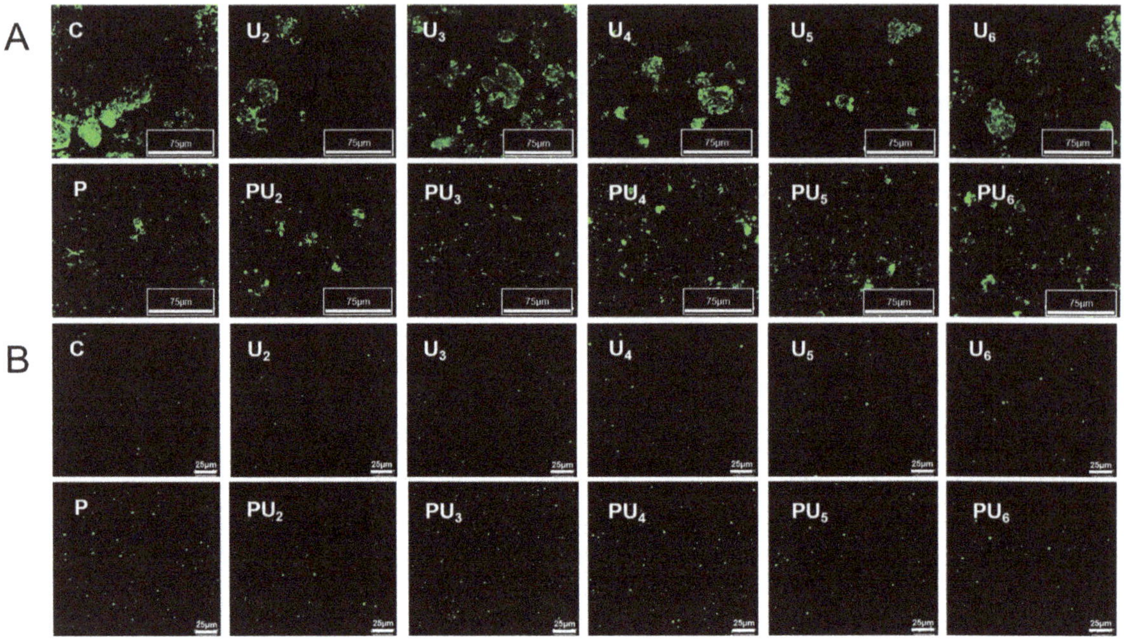

Figure 2. Confocal laser scanning microscopy (CLSM) micrographs of (**A**) the suspensions and (**B**) the supernatant of native (C) and treated PP with high-intensity ultrasound (HIU, 200–600 W) and pH-shifting (pH 12), individually (U_2, U_3, U_4, U_5, U_6, and P) or in combination (PU_2, PU_3, PU_4, PU_5, and PU_6).

3.4. Sodium Dodecyl Sulfate-Polyacrylamide Gel Electrophoresis (SDS-PAGE)

The SDS-PAGE results of the PP solution and supernatant are shown in Figure 3. Approximately 90 wt% of PP are globular proteins, which can be categorized into three primary groups: arachin (66 wt% of the PP), conarachin I (13 wt%), and conarachin II (16 wt%) [49]. From Figure 3A, in the absence of β-mercaptoethanol, all samples displayed the typical electrophoretic band pattern of PP, with visible bands characteristic of conarachin II (67.0 kDa), acid arachin (35.5 kDa), basic arachin (21.6 kDa), and conarachin I (18.0, 17.0, and 15.5 kDa). The electrophoretic bands of PP remained largely unchanged after HIU alone, indicating that HIU did not alter the subunit composition of PP. However, under pH-shifting treatment alone or in combination with HIU, the bands corresponding to acid arachin intensified, and new bands appeared in the range of 15.5 to 18.0 kDa. suggesting that alkaline pH-shifting can lead to partial degradation of PP. Conarachin II and conarachin I are trimers, which can be dissociated into smaller subunits upon the addition of a reducing agent [50]. Thus, in the SDS-PAGE pattern of PP under reducing conditions, conarachin II dissociates into acid arachin (40.5, 37.5, and 35.5 kDa), as confirmed by Figure 3C.

Based on the SDS-PAGE analysis of the supernatant (Figure 3B), the conarachin I band intensity was notably darker after pH-shifting treatment, whether individual or combined. This demonstrates that the partial cleavage of peptide bonds is caused by pH-shifting rather than HIU, and the former can alter the structure of PP to a greater extent than the latter. Concurrently, samples treated with pH-shifting alone or in combination exhibited high molecular weight bands located at the upper part of the gel, with the intensity varying with HIU power. This indicates that some insoluble aggregates have been converted into soluble oligomers [51]. These may be connected by hydrogen bonds, as hydrophobic interactions and S-S bonds have been disrupted by SDS and β-mercaptoethanol (Figure 3D).

Sun et al. [52] demonstrated that during the transformation of proteins into a molten globule state by pH-shifting, the cleavage of S-S bonds occurs, prompting the disassembly of protein aggregates into soluble oligomers.

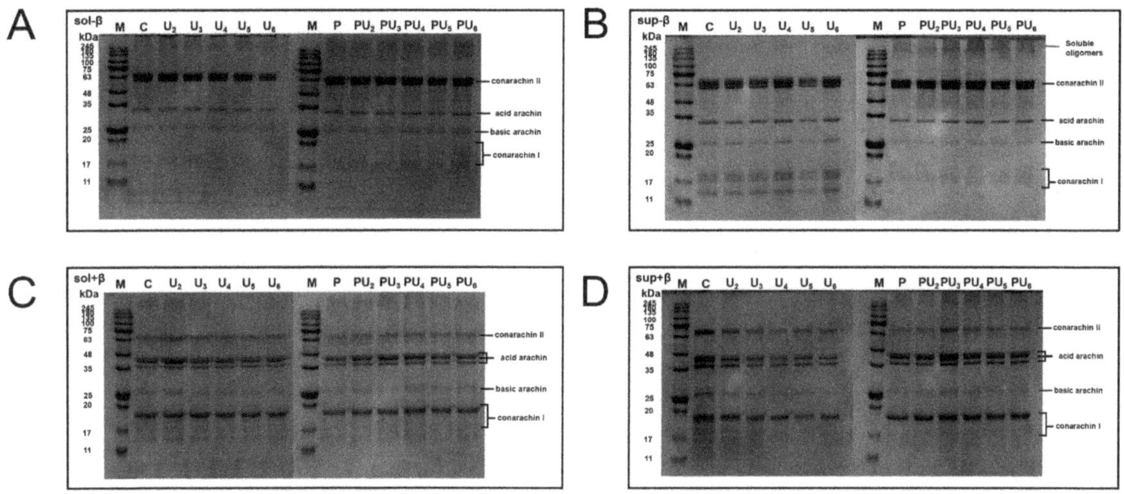

Figure 3. Sodium dodecyl sulfate-polyacrylamide gel electrophoresis (SDS-PAGE) pattern of (**A**,**C**) solutions and (**B**,**D**) supernatants for native (C) and treated PP with high-intensity ultrasound (HIU, 200–600 W) and pH-shifting (pH 12), alone (U_2, U_3, U_4, U_5, U_6, and P) or in combination (PU_2, PU_3, PU_4, PU_5, and PU_6) under (**A**,**B**) nonreducing and (**C**,**D**) reducing conditions. M indicates the marker.

3.5. Structural Characterization of PP

3.5.1. Fourier Transform Infrared Spectroscopy (FTIR)

Building on our evidence suggesting that the structure of PP may have undergone alterations during the combined pH-shifting and HIU treatment, our immediate focus was on validating the impact of individual and combined treatments on the PP structure. FTIR spectroscopy is a reliable tool for analyzing molecular composition, structure, and intermolecular interactions, offering valuable insights into the secondary structure of proteins [53]. From Figure 4A,B, there were minimal changes in the FTIR spectra of PP after various treatments, suggesting the absence of notable alterations in the protein polypeptide backbone.

The amide I band (1700~1600 cm^{-1}), corresponding to C=O stretching vibrations, exhibits sensitivity to alterations in the secondary structure of the protein [20]. Consequently, we performed deconvolution of the amide I band (Figure 4C,D) and quantified the relative abundance of secondary structures in PP following various treatments (Table 1). The results revealed varying extents of reduction in the α-helix and β-sheet contents, with an increase in β-turn and random coil structures. The combined treatments exhibited the most significant structural alterations ($p < 0.05$). Compared to native PP, the α-helix and β-sheet contents of combined-treated PP initially increased with increasing HIU power up to 300 W (PU_3), while the β-turn and random coil contents decreased.

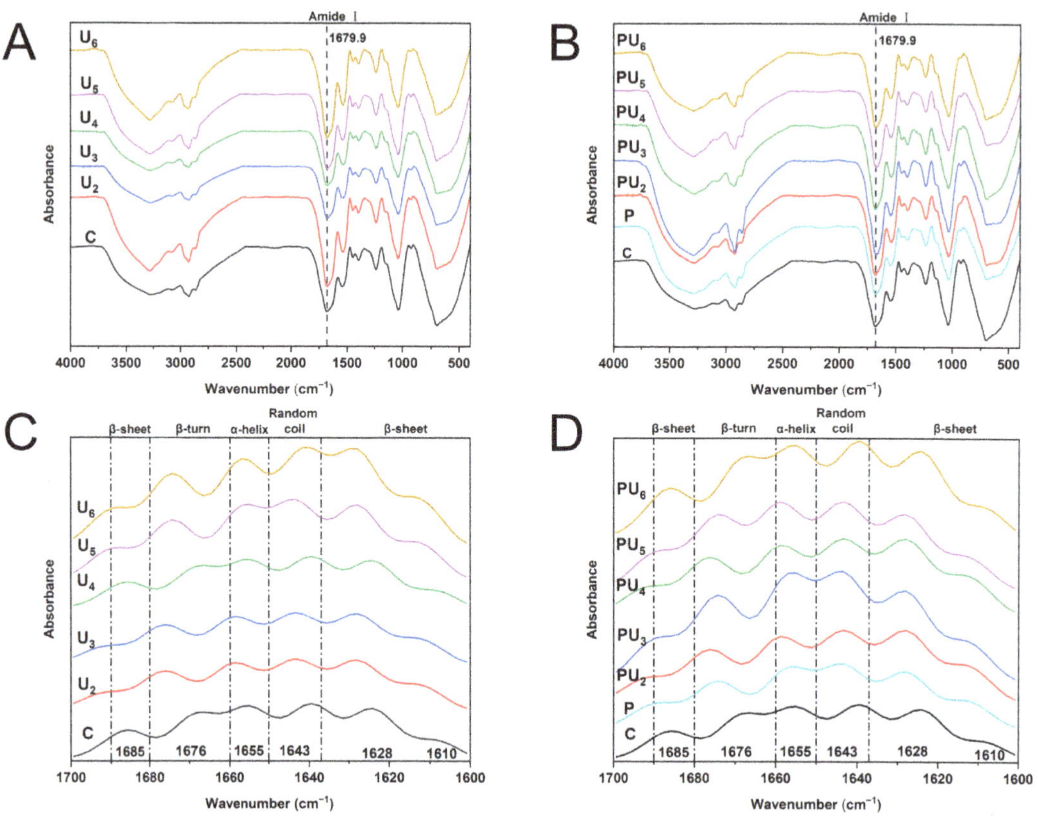

Figure 4. (**A**,**B**) Fourier transform infrared (FTIR) spectra and (**C**,**D**) deconvolution diagram of amide I band of native (C) and treated PP with high-intensity ultrasound (HIU, 200–600 W) and pH-shifting (pH 12), individually (U_2, U_3, U_4, U_5, U_6, and P) or in combination (PU_2, PU_3, PU_4, PU_5, and PU_6).

Table 1. Secondary structure contents of PP after high-intensity ultrasound, pH-shifting, and combined treatments.

HIU (W)	α-Helices (%)		β-Sheets (%)		β-Turns (%)		Random Coils (%)	
	U	U-pH	U	U-pH	U	U-pH	U	U-pH
0 (C)	24.43 ± 0.07^{a}	21.09 ± 0.09^{A}	42.54 ± 0.12^{e}	43.22 ± 0.09^{E}	15.25 ± 0.06^{a}	13.70 ± 0.15^{B}	17.79 ± 0.12^{e}	21.99 ± 0.12^{E}
200	22.66 ± 0.07^{b}	19.97 ± 0.12^{B}	42.73 ± 0.14^{d}	44.27 ± 0.05^{C}	14.88 ± 0.04^{b}	13.08 ± 0.08^{D}	19.73 ± 0.09^{d}	22.68 ± 0.16^{D}
300	22.31 ± 0.13^{c}	18.17 ± 0.08^{E}	42.84 ± 0.12^{a}	45.64 ± 0.03^{A}	14.79 ± 0.13^{b}	12.24 ± 0.05^{F}	20.06 ± 0.06^{c}	23.95 ± 0.11^{A}
400	21.82 ± 0.07^{d}	18.91 ± 0.10^{D}	42.96 ± 0.05^{c}	44.82 ± 0.09^{D}	13.83 ± 0.08^{c}	12.90 ± 0.04^{E}	21.39 ± 0.05^{b}	23.37 ± 0.11^{B}
500	21.75 ± 0.02^{d}	19.71 ± 0.13^{C}	43.41 ± 0.05^{b}	43.92 ± 0.09^{B}	13.48 ± 0.11^{d}	13.22 ± 0.11^{C}	21.36 ± 0.06^{b}	23.15 ± 0.07^{C}
600	20.77 ± 0.03^{e}	19.60 ± 0.06^{C}	43.97 ± 0.14^{a}	43.18 ± 0.16^{E}	13.43 ± 0.15^{d}	14.05 ± 0.06^{A}	21.83 ± 0.06^{a}	23.18 ± 0.17^{BC}

Data are presented as the mean ± standard deviation (SD) of triplicate experiments. U and U-pH represent HIU treatment alone and the combined treatment of HIU and pH-shifting, respectively. Different uppercase (A–E) and lowercase (a–e) letters indicate markedly differences in the secondary structure contents of PP with HIU treatment alone and with a combination of pH-shifting and HIU, respectively ($p < 0.05$).

Generally, α-helices and β-turns constitute highly ordered and relatively stable structural elements [54]. A reduction in these structures could possibly be linked to the disruption of intrachain hydrogen bonds involving amino hydrogen and carbonyl oxygen (C=O) [49]. The β-sheet and random coil structures exhibit flexibility, with the β-sheet being relatively stretched and the random coil possessing non-repetitive and indeterminate characteristics [55]. Consequently, when the ultrasonic power is below 300 W, the PP structure transitions from a regular, rigid configuration to a disordered, flexible state as the power increases. These findings align with those presented by Kong et al. [56] for soy protein and Rafique et al. [57] for oat protein. The notable increase in β-sheet content and the accompanying decrease in β-turn and random coil structures observed when the HIU power surpassed 300 W may be related to the overprocessing effect caused by excessive HIU power, which induces protein refolding and even aggregation, thereby decreasing the proportion of flexible structures [58].

3.5.2. Intrinsic Fluorescence Spectroscopy

Intrinsic fluorescence spectroscopy is considered a robust indicator to evaluate changes in protein tertiary structure, indicating the extent of exposure of aromatic amino acids (Trp, Tyr, and Phe) to the polar environment [59]. The intrinsic fluorescence (Figure 5A,B) of PP increased with the increase in HIU power ($p < 0.05$). In comparison to individual HIU and pH-shifting treatments, the combined treatment demonstrated a more significant enhancement, reaching its maximum value in the PU_3 treatment. Typically, the increased fluorescence intensity can be attributed to more extensive protein unfolding and the increased exposure of Tyr and Trp residues [60]. No red or blue shift was observed in any samples, indicating that the maximum emission wavelength remained unchanged.

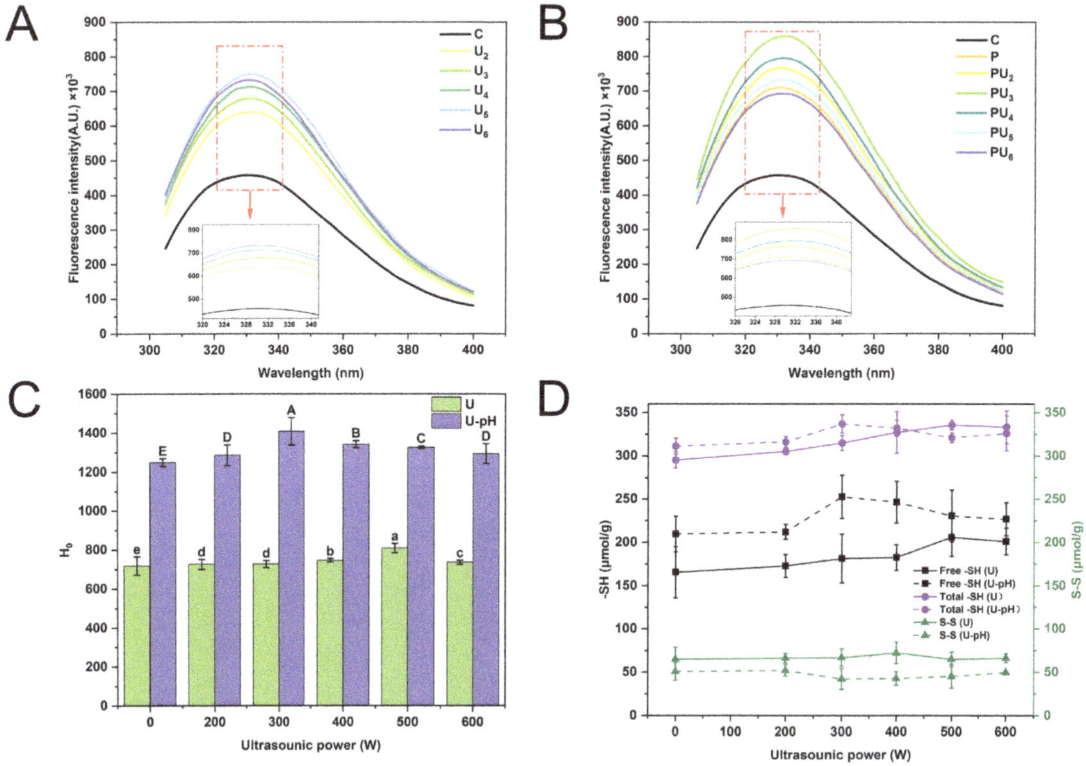

Figure 5. (**A**,**B**) intrinsic emission fluorescence spectra, (**C**) surface hydrophobicity (H_0), and (**D**) sulfhydryl groups (-SH) and disulfide bonds (S-S) of native (C) and treated PP with high-intensity ultrasound (HIU, 200~600 W) and pH-shifting (pH 12), individually (U_2, U_3, U_4, U_5, U_6, and P) or in combination (PU_2, PU_3, PU_4, PU_5, and PU_6). U and U-pH represent HIU treatment alone and the combined treatment of HIU and pH-shifting, respectively. Different upper (A–E) and lower (a–e) case letters denote marked differences in the H_0 of PP with HIU treatment alone and with a combination of pH-shifting and HIU, respectively ($p < 0.05$).

However, the fluorescence intensity decreased after the U_6 treatment, and the combined treatment showed a reduction at 400 W HIU power. This phenomenon may be attributed to overprocessing-induced overheating, leading to protein aggregation that obscures the exposed Tyr and Trp residues. Alternatively, excessive cavitation effects may quench the exposed fluorescent chromophores [61]. When protein is treated by pH-shifting during HIU it is more sensitive to the cavitation effects of HIU than either treatment alone, resulting in an earlier decline in fluorescence intensity [62].

3.5.3. Surface Hydrophobicity (H_0)

The H_0 serves as an indicator of changes in the protein tertiary structure [63]. As depicted in Figure 5C, the H_0 of PP initially increased markedly as the HIU power was increased ($p < 0.05$). The H_0 of PU_3 (14,083.00) was nearly twice as much as that of C (7181.43) ($p < 0.05$). Because the H_0 is contingent upon the quantity of exposed hydrophobic groups on the protein surface that are accessible for binding [64], the increase in H_0 is likely due to protein unfolding, exposing initially buried hydrophobic groups [65]. Higher H_0 of PP correlates with increased solubility (Figure 1A), suggesting a reduction in interactions between hydrophobic patches on the protein surface, leading to a decrease in PP particle size. Similar conclusions were reached by Mozafarpour et al. [66] in their research. Furthermore, in line with the intrinsic fluorescence intensity, H_0 decreased at excessive HIU power. This can be attributed to overheating or oxidation caused by excessive HIU, triggering protein denaturation and leading to the reburial of initially exposed hydrophobic groups.

3.5.4. Sulfhydryl (-SH) and Disulfide Bonds (S-S)

The -SH and S-S contents provide further insight into protein structural changes. The contents of total -SH, free -SH, and S-S in PP are illustrated in Figure 5D. With increasing HIU power, a marked increase in the total -SH and free -SH contents was observed initially, accompanied by a decreasing trend in the S-S content, with combined treatments showing more pronounced alterations compared to individual treatments ($p < 0.05$). The initial increase in free -SH content can be attributed to two main reasons. The first is the cleavage of S-S bonds, leading to the formation of new -SH groups [67]. A concurrent increase in total -SH and a decrease in S-S also occur. The second is protein structural unfolding, leading to the exposure of internal -SH groups, which are then converted into free -SH groups [68]. Moreover, as shown in the SDS-PAGE results (Figure 3), conarachin II is a trimer formed by S-S bonds. It contains sulfur-containing amino acids, such as Cys and Met residues, whose sulfur atoms are highly reactive with oxygen-free radicals produced during oxidation [69,70]. This leads to the interchange reaction between -SH and S-S, which may also result in changes in the content of free -SH. In contrast, at relatively high HIU power (PU_4), a reduction in free -SH content was observed. This may be attributed to the oxidation of increased -SH groups by radicals such as H• and •OH generated from cavitation effects, facilitating the formation of S-S bonds and thus rigidifying the previously loosened protein structure [33,71].

3.6. Emulsifying Properties

3.6.1. Particle Size, Zeta Potential, and Storage Stability of PP Emulsion

Proteins stabilize emulsions through two primary mechanisms: (1) adsorption at the oil–water interface to form a protective interfacial film, and (2) increasing the viscosity of the continuous phase, which retards droplet coalescence and phase separation [72,73]. The emulsifying property of proteins is closely related to their solubility, hydrophobicity, and conformational flexibility, as these factors govern the adsorption kinetics and interfacial film strength. In addition, the emulsifying property is also related to the protein concentration, which domains the viscosity and molecular interactions within the aqueous phase in emulsions [74,75]. Consequently, in light of the findings above, we chose C, U_3, P, PU_2, PU_3, and PU_4 for emulsion preparation, and we measured the particle size and droplet size distribution of emulsions in both fresh and 7-day stored conditions. From Figure 6B it can be noticed, although all fresh emulsions exhibit a uniform opalescent appearance, that the emulsion formulated with C has a larger particle size (50.09 μm) and a broader droplet size distribution (Figure 6A,E). Emulsions prepared from PP treated by HIU and pH-shifting individually (U_3 and P) exhibited a significant reduction in particle size, with the droplet size distribution shifting to the left to varying degrees ($p < 0.05$). Notably, the combined treatment (PU_3) resulted in the smallest particle size (15.68 μm, $p < 0.05$), demonstrating the synergistic effect of HIU and pH-shifting on enhancing the emulsifying properties of PP. This improvement can be attributed to the disruption of protein conformational structures by HIU and pH-shifting, which increases protein solubility and promotes protein adsorption at the oil–water interface. The reduced interfacial tension and the formation of a stable interfacial film contribute to the enhanced emulsion stability. Specifically, HIU and pH-shifting increase the surface hydrophobicity (H_0) of proteins (Figure 5C), achieving a balanced hydrophobic-to-hydrophilic ratio essential for emulsification, which in turn promotes rapid adsorption of proteins at the oil–water interface. Additionally, the mechanical shear forces generated by HIU and the conformational changes induced by pH-shifting create a rigid and dense interfacial film, further stabilizing the emulsion [76,77].

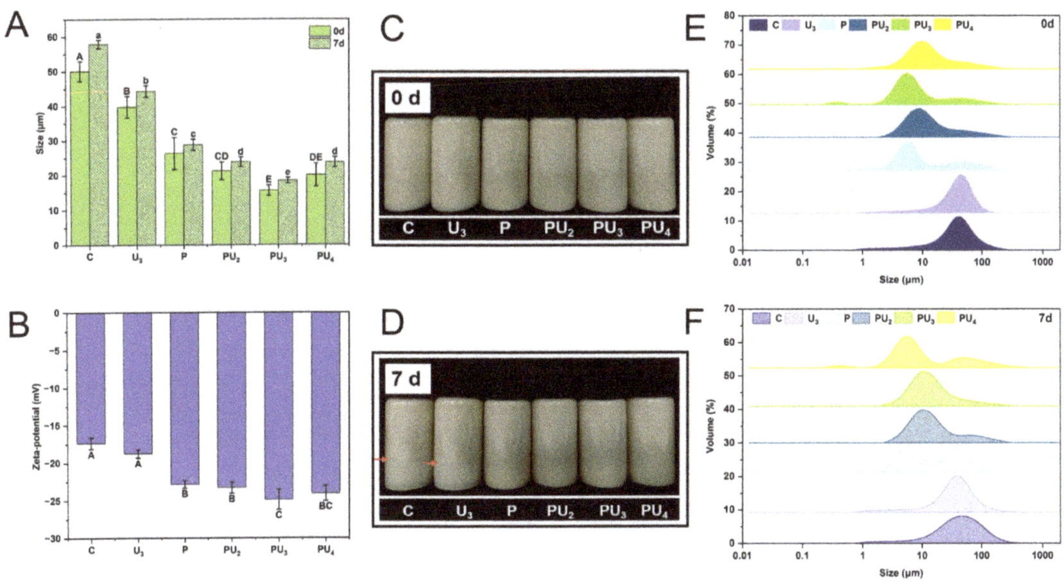

Figure 6. (**A**) Mean particle size, (**B**) zeta potential, (**C**,**D**) appearance, and (**E**,**F**) particle size distribution of fresh and 7-day-stored (**D**) emulsions stabilized by native (C) and treated PP with high-intensity ultrasound (HIU, 200–600 W) and pH-shifting (pH 12), individually (U_3 and P) or in combination (PU_2, PU_3, and PU_4). Different upper (A–E) and lower (a–e) case letters denote marked differences in the size and zeta potential of varying treatments, respectively ($p < 0.05$).

After 7 days of storage, samples prepared from C and HIU-treated PP (individual treatment) demonstrated different levels of phase separation, with C showing a clearer lower liquid phase, and HIU-treated PP displaying a slight difference in clarity between the lower and upper layers. In contrast, the emulsions prepared from combined-treated

PP maintained a uniform opalescent appearance (Figure 6D). Meanwhile, an increase in particle size led to a rightward shift or broadening of the droplet size distribution (Figure 6F). Nevertheless, the trend in particle size was consistent with that of the fresh emulsion, with PU_3 remaining the smallest (18.53 μm, $p < 0.05$), indicating that the combined treatment of PP led to stabilized emulsions with superior storage stability (Figure 6A, 7d). The improved stability can be attributed to several factors. First, the reduction in protein aggregation and enhanced protein–water interactions increase the exposure of hydrophobic groups on the protein surface, augmenting steric hindrance and molecular flexibility. Smaller protein particles adsorb more readily at the oil–water interface, reducing interfacial tension and promoting better emulsification. Additionally, higher solubility accelerates the protein interfacial diffusion rate, further stabilizing the system [78,79].

The zeta potential, which reflects the overall surface charge of emulsion droplets, is a critical factor influencing emulsion stability. A higher absolute zeta potential value indicates stronger electrostatic repulsion between droplets, which prevents aggregation and enhances stability [80]. As shown in Figure 6B, emulsions prepared from PP subjected to individual HIU and pH-shifting treatments exhibit markedly greater absolute zeta potential values than the control emulsion ($p < 0.05$). The maximum value was observed in PU_3 (−24.87 mV, $p < 0.05$), demonstrating the strongest electrostatic repulsion and the highest stability among all samples [81]. In PU_4, however, the emulsion exhibits an enlarged particle size and a diminished absolute zeta potential, possibly owing to excessive exposure of hydrophobic groups, ultimately failing to form a continuous and uniform protein film around the droplets, leading to emulsion flocculation [82]. Similar observations were reported by Yu et al. [76], who found that overprocessing could result in a slight increase in emulsion droplet size due to protein reaggregation.

3.6.2. PP Emulsion Microstructure Characterization

CLSM was used to inspect the emulsion's microstructure, thereby facilitating the observation of the distribution of oil droplets and proteins within the emulsion. From Figure 7, it can be observed that the oil droplets in the control emulsion are relatively larger and mostly exhibit an irregular spherical shape. This could be ascribed to the rigid structure of native protein aggregates, which possess poor surface characteristics, leading to slow diffusion rates and weak affinity for the oil–water interface [83]. Additionally, it can be observed from the images that although the control emulsion's protein coats the oil droplets, the resulting interfacial film is discontinuous and non-uniform, with some droplets not being fully surrounded by the protein. Thus, we hypothesize that the instability of emulsions prepared with native PP is due to the variable size of oil droplets and the extent of protein aggregation, which together prevent the formation of a homogeneous and continuous oil-in-water structure. Similarly, Yang et al. [16] have proposed that the aggregation of EYP particles induces a non-uniform interfacial film on emulsion droplets, which in turn leads to a reduction in the film's mechanical robustness and an increased likelihood of emulsion destabilization, characterized by enlarged droplet sizes. As observed in Figure 7, individual treatment with HIU or pH-shifting leads to a reduction in droplet size and a more homogeneous distribution. The combined treatment, PU_3, which contrasts with the samples treated with HIU or pH-shifting individually, yields a diminished droplet size and fosters a more uniform dispersion. Concurrently, as evidenced by the protein phase image, the protein aggregates in PU_3 are well-dispersed, allowing for homogeneous adsorption at the oil–water interface, which is more conducive to emulsion stability.

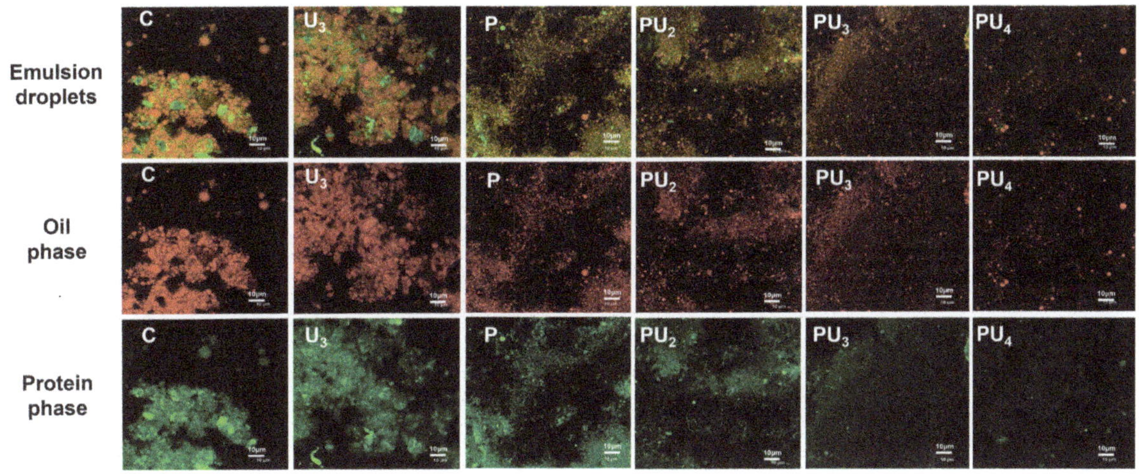

Figure 7. Confocal laser scanning microscopy (CLSM) micrographs of the emulsions stabilized by native (C) and treated PP with high-intensity ultrasound (HIU, 200–600 W) and pH-shifting (pH 12), individually (U_3 and P) or in combination (PU_2, PU_3, and PU_4).

Figure 7 indicates that pH-shifting of PP appears to significantly enhance its emulsification property. This may be ascribed to the molten globule structure induced by pH-shifting. The increased hydrophobic surface area within the molten side chain structure leads to the formation of a stronger viscoelastic protein film at the oil–water interface. Proteins in the molten globule form lose tertiary interactions and fully packed amino acid side chains, exhibiting a disordered tertiary structure in a partially folded conformation, which can promote interactions among proteins, thereby augmenting the mechanical strength of the interfacial adsorption film [84,85]. Due to the increased flexibility of the protein structure, HIU treatment results in faster protein diffusion on the oil–water interface, as proteins with more exposed hydrophobic regions more readily interact with the oil phase. However, with increasing power (PU_4), structural changes induced by HIU may hinder the interaction between exposed hydrophobic groups and oil droplets, adversely affecting emulsifying properties and hindering the development of a continuous and homogenous protein film surrounding the droplets, resulting in emulsion flocculation [82,86].

3.6.3. PP Emulsion Rheological Properties

Rheological properties are crucial indicators of emulsion performance and significantly influence emulsion stability. As shown in Figure 8A, the apparent viscosity of all samples sharply decreases as the shear rate increases, indicating shear-thinning behavior. This result may be attributed to the gradual stretching, deformation, or disintegration and rearrangement of emulsion structures under shear stress [87]. Alternatively, an increase in shear rate may lead to flocculation within the emulsion system, reducing the dispersive resistance between droplets, which in turn leads to a decrease in viscosity [88]. Emulsions prepared from PP treated with pH-shifting alone (P) exhibit higher apparent viscosity compared to those treated with HIU alone (U_3) due to the stronger effect of pH-shifting on PP, which exposes more hydrophobic groups (Figure 5C), thereby improving interfacial properties. The emulsion prepared from combined-treated PP (PU_3) exhibited the highest apparent viscosity among all samples. This is attributed to its smallest droplet size and most uniform distribution (Figure 7), which reduce droplet collision frequency and enhance emulsion stability, as supported by classical colloidal theory [89,90]. In accordance with

Stokes' law, an increase in the viscosity of the emulsion leads to a decrease in the droplet settling or floating rate, thereby enhancing the stability of the emulsion [91].

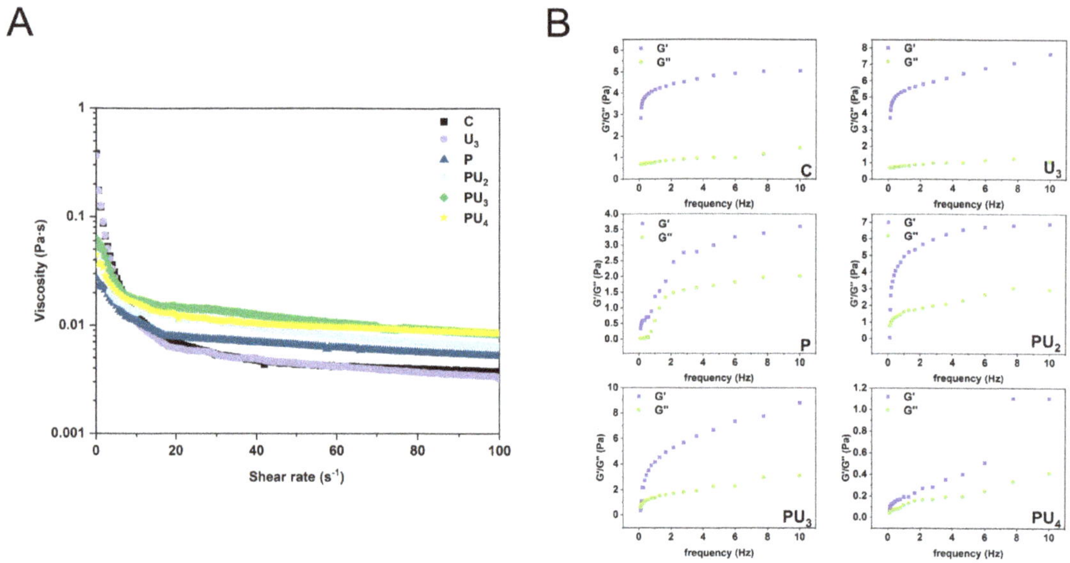

Figure 8. (**A**) Apparent viscosity and (**B**) frequency sweep of the emulsions stabilized by native (C) and treated PP with high-intensity ultrasound (HIU, 200–600 W) and pH-shifting (pH 12), individually (U_3 and P) or in combination (PU_2, PU_3, and PU_4).

Figure 8B illustrates the frequency scan results for each emulsion. Both G′ and G″ of the emulsions show an increasing trend, indicating that PP treated with HIU and pH-shifting favors the development of a stable, viscoelastic film at the oil–water interface [26]. Within the linear viscoelastic region, G′ is greater than G″, indicating that the emulsion is more inclined to exhibit elastic behavior. With the combined treatment, G′ and G″ first increase and then decrease with increasing HIU power, reaching a maximum at 300 W (PU_3). This is because pH-shifting loosens the PP structure, making it more sensitive to subsequent HIU treatment. By exposing more reactive groups and improving interactions between aggregated proteins, a thicker and denser interfacial film is formed, enhancing the viscoelasticity of the emulsion [92]. At 400 W, the excessive HIU treatment promotes protein aggregation, which weakens the strength and density of the protein network structure, thereby reducing the stability of the emulsion and leading to a decrease in viscosity, G′, and G″ [33].

4. Conclusions

The study substantiated that the combination of HIU and pH-shifting treatment exerts a synergistic effect on augmenting the solubility and emulsification properties of PP. Optimal modification parameters were identified as a pH_{12}-shifting for 2 h in conjunction with 300 W power HIU treatment for 5 min. The solubility of PP increased significantly from 9.95% to 54.37%, while the emulsifying properties were markedly enhanced, as evidenced by the reduction in emulsion particle size to 15.68 μm in PU_3. These improvements are attributed to the disruption of the rigid spherical structure of PP by pH-shifting, followed by the mechanical shear forces of HIU, which further unfold the protein and expose reactive groups, thereby enhancing solubility and interfacial adsorption. The combined treatment enables PP to adsorb more readily at the oil–water interface, generating a more robust viscoelastic interfacial film during emulsification. Future studies should explore how modified peanut proteins perform in formulated products like plant-based milk or dressings to evaluate their sensory attributes and functionality in real food matrices. The combination of pH-shifting and HIU not only improves the utilization of PP as a byproduct of peanut oil extraction but also broadens its application in the food field, contributing to resource sustainability.

Author Contributions: Conceptualization, Z.J.; Methodology, Z.J., S.Z. and Q.C.; Software, Z.J., Z.F. and Y.W.; Formal analysis, Z.F. and M.F.; Investigation, S.Z., Y.W., M.F. and Q.C.; Resources, B.K.; Writing—original draft, Z.J.; Writing—review and editing, H.L.; Visualization, B.K.; Supervision, Q.C., B.K. and H.L.; Project administration, H.L.; Funding acquisition, H.L. All authors have read and agreed to the published version of the manuscript.

Funding: This study was supported by the National Key Research and Development Program during the 14th Five-Year Plan in China (2024YFF1106603) and the National Natural Science Foundation of China (32202088).

Data Availability Statement: Data are contained within the article.

Conflicts of Interest: The authors declare no conflicts of interest.

Abbreviations

The following abbreviations are used in this manuscript:

PP	Peanut protein
HIU	High-intensity ultrasound
CLSM	Confocal laser scanning electron microscopy
SDS-PAGE	Sodium dodecyl sulfate-polyacrylamide gel electrophoresis
FTIR	Fourier transform infrared
H0	Surface hydrophobicity
-SH	Sulfhydryl groups
S-S	Disulfide bonds

References

1. Cui, S.; McClements, D.J.; Xu, X.; Jiao, B.; Zhou, L.; Zhou, H.; Xiong, L.; Wang, Q.; Sun, Q.; Dai, L. Peanut proteins: Extraction, modifications, and applications: A comprehensive review. *Grain Oil Sci. Technol.* **2023**, *6*, 135–147. [CrossRef]
2. Wang, X.; Zhao, Y.; Wang, T.; Li, Y.; Wang, C.; Zhao, X.; Liang, Y.; Wang, H. The structural and functional properties changing of peanut protein isolate via different treatments and their application in pork meatballs production. *LWT* **2023**, *184*, 115046. [CrossRef]
3. Zhao, Y.; Chen, F. Effects of simultaneous water-enzyme separation on the physical and chemical properties of peanut oil and protein. *J. Food Compos. Anal.* **2024**, *126*, 105858. [CrossRef]
4. Sánchez-Reséndiz, A.; Rodríguez-Barrientos, S.; Rodríguez-Rodríguez, J.; Barba-Dávila, B.; Serna-Saldívar, S.O.; Chuck-Hernández, C. Phosphoesterification of soybean and peanut proteins with sodium trimetaphosphate (STMP): Changes in structure to improve functionality for food applications. *Food Chem.* **2018**, *260*, 299–305. [CrossRef]
5. Ravindran, N.; Kumar Singh, S.; Singha, P. A comprehensive review on the recent trends in extractions, pretreatments and modifications of plant-based proteins. *Food Res. Int.* **2024**, *190*, 114575. [CrossRef]
6. Tang, J.; Yao, D.; Xia, S.; Cheong, L.; Tu, M. Recent progress in plant-based proteins: From extraction and modification methods to applications in the food industry. *Food Chem. X* **2024**, *23*, 101540. [CrossRef]
7. Wang, M.; Bu, G.; Zhu, T.; Liu, J.; Li, M.; Rashid, M.T.; Han, M. Effects of enzymatic hydrolysis combined with glycation on the emulsification characteristics and emulsion stability of peanut protein isolate. *Food Res. Int.* **2024**, *192*, 114722. [CrossRef]
8. Chen, L.; Chen, J.; Yu, L.; Wu, K.; Zhao, M. Emulsification performance and interfacial properties of enzymically hydrolyzed peanut protein isolate pretreated by extrusion cooking. *Food Hydrocoll.* **2018**, *77*, 607–616. [CrossRef]
9. Zhu, Q.; Xu, W.; Zhang, C.; Gong, J.; Qin, X.; Zhang, H.; Liu, G. Transglutaminase-mediated glycosylation enhances the physicochemical and functional properties of ovalbumin. *Food Hydrocoll.* **2024**, *153*, 109992. [CrossRef]

10. Magalhães, I.S.; Guimarães, A.D.B.; Tribst, A.A.L.; Oliveira, E.B.d.; Leite Júnior, B.R.d.C. Ultrasound-assisted enzymatic hydrolysis of goat milk casein: Effects on hydrolysis kinetics and on the solubility and antioxidant activity of hydrolysates. *Food Res. Int.* **2022**, *157*, 111310. [CrossRef]
11. Chen, L.; Zhang, S.-B. Structural and functional properties of self-assembled peanut protein nanoparticles prepared by ultrasonic treatment: Effects of ultrasound intensity and protein concentration. *Food Chem.* **2023**, *413*, 135626. [CrossRef]
12. Liu, Q.; Liu, Y.; Huang, H.; Xiong, M.; Yang, Y.; Lin, C.; Yang, F.; Xie, Y.; Yuan, Y. Improvement of the emulsifying properties of Zanthoxylum seed protein by ultrasonic modification. *Ultrason. Sonochem.* **2023**, *100*, 106638. [CrossRef]
13. Yu, Z.; Gao, Y.; Jia, X.; Cui, S.; Ma, L.; Zheng, D.; Li, X.; Li, L.; Zhang, L.; Chen, Y. Recent advance in high-intensity ultrasound modification of blue food protein: Mechanisms, functional properties and structural alterations. *Trends Food Sci. Technol.* **2024**, *143*, 104271. [CrossRef]
14. Zhang, J.; Liu, Q.; Chen, Q.; Sun, F.; Liu, H.; Kong, B. Synergistic modification of pea protein structure using high-intensity ultrasound and pH-shifting technology to improve solubility and emulsification. *Ultrason. Sonochem.* **2022**, *88*, 106099. [CrossRef]
15. Liu, Z.; Li, X.; Guan, Z.; Jia, Z.; Zhang, Z.; Yang, C.; Wang, J. Transglutaminase-crosslinked cold-set pea protein isolate gels modified by pH shifting: Properties, structure and formation mechanisms. *Food Hydrocoll.* **2024**, *154*, 110158. [CrossRef]
16. Yang, Y.; Jin, H.; Chen, B.; Zhang, Y.; Cai, Z.; Sheng, L. Disturbing egg yolk protein structure via pH-shifting treatment for interface reorganization: Improving solubility to enhance oil-water interface adsorption and emulsification properties. *Food Chem.* **2025**, *468*, 142516. [CrossRef]
17. Liu, C.; Deng, Z.; Wang, L.; Zhang, M.; Liu, J. Complexation between curcumin and walnut protein isolate modified by pH shifting combined with protein-glutaminase. *Food Chem.* **2025**, *464*, 141693. [CrossRef]
18. Ma, C.; Wan, Q.; Song, J.; Hao, T.; Xia, S.; Shen, S.; Li, K.; Xue, C.; Jiang, X. Ultrasound-assisted pH shift-induced interfacial remodeling for enhancing soluble yeast protein content: Effects on structure and interfacial properties of proteins under different treatment conditions. *Food Hydrocoll.* **2024**, *149*, 109521. [CrossRef]
19. Han, G.; Zhao, S.; Liu, Q.; Xia, X.; Chen, Q.; Liu, H.; Kong, B. High-intensity ultrasound combined with glycation enhances the thermal stability and in vitro digestion behaviors of myofibrillar protein aqueous solution. *Int. J. Biol. Macromol.* **2023**, *251*, 126301. [CrossRef]
20. Ji, X.; Xiong, Y.L.; Jiang, J. Tunable rice protein–starch composite soft gels: Structural role of ultrasound-modified protein. *Food Hydrocoll.* **2024**, *148*, 109462. [CrossRef]
21. Huang, X.; Li, Y.; Cui, C.; Sun-Waterhouse, D. Structural, functional properties, and in vitro digestibility of sunflower protein concentrate as affected by extraction method: Isoelectric precipitation vs ultrafiltration. *Food Chem.* **2024**, *439*, 138090. [CrossRef]
22. Wang, K.; Zhang, J.; Fu, Z.; Luo, Y.; Pu, C.; Tang, W.; Sun, Q. Oleogels based on peanut protein isolate fibrils: Structural characterization dependent on induction time and suitability in marguerite biscuits. *Food Hydrocoll.* **2024**, *154*, 110106. [CrossRef]
23. Yang, J.; Duan, Y.; Zhang, H.; Huang, F.; Wan, C.; Cheng, C.; Wang, L.; Peng, D.; Deng, Q. Ultrasound coupled with weak alkali cycling-induced exchange of free sulfhydryl-disulfide bond for remodeling interfacial flexibility of flaxseed protein isolates. *Food Hydrocoll.* **2023**, *140*, 108597. [CrossRef]
24. Gao, Y.; Chen, L.; Chi, H.; Li, L.; Teng, F. Insights into the soybean protein isolate hydrolysates: Performance characterization, emulsion construction and in vitro digestive behavior. *Int. J. Biol. Macromol.* **2024**, *279*, 135372. [CrossRef]
25. Li, R.; Xiong, Y.L. Disulfide cleavage to improve interfacial behavior and emulsification properties of oat protein. *Food Chem.* **2023**, *404*, 134511. [CrossRef]
26. Wang, N.; Wang, D.; Xing, K.; Han, X.; Gao, S.; Wang, T.; Yu, D.; Elfalleh, W. Ultrasonic treatment of rice bran protein-tannic acid stabilized oil-in-water emulsions: Focus on microstructure, rheological properties and emulsion stability. *Ultrason. Sonochem.* **2023**, *99*, 106577. [CrossRef]
27. Grossmann, L.; McClements, D.J. Current insights into protein solubility: A review of its importance for alternative proteins. *Food Hydrocoll.* **2023**, *137*, 108416. [CrossRef]
28. Xia, B.; Liu, Y.; Dong, C.; Shen, Y.; Wang, C. Enhancing the usability of pea protein in emulsion applications through modification by various approaches: A comparative study. *Food Res. Int.* **2024**, *188*, 114477. [CrossRef]
29. Huang, D.; Xu, Y.; Zhang, W.; Liu, Y.; Zhang, T.; Liu, H.; Jiang, Y.; Li, D. Enhancement of foaming property of ormosia protein: Insights into the effect of high-intensity ultrasound on physicochemical properties and structure analysis. *Food Hydrocoll.* **2024**, *152*, 109902. [CrossRef]
30. Gao, K.; Zha, F.; Yang, Z.; Rao, J.; Chen, B. Structure characteristics and functionality of water-soluble fraction from high-intensity ultrasound treated pea protein isolate. *Food Hydrocoll.* **2022**, *125*, 107409. [CrossRef]
31. Jeong, M.-S.; Cho, S.-J. Effect of pH-shifting on the water holding capacity and gelation properties of mung bean protein isolate. *Food Res. Int.* **2024**, *177*, 113912. [CrossRef]
32. Chen, X.W.; Zhao, H.Q.; Wang, H.; Xu, P.J.; Chen, M.L.; Xu, Z.; Wen, L.; Cui, B.; Yu, B.; Zhao, H.B.; et al. Preparation of high-solubility rice protein using an ultrasound-assisted glycation reaction. *Food Res. Int.* **2022**, *161*, 111737. [CrossRef]

33. Xu, Y.; Yang, Y.; Ma, C.-m.; Bian, X.; Ren, L.-K.; Liu, B.-x.; Ai, L.-z.; Zhang, N. The improvement of the oxidative oat (Avena sativa L.) protein based on ultrasound treatment: Study of structural, emulsifying, and rheological properties. *Food Hydrocoll.* **2023**, *144*, 109047. [CrossRef]
34. Cao, H.; Sun, R.; Shi, J.; Li, M.; Guan, X.; Liu, J.; Huang, K.; Zhang, Y. Effect of ultrasonic on the structure and quality characteristics of quinoa protein oxidation aggregates. *Ultrason. Sonochem.* **2021**, *77*, 105685. [CrossRef]
35. Wang, J.; Wang, X.; Wang, W.; Zhang, L.; Zhao, Y. Functionalization of pine kernel protein by pH-shifting combined with ultrasound treatments: Further improvement with increasing acidity. *Int. J. Biol. Macromol.* **2023**, *248*, 125884. [CrossRef]
36. Yang, J.; Duan, Y.; Geng, F.; Cheng, C.; Wang, L.; Ye, J.; Zhang, H.; Peng, D.; Deng, Q. Ultrasonic-assisted pH shift-induced interfacial remodeling for enhancing the emulsifying and foaming properties of perilla protein isolate. *Ultrason. Sonochem.* **2022**, *89*, 106108. [CrossRef]
37. Jiang, S.; Ding, J.; Andrade, J.; Rababah, T.M.; Almajwal, A.; Abulmeaty, M.M.; Feng, H. Modifying the physicochemical properties of pea protein by pH-shifting and ultrasound combined treatments. *Ultrason. Sonochem.* **2017**, *38*, 835–842. [CrossRef]
38. Li, Y.; Wang, J.; Zhu, X.; Tan, L.; Xie, D.; Xu, W.; Gui, Y.; Zhao, Y.; Wang, J.J. Basic electrolyzed water coupled with ultrasonic treatment improves the functional properties and digestibility of Antarctic krill proteins. *Food Res. Int.* **2022**, *162*, 112201. [CrossRef]
39. Yang, C.; Zhu, X.; Huang, J.; Wei, Y.; Wen, L.; Yang, F.; Yang, F.; Liu, W. Harnessing ultrasonic power to optimize quinoa byproduct protein for sustainable utilization. *LWT* **2024**, *207*, 116629. [CrossRef]
40. Pan, J.Y.; Xu, H.N.; Dabbour, M.; Mintah, B.K.; Chen, W.; Yang, F.; Zhang, Z.L.; Cheng, Y.; Dai, C.H.; He, R.H.; et al. Effect of alkaline pH-shifting process on extraction rate, structural, and functional properties of black soldier fly larvae protein. *LWT-Food Sci. Technol.* **2023**, *185*, 115180. [CrossRef]
41. Tang, L.; Liu, X.; Bai, S.; Zhao, D.; Guo, X.; Zhu, D.; Su, G.; Fan, B.; Wang, B.; Zhang, L.; et al. Okara protein extracted by alternating ultrasonic/alkali treatment and its improved physicochemical and functional properties. *Ultrason. Sonochem.* **2024**, *111*, 107129. [CrossRef]
42. Wang, K.; Zhang, Y.; Sun, J. Synergistic effect of high-intensity ultrasound and pH-shifting on the functionalities of chicken wooden breast myofibrillar protein: Reveal the mechanism of protein structure change. *LWT* **2023**, *181*, 114743. [CrossRef]
43. Hussain Badar, I.; Wang, Z.; Chen, Q.; Liu, Q.; Ma, J.; Liu, H.; Kong, B. Ultrasonic enhancement of structural and emulsifying properties of heat-treated soy protein isolate nanoparticles to fabricate flaxseed-derived diglyceride-based pickering emulsions. *Food Chem.* **2024**, *442*, 138469. [CrossRef]
44. Xiang, F.; Li, Z.; Zheng, Y.; Ding, C.; Benu, A.; Ma, X.; Xu, X.; Zhu, J.; Abubakar, B.Z.; Shi, A.; et al. Characterization and correlation of engineering properties and microstructure of peanuts: From microscopic to macroscopic description. *J. Integr. Agric.* **2024**, *24*, 339–352. [CrossRef]
45. Ozkan, G.; Tataroglu, P.; Gulec, S.; Capanoglu, E. Modification of pea protein isolates by high-intensity ultrasonication: Functional, structural and nutritional properties. *Food Chem. Adv.* **2024**, *5*, 100793. [CrossRef]
46. Sun, X.; Zhang, W.; Zhang, L.; Tian, S.; Chen, F. Molecular and emulsifying properties of arachin and conarachin of peanut protein isolate from ultrasound-assisted extraction. *LWT* **2020**, *132*, 109790. [CrossRef]
47. Dabbour, M.; Hamoda, A.; Xu, H.; Mintah, B.K.; Wahia, H.; Betchem, G.; Yolandani; He, R.; Ma, H.; Fikry, M. pH-shifting and sonication synergistically altered cottonseed protein: Correlating the conformational and functional characteristics. *Ind. Crops Prod.* **2024**, *222*, 120043. [CrossRef]
48. Su, C.; Huang, Y.; Chen, J.; Li, H.; Zhang, D.; Tang, Y. Effect of ultrasound-assisted phosphates treatment on solubilization and stable dispersion of rabbit Myofibrillar proteins at low ionic strength. *Food Chem.* **2025**, *472*, 142898. [CrossRef]
49. Basse, B.; Bosc, V.; Saiter, J.-M.; Chan-Huot, M.; Dupas, J.-P.; Maillard, M.-N.; Menut, P. Combined effects of ionic strength and enzymatic pre-treatment in thermal gelation of peanut proteins extracts. *Food Res. Int.* **2020**, *137*, 109362. [CrossRef]
50. Zhao, C.; Wang, F.; Yang, X.; Mao, Y.; Qi, Q.; Zheng, M.; Xu, X.; Cao, Y.; Wu, Y.; Liu, J. Synergistic influence of ultrasound and dietary fiber addition on transglutaminase-induced peanut protein gel and its application for encapsulation of lutein. *Food Hydrocoll.* **2023**, *137*, 108374. [CrossRef]
51. Zhao, R.; Fu, W.; Li, D.; Dong, C.; Bao, Z.; Wang, C. Structure and functionality of whey protein, pea protein, and mixed whey and pea proteins treated by pH shift or high-intensity ultrasound. *J. Dairy Sci.* **2024**, *107*, 726–741. [CrossRef]
52. Sun, Y.; Wang, L.; Wang, H.; Zhou, B.; Jiang, L.; Zhu, X. Effect of pH-shifting and ultrasound on soy/potato protein structure and gelation. *Food Hydrocoll.* **2025**, *159*, 110672. [CrossRef]
53. Iscimen, E.M.; Dursun Capar, T.; McClements, D.J.; Yalcin, H.; Hayta, M. Ultrasound-assisted preparation of faba bean protein isolate-*Vitis vinifera* L. polyphenol extract conjugates: Structural and functional characterization. *Food Biosci.* **2023**, *55*, 103041. [CrossRef]
54. Zhang, W.; Boateng, I.D.; Wang, Y.; Lin, M.; Vardhanabhuti, B. High-intensity ultrasound-assisted alkaline extraction of soy protein: Optimization, modeling, physicochemical and functional properties. *Int. J. Biol. Macromol.* **2024**, *283*, 137494. [CrossRef]

55. Jiang, F.; Zhu, Y.; Hu, W.-X.; Li, M.; Liu, Y.; Feng, J.; Lv, X.; Yu, X.; Du, S.-k. Characterization of quinoa starch nanoparticles as a stabilizer for oil in water Pickering emulsion. *Food Chem.* **2023**, *427*, 136697. [CrossRef]
56. Kong, Y.; Sun, L.; Wu, Z.; Li, Y.; Kang, Z.; Xie, F.; Yu, D. Effects of ultrasonic treatment on the structural, functional properties and beany flavor of soy protein isolate: Comparison with traditional thermal treatment. *Ultrason. Sonochem.* **2023**, *101*, 106675. [CrossRef]
57. Rafique, H.; Peng, P.; Hu, X.; Saeed, K.; Khalid, M.Z.; Khalid, W.; Morya, S.; Alsulami, T.; Mugabi, R.; Nayik, G.A. Ultrasound-assisted modification of oat protein isolates: Structural and functional enhancements. *Ultrason. Sonochem.* **2025**, *112*, 107204. [CrossRef]
58. Sun, P.; Zhang, Q.; Zhao, Y.; Zhao, D.; Zhao, X.; Jiang, L.; Zhang, Y.; Wu, F.; Sui, X. Improving gel properties of soy protein isolate through alkaline pH-shifting, mild heat treatment, and TGase cross-linking. *Food Hydrocoll.* **2023**, *144*, 108924. [CrossRef]
59. Li, R.; True, A.D.; Sha, L.; Xiong, Y.L. Structural modification of oat protein by thermosonication combined with high pressure for O/W emulsion and model salad dressing production. *Int. J. Biol. Macromol.* **2024**, *255*, 128109. [CrossRef]
60. Chen, W.; Ma, H.; Wang, Y.-Y. Recent advances in modified food proteins by high intensity ultrasound for enhancing functionality: Potential mechanisms, combination with other methods, equipment innovations and future directions. *Ultrason. Sonochem.* **2022**, *85*, 105993. [CrossRef]
61. Zhao, Y.; Ma, Q.; Zhou, T.; Liu, L.; Wang, Y.; Li, X.; Zhang, X.; Dang, X.; Jean Eric-Parfait Kouame, K. Ultrasound-induced structural changes of different milk fat globule membrane protein-phospholipids complexes and their effects on physicochemical and functional properties of emulsions. *Ultrason. Sonochem.* **2024**, *103*, 106799. [CrossRef]
62. Ding, S.; Ye, X.; Qu, L.; Mu, J.; Huang, L.; Dai, C. Modification of whey protein isolate by ultrasound-assisted pH shift for complexation with carboxymethylcellulose: Structure and interfacial properties. *Int. J. Biol. Macromol.* **2023**, *252*, 126479. [CrossRef]
63. Wang, Y.; Liu, X.; Luo, S.; Zhong, C.; Ye, J.; Liu, C. The impact of pH shifting combined high-pressure homogenization on structural and functional properties of rice dreg protein. *Innov. Food Sci. Emerg. Technol.* **2024**, *91*, 103520. [CrossRef]
64. Mir, N.A.; Riar, C.S.; Singh, S. Physicochemical, molecular and thermal properties of high-intensity ultrasound (HIUS) treated protein isolates from album (Chenopodium album) seed. *Food Hydrocoll.* **2019**, *96*, 433–441. [CrossRef]
65. Bi, C.-h.; Chi, S.-y.; Zhou, T.; Zhang, J.-y.; Wang, X.-y.; Li, J.; Shi, W.-t.; Tian, B.; Huang, Z.-g.; Liu, Y. Effect of low-frequency high-intensity ultrasound (HIU) on the physicochemical properties of chickpea protein. *Food Res. Int.* **2022**, *159*, 111474. [CrossRef]
66. Mozafarpour, R.; Koocheki, A.; Nicolai, T. Modification of grass pea protein isolate (Lathyrus sativus L.) using high intensity ultrasound treatment: Structure and functional properties. *Food Res. Int.* **2022**, *158*, 111520. [CrossRef]
67. Liu, G.; Hu, M.; Du, X.; Liao, Y.; Yan, S.; Zhang, S.; Qi, B.; Li, Y. Correlating structure and emulsification of soybean protein isolate: Synergism between low-pH-shifting treatment and ultrasonication improves emulsifying properties. *Colloids Surf. A: Physicochem. Eng. Asp.* **2022**, *646*, 128963. [CrossRef]
68. Chen, H.; Guo, Z.; Wang, Z.; Yang, B.; Chen, X.; Wen, L.; Yang, Q.; Kan, J. Structural and physicochemical properties of the different ultrasound frequency modified Qingke protein. *Ultrason. Sonochem.* **2023**, *94*, 106338. [CrossRef]
69. Chen, Y.; Ye, S.; Liu, L.; Ren, Y.; Li, Q.; Zhang, C.; Qian, J.-Y. Influence of ohmic heating on structure, texture and flavor of peanut protein isolate. *Innov. Food Sci. Emerg. Technol.* **2023**, *90*, 103512. [CrossRef]
70. Li, Q.; Shen, F.; He, X.; Xing, C.; Yan, W.; Fang, Y.; Hu, Q. Modification of soy protein isolate using dielectric barrier discharge cold plasma assisted by modified atmosphere packaging. *Food Chem.* **2023**, *401*, 134158. [CrossRef]
71. Ni, X.; Chen, C.; Li, R.; Liu, Q.; Duan, C.; Wang, X.; Xu, M. Effects of ultrasonic treatment on the structure and functional characteristics of myofibrillar proteins from black soldier fly. *Int. J. Biol. Macromol.* **2024**, *278*, 135057. [CrossRef]
72. Liu, Y.; Wu, Q.; Zhang, J.; Yan, W.; Mao, X. Food emulsions stabilized by proteins and emulsifiers: A review of the mechanistic explorations. *Int. J. Biol. Macromol.* **2024**, *261*, 129795. [CrossRef]
73. Kenar, J.A. Review of Food Emulsions: Principles, Practices, and Techniques, Second Edition by D.J. McClements. *Int. News Fats Oils Relat. Mater.* **2005**, *12*, 760.
74. Shi, L.S.; Yang, X.Y.; Gong, T.; Hu, C.Y.; Shen, Y.H.; Meng, Y.H. Ultrasonic treatment improves physical and oxidative stabilities of walnut protein isolate-based emulsion by changing protein structure. *LWT* **2023**, *173*, 114269. [CrossRef]
75. Dickinson, E. Food emulsions and foams: Stabilization by particles. *Curr. Opin. Colloid Interface Sci.* **2010**, *15*, 40–49. [CrossRef]
76. Yu, C.; Li, S.; Sun, S.; Yan, H.; Zou, H. Yellow horn as an alternative source of plant-based protein: The effects of high-intensity ultrasonication treatment on its physicochemical properties and emulsifying properties. *LWT* **2022**, *167*, 113820. [CrossRef]
77. Zhao, Q.; Xie, T.; Hong, X.; Zhou, Y.; Fan, L.; Liu, Y.; Li, J. Modification of functional properties of perilla protein isolate by high-intensity ultrasonic treatment and the stability of o/w emulsion. *Food Chem.* **2022**, *368*, 130848. [CrossRef]
78. Huang, D.; Hao, R.; Zhang, W.; Liu, Y.; Lin, X.; Song, W.; Jiang, Y.; Sun-Waterhouse, D.; Li, D. High-intensity ultrasound-modified Jerusalem artichoke leaf protein for stabilizing corn oi-in-water emulsion and Enhacing curcumin delivery. *Food Chem.* **2025**, *463*, 141240. [CrossRef]

79. Zou, H.; Zhao, N.; Sun, S.; Dong, X.; Yu, C. High-intensity ultrasonication treatment improved physicochemical and functional properties of mussel sarcoplasmic proteins and enhanced the stability of oil-in-water emulsion. *Colloids Surf. A Physicochem. Eng. Asp.* **2020**, *589*, 124463. [CrossRef]
80. Chen, Y.; Li, L.; Zhao, X.; Zeng, X.; Xu, X. How environmental stresses affect the physical stability of oil in water emulsion prepared using pH-shifted myofibrillar protein? *LWT* **2023**, *186*, 115200. [CrossRef]
81. Gao, Y.; Chen, L.; Li, L.; Chi, H.; Teng, F. High-pressure homogenization assisted pH-shifting modified soybean lipophilic protein interacting with chitosan hydrochloride: Double emulsion construction, physicochemical properties, stability, and in vitro digestion analysis. *Food Hydrocoll.* **2025**, *160*, 110834. [CrossRef]
82. Wang, K.; Liu, H.; Sun, J. Improved gelling and emulsifying properties of chicken wooden breast myofibrillar protein by high-intensity ultrasound combination with pH-shifting. *Poult. Sci.* **2023**, *102*, 103063. [CrossRef]
83. Zhang, R.; Cheng, L.; Luo, L.; Hemar, Y.; Yang, Z. Formation and characterisation of high-internal-phase emulsions stabilised by high-pressure homogenised quinoa protein isolate. *Colloids Surf. A Physicochem. Eng. Asp.* **2021**, *631*, 127688. [CrossRef]
84. Karabulut, G.; Kapoor, R.; Yemis, O.; Feng, H. Manothermosonication, high-pressure homogenization, and their combinations with pH-shifting improve the techno-functionality and digestibility of hemp protein. *Food Hydrocoll.* **2024**, *150*, 109661. [CrossRef]
85. Zhou, B.; Cao, X.; Rong, Y.; Wu, C.; Hu, Y.; Li, B.; Cui, B. Emulsifying and interfacial properties of glutenin processed by pH-shifting treatment. *J. Mol. Liq.* **2024**, *403*, 124889. [CrossRef]
86. Choe, U.; Chang, L.; Ohm, J.-B.; Chen, B.; Rao, J. Structure modification, functionality and interfacial properties of kidney bean (*Phaseolus vulgaris* L.) protein concentrate as affected by post-extraction treatments. *Food Hydrocoll.* **2022**, *133*, 108000. [CrossRef]
87. Zhou, L.; Kang, D.; Wang, J.; Cai, J.; Xing, L.; Zhang, W. Effects of ultrasound-assisted emulsification and carboxymethyl cellulose addition on the rheological and microstructure properties of myofibrillar protein-soybean oil emulsion gel. *Food Bioprod. Process.* **2024**, *144*, 203–213. [CrossRef]
88. Wang, N.; Wang, T.; Yu, Y.; Xing, K.; Qin, L.; Yu, D. Dynamic high-pressure microfluidization assist in stabilizing hemp seed protein-gum Arabic bilayer emulsions: Rheological properties and oxidation kinetic model. *Ind. Crops Prod.* **2023**, *203*, 117201. [CrossRef]
89. Niu, H.; Wang, W.; Dou, Z.; Chen, X.; Chen, X.; Chen, H.; Fu, X. Multiscale combined techniques for evaluating emulsion stability: A critical review. *Adv. Colloid Interface Sci.* **2023**, *311*, 102813. [CrossRef]
90. Mcclements, D.J. *Food Emulsions: Principles, Practices, and Techniques*, 3rd ed.; CRC Press: Boca Raton, FL, USA, 2015. [CrossRef]
91. Li, D.; Zhao, Y.; Wang, X.; Tang, H.; Wu, N.; Wu, F.; Yu, D.; Elfalleh, W. Effects of (+)-catechin on a rice bran protein oil-in-water emulsion: Droplet size, zeta-potential, emulsifying properties, and rheological behavior. *Food Hydrocoll.* **2020**, *98*, 105306. [CrossRef]
92. Li, H.; Liu, Y.; Tan, H.; Wu, X.; Wu, W. Effect of ultrasonic pretreatment on the emulsion rheological properties and interface protein structure of epigallocatechin-3-gallate and rice bran protein complex. *Food Chem.* **2025**, *463*, 141406. [CrossRef]

Disclaimer/Publisher's Note: The statements, opinions and data contained in all publications are solely those of the individual author(s) and contributor(s) and not of MDPI and/or the editor(s). MDPI and/or the editor(s) disclaim responsibility for any injury to people or property resulting from any ideas, methods, instructions or products referred to in the content.

foods

MDPI

Article

Betalain–Chickpea Protein Particles Produced by Freeze Drying and Spray Drying: Physicochemical Aspects, Storage Stability, and In Vitro Digestion

Mary H. Grace [1], Roberta Targino Hoskin [1], Malak Alghamdi [1], Mary Ann Lila [1] and Vesela I. Chalova [1,2,*]

[1] Plants for Human Health Institute, Department of Food, Bioprocessing and Nutrition Sciences, North Carolina State University, 600 Laureate Way, Kannapolis, NC 28081, USA; mhgrace@ncsu.edu (M.H.G.); rtcorrei@ncsu.edu (R.T.H.); malgham@ncsu.edu (M.A.); mlila@ncsu.edu (M.A.L.)

[2] Department of Biochemistry and Nutrition, University of Food Technologies, 4002 Plovdiv, Bulgaria

* Correspondence: v_chalova@uft-plovdiv.bg

Abstract: Beetroots are one of the primary sources of betalains, nitrogenous pigments with anti-inflammatory and antioxidant properties. However, due to their chemical instability, betalains have limited use in food applications. This work investigated whether betalains encapsulated in chickpea protein could be stabilized and delivered in a shelf-stable format. Freeze-dried (CB-FD) and spray-dried (CB-SD) protein–betalain particles encapsulated in chickpea protein isolate (6% w/v) were prepared. The encapsulation method affected particles' morphology, water activity, hygroscopicity, solubility, and color. Particles captured total betalains of 9.30 ± 0.61 and 4.40 ± 0.92 mg/g for CB-SD and CB-FD, respectively. LC-MS identified 12 betacyanins and 6 betaxanthins. The stability of betalains revealed that encapsulation efficiently preserved betalain integrity of over 6 weeks of storage at 4, 22, and 40 °C compared to dry beetroot extract. CB-SD particles were stable with no significant changes, while CB-FD showed slight degradation after 4 weeks due to increased A_w. Antioxidant activity correlated well with betalain concentration. In vitro digestion resulted in only 25% bioaccessibility of betacyanins, while betaxanthins were more stable with 100% recovery. Encapsulation with chickpea protein isolate is an efficient and straightforward strategy for expanding and diversifying applications of phytochemical-rich beetroot extracts for the food industry.

Keywords: plant based; pigments; protein-rich particles; valorization; delivery method

1. Introduction

Beetroots (*Beta vulgaris*, family *Chenopodiaceae*) are vegetables produced worldwide throughout the seasons and are one of the main natural sources of betalains, as well as phenolic compounds, vitamins, carotenoids, and minerals [1]. Betalains are water-soluble nitrogenous pigments composed of reddish-violet betacyanins and yellowish-orange betaxanthin pigments with documented biological effects, particularly in the prevention and management of chronic conditions such as diabetes, obesity, and cardiovascular diseases [1].

However, the use of beetroots in food formulations is challenged by their earthy smell and taste and natural perishability [2]. Moreover, betalains are heat-sensitive molecules and susceptible to degradation caused by pH fluctuations, oxygen and light, and the presence of metal cations [3]. Therefore, creative solutions to enable longer shelf life, expand the portfolio of products, and provide value-added methods for the use of beetroots in the food industry are needed [4].

Encapsulation techniques have been intensively studied as technologically sound methods for improving stability, preserving biological activity, and ensuring easier handling and transportation of food products. Spray drying is one of the most popular encapsulation techniques in the food industry today [5–7], while freeze drying, although generally more expensive than spray drying, is known for yielding well preserved, high-quality compounds [8].

Our group explored encapsulation techniques to obtain protein–polyphenol particles with preserved phytochemicals, enhanced storability, and desirable technological attributes [9,10], for multiple food applications. In this work, we used our well-established processing strategy to investigate and compare spray drying and freeze drying techniques as technological routes to produce protein–betalain ingredients using chickpea protein isolate and betalain-rich extracts from beetroots. The resulting particles produced by spray drying and freeze drying were comprehensively analyzed in terms of physicochemical attributes, betalain content, profile, storage stability, and in vitro bioaccessibility. The outcomes of this study provide practical information on the processing of abundant yet underexplored phytochemicals from beetroots to deliver safe, stable, and marketable food products.

2. Materials and Methods

2.1. Materials

Fresh organic red beetroots (Merlin F1-hybrid, *Beta vulgaris* L. subsp. vulgaris var. conditiva Alef.) provided by a local grower were stored in a cold room (4–6 °C) and processed within two weeks. Chickpea protein isolate (ChickP S930) was obtained from ChickP Protein Ltd. (Rehovot, Israel) and contained 89.7% protein, 7.0% total carbohydrates, and less than 0.1% fat (according to the manufacturer's specifications). Solvents used in high-performance liquid chromatography (HPLC) or liquid chromatography-mass spectrometry (LC-MS) studies were of ACS or HPLC grade. All remaining reagents were of analytical grade. Distilled or double-distilled water was used to prepare the solutions.

2.2. Preparation of Beetroot Extract

Beetroots were washed and cut into small pieces that were mixed with 80% aqueous ethanol solution in a laboratory blender 100:400 *w*/*v* (Vitamix, Vita-Mix Corporation, Cleveland, OH, USA), then processed under vacuum for 5 min at room temperature, followed by vacuum filtration through cheesecloth and centrifugation (Sorvall Legend RT, Thermo Electron Corporation, Osterode, Germany) at 4000 rpm, 10 °C for 20 min. The supernatant was collected and vacuum concentrated in a rotary evaporator at 46 °C (Rotavapor R-210, BUCHI Corporation, New Castle, DE, USA) until complete evaporation of ethanol and concentration to yield an aqueous concentrated beetroot extract (°Brix 12.5 ± 0.5, pH 5.8 ± 0.5). The concentrated extract was collected, stored in dark containers at 4 °C, and used to prepare chickpea protein–betalain particles within two days.

2.3. Freeze Drying (FD) and Spray Drying (SD) Processes

Before drying, the concentrated beetroot extract was mixed with water in a 2.5-fold dilution, resulting in a solution with a total betalain concentration of 9.075 mg/g dw. To identify the optimal combination of chickpea protein and beetroot extract that maximized betalain content, preliminary experiments were conducted (Tables S1 and S2). A 6% *w*/*v* chickpea solution, determined during these initial tests, was combined with beetroot solution and stirred for 30 min at room temperature until complete homogenization was achieved. For FD, the beetroot extract-chickpea protein suspension was centrifuged for 20 min at 4000 rpm. The supernatant containing the excess beetroot extract was discarded, and the pelleted material was collected, and subjected to −20 °C for 12 h before being FD.

The FD procedure lasted 48 h (Labconco, Labconco Corporation, Kansas City, MO, USA). The condenser temperature was −55 °C, and the vacuum pressure was 0.105 Torr. The dried material was pulverized using an IKA A11 analytical mill (Wilmington, NC, USA), placed in tightly closed containers, and stored in a −20 °C freezer for further analysis.

The SD was carried out based on our protocol [11] using a lab-scale spray dryer (model B-290, Buchi Labortechnik AG, Flawil, Switzerland) using a 0.7 mm nozzle, 7.5 mL/min of feed flow controlled by a peristaltic pump, aspirator rate at 100%, with air in co-current flow. The feed solutions prepared by mixing chickpea protein and beetroot extracts were kept under constant stirring at room temperature during the spray drying process. The drying inlet air temperature was set to 125 °C, and the outlet air temperature was 64–66 °C. Spray-dried protein–betalain particles were collected from the powder collection vessel only, weighed, and stored in sealed plastic containers at −20 °C until further use. The SD yield was calculated according to Hoskin et al. [10]. Freeze-dried and spray-dried chickpea protein–betalain particles were labeled as CB-FD and CB-SD, respectively.

2.4. *Characterization of Chickpea Protein–Betalain Particles*

2.4.1. Water Activity (A_w)

The water activity (A_w) was determined by using a water activity meter (Aqualab 4TE, Meter Group, Inc., Pulman, WA, USA).

2.4.2. Hygroscopicity and Solubility

Hygroscopicity and solubility analyses were determined according to Correia et al. [12]. For hygroscopicity, samples (0.5 g) were placed in a desiccator containing a saturated solution of NaCl (RH 75.3%). Results were expressed as the mass of water absorbed per 100 g of sample after 7 days of storage. For solubility, samples (0.5 g) were mixed with 50 mL of distilled water and vigorously homogenized at high velocity for 5 min and centrifuged at 4000 rpm for 5 min. An aliquot (25 mL) of the supernatant was removed, transferred to aluminum dishes, and dried to constant weight in an oven at 105 °C. Results expressed as a percentage (%) were calculated as the ratio between the weight of the supernatant (soluble solids in solution) and the weight of the sample.

2.4.3. Instrumental Color Evaluation

Color measurements were performed using a CR-5 colorimeter (Konica Minolta, Tokyo, Japan). The CIELAB L* (lightness), a* (redness-greenness), and b* (yellowness-blueness) parameters were recorded and used to calculate total color difference (ΔE) [13]. The colorimetric analyses were performed in triplicate.

2.4.4. Morphology by Scanning Electron Microscopy (SEM)

The morphological analysis was conducted using a scanning electron microscope (SU3900, Hitachi, Tokyo, Japan). Samples were placed on a conductive carbon tape and analyzed under a vacuum. The morphological analysis was conducted at an accelerating voltage of 20 kV and 250, 500, and 1000× magnifications.

2.5. *Phytochemical Analyses*

2.5.1. Elution of Phytochemicals

Protein–betalain particles were sampled (0.1 g) and mixed with 4 mL 80% aqueous methanol solution. The extraction of phytochemicals was performed by sonication for 5 min at 30 °C followed by centrifugation (12,000 rpm, 10 min). The procedure was repeated 2 times. The supernatants were pooled together and transferred to a 10 mL volumetric flask. The volume was leveled with extraction solvent. The eluents were used for betalain and antioxidant capacity assessments.

2.5.2. Betalain Content by Spectrophotometry

Spectrophotometric evaluation of betalain content was performed as described by Šaponjac et al. [14]. Briefly, absorbance readings of liquid samples were taken by a microplate reader (SpectaMax Plus 384, Molecular Devices, LLC., San Jose, CA, USA) at 538 (a) and 476 nm (b) for evaluation of betacyanin and betaxanthin, respectively. A third measurement at 600 nm (c) was taken to correct for colored impurities. The absorbance of betanin (x) and vulgaxanthin-I (y) corrected for colored impurities were calculated using the following equations:

$$x = 1.095 \times (a - c)$$

$$y = b - z - x/3.1$$

$$z = a - x$$

where z is the absorbance of impurities. The concentrations of betacyanin (in terms of betanin) and betaxanthin (in terms of vulgaxanthin-I) were calculated by using the following equation:

$$C\ (mg/100\ mL) = x(y) \times F \times 1000/A^{1\%},$$

where F is the dilution factor, and $A^{1\%}$ is the absorbance coefficient for betanin (1120) or violaxanthin-I (750). The contents of betacyanin or betaxanthin in the protein–betalain particles were expressed as betanin or vulgaxanthin-I equivalents (mg BE or VE/g sample).

2.5.3. HPLC Profile for Betalains

The analysis of betacyanins was conducted using an Agilent 1200 HPLC (Agilent Technologies, Santa Clara, CA, USA) equipped with a photodiode array detector (DAD). Separation of the compounds was carried out on Phenomenex Synergi 4 µm hydro-RP 80A column (250 mm × 4.6 mm × 5 µm, Torrance, CA, USA) with a constant flow rate of 1 mL/min at 30 °C. The elution was conducted using a binary solvent system consisting of solvent A (5% aqueous formic acid) and solvent B (100% acetonitrile) and a gradient as follows: 0% B (0–20 min), 30% B (20–25 min), 70% B (25–27 min), 80% B (27–29 min), 0% B (29–35 min). Spectra were collected at 532 nm for betacyanins and at 480 nm for betaxanthins. Betacyanins in the samples were identified by matching the retention time and spectral characteristics of betanin and isobetanin in red beet extract diluted with dextrin (dissolved in 50% methanol).

2.6. Antioxidant Activity—DPPH Assay

The antioxidant activity was evaluated using the 2,2-diphenyl-1-picrylhydrazyl radical scavenging method (DPPH assay) [15]. Briefly, a 20 µL sample was mixed with 180 µL DPPH solution (150 µM in methanol 80%) in a microplate. Incubation for 40 min in the dark (room temperature) was followed by an absorbance measurement at 515 nm. Trolox (6-hydroxy-2,5,7,8-tetramethylchroman-2-carboxylic acid) was used to generate a standard curve. The results were expressed in Trolox equivalents (µmol TE).

2.7. Storage Stability

Protein–betalain particles and FD BEx were evaluated during a 6-week storage period. For this, samples were placed in opaque vials and stored in the dark at 5, 22, and 40 °C. Samples were analyzed right after production (week 0) and after 2 (week 2), 4 weeks (week 4), and 6 weeks (week 6) of storage for the determination of betalain content and antioxidant activity (measured as DPPH radical scavenging activity). Water activity and color evaluation, including L*, a*, b*, and total color difference (ΔE), were analyzed at week 4 and calculated according to Zhang et al. [16].

2.8. In Vitro Gastrointestinal Digestion Bioaccessibility Assay

This study's in vitro gastrointestinal digestion (GID) method was adapted from [17]. Initially, simulated salivary fluid (SSF), simulated gastric fluid (SGF), and simulated intestinal fluid (SIF) electrolyte stock solutions were prepared with the corresponding electrolytes. CB-FD and CB-SD particles (n = 3 replicates) were normalized to a concentration of 10 mg/g total betalains. For the oral phase, samples were suspended in 1 mL of distilled water, mixed with 0.7 mL of SSF, followed by sequential addition of 100 µL of 1500 U/mL porcine pancreas α-amylase solution, 50 µL of 0.03 M $CaCl_2$, and 150 µL of water (pH 7.0), homogenized for 2 min. The gastric phase started with mixing the 2 mL oral bolus with 0.64 mL of SGF, 160 µL porcine pepsin solution (25,000 U/mL), 5 µL of 0.03 M $CaCl_2$, and the pH was adjusted to 3.0 with HCl. Distilled water was added to achieve a total volume of 4 mL followed by agitation at 37 °C for 2 h (Innova 44r Incubator Shaker Series, New Brunswick Scientific, Edison, NJ, USA). For the intestinal stage, 4 mL of gastric chyme was mixed with 2.2 mL of SIF, 1 mL of a pancreatin solution (800 U/mL), 500 µL fresh bile (160 mM in fresh bile), 80 µL of 0.03 M $CaCl_2$, and the solution was adjusted to pH 7.0 with NaOH. Distilled water was added to achieve a total volume of 8 mL, and the mixture was shaken for an additional 2 h at 37 °C. Samples were centrifuged to obtain the soluble fraction (supernatant) and the residual fraction, which were immediately frozen and freeze-dried. The bioaccessibility index (BI) for betacyanidins and betaxanthins was calculated according to

$$BI\% = \left(\frac{A}{B}\right) * 100$$

where A corresponds to the metabolite (betalain) contents (mg/g) quantified in the intestinal supernatant residues for CB-FD or CB-SD, and B is for the initial mg/g betalain content for CB-FD, or CB-SD, quantified pre in vitro digestion.

2.9. Statistical Analysis

Data are presented as mean values and standard deviation of experiments performed in triplicates unless specified otherwise. Prism 10.0.3 (GraphPad Software, Boston, MA, USA) was used to perform the analysis of variance (ANOVA) with $p < 0.05$ using Tukey's or Dunnett's multiple comparisons tests. An unpaired t-test was used for comparison between two groups.

3. Results and Discussion

3.1. Characterization of Protein–Betalain Particles

3.1.1. Visual Aspect and Morphology

The beetroot extract had a bright red color, while chickpea protein is a whitish fine powder. CB-SD particles showed a bright reddish color and the aspect of a fine powder, which is typical for spray-dried powders. In contrast, CB-FD particles had a lighter pinkish-red color than CB-SD, and particles had a rough, granular aspect (Figure 1).

SEM results show the impact of each encapsulation technique on particle structure. While CB-SD produced spherical particles of various sizes with some shrinkage and concavities, CB-FD particles were mostly characterized by irregular shapes with visible pores and edges (Figure 2). In addition, CB-FD particles showed plate-shaped, fractured structures resembling the aspect of broken glass, while CB-SD particles had smoother surfaces. This is explained by the forced passage through the atomization nozzle of the spray dryer, which induces the formation of more regular spherical shapes for spray-dried particles [12]. On the other hand, FD induces the formation of porous structures due to ice crystal formation during freezing and the sublimation process during drying [18]. When comparing images with the same magnification, CB-SD particles are visibly smaller than

CB-FD. Similar morphological observations were made when comparing spray-dried and freeze-dried particles derived from blackberry juice [19], beetroot juice [20], and moringa extracts [13]. These morphological differences may affect the stability of food particles. For example, particles with significantly fractured surfaces might lead to higher oxygen permeability during storage and consequent higher core material degradation [21].

Figure 1. Concentrated beetroot extract (**A**), chickpea protein isolate (**B**), freeze-dried chickpea protein–betalain particles (**C**) and spray-dried chickpea protein–betalain particles (**D**).

3.1.2. A_w, Solubility, and Hygroscopicity

CB-SD samples had an A_w value (0.2868) typically reported in the literature for spray-dried samples. In contrast, CB-FD had a significantly lower A_w (0.0436; $p < 0.05$; Table 1), which agrees with the observed granular, crispy texture of freeze-dried samples. Water activity depends on the applied operational conditions, and it is strongly correlated with food texture. A similar trend was observed for spray-dried and freeze-dried beetroot juice with pumpkin protein [22]. Despite the observed differences, both treatments are within the microbiologically safe range ($A_w < 0.65$) for food products [21].

Table 1. Betalain retention and physicochemical properties of protein–betalain particles produced by freeze drying and spray drying.

Parameters	CB-FD	CB-SD
Moisture, %	1.66 ± 0.12 [b]	4.65 ± 0.05 [a]
Water activity (A_w)	0.0436 ± 0.0043 [b]	0.2868 ± 0.0024 [a]
Hygroscopicity (% w/w)	14.90 ± 0.20 [b]	18.85 ± 0.03 [a]
Solubility (% w/w)	52.41 ± 0.94 [b]	78.25 ± 0.58 [a]
Instrumental color	L* = 44.33 ± 1.13 a* = 33.31 ± 0.82 b* = -2.63 ± 0.41	L* = 41.22 ± 1.09 a* = 42.83 ± 2.84 b* = -6.16 ± 0.23
Betalain Content		
Betacyanins (mg/g)	2.59 ± 0.12	5.48 ± 0.36
Betaxanthins (mg/g)	1.84 ± 0.19	3.84 ± 0.29
Total Betalains (mg/g)	4.40 ± 0.29	9.32 ± 1.51
DPPH Antioxidant capacity (μMol TE)	64.75 ± 4.2	82.95 ± 10.0

Results are shown as mean ± standard deviation. Sample identification: freeze-dried chickpea protein–betalain particles (CB-FD) and spray-dried chickpea protein–betalain particles (CB-SD). Different letters ([a],[b]) on the same line indicate a significant difference using an unpaired t-test ($p < 0.05$).

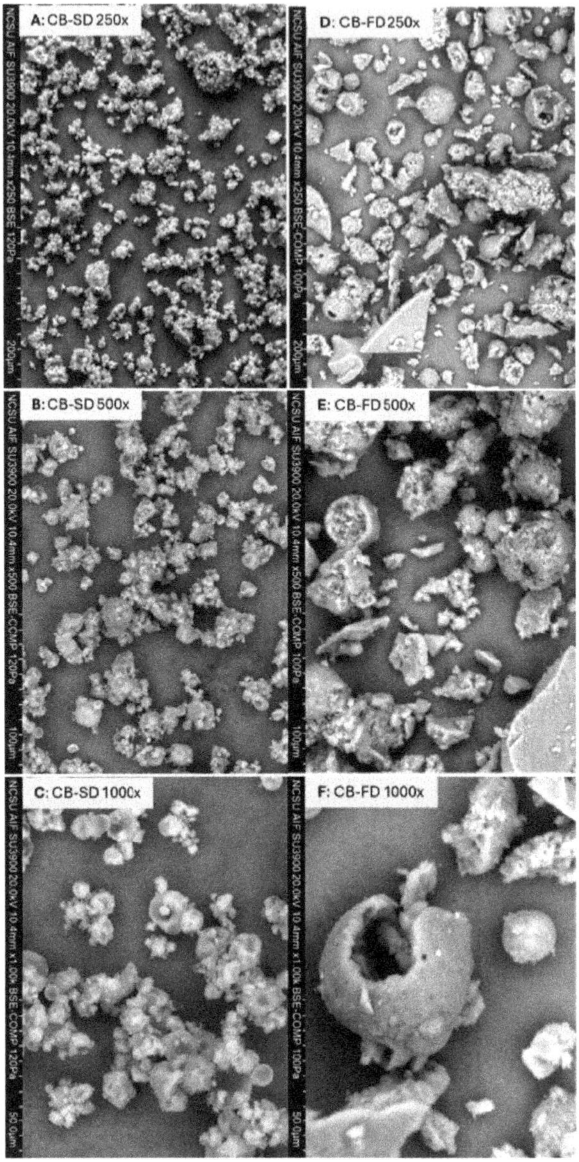

Figure 2. Scanning electron micrographs of CB-SD: spray-dried chickpea protein–betalain particles (**A–C**) and CB-SD: freeze-dried chickpea protein–betalain particles (**D–F**). Images (**A**,**D**): magnification 250×; images (**B**,**E**): magnification 500×; images (**C**,**E**): magnification 1000×.

A fast and complete powder reconstitution is a critical quality aspect. Indeed, when it comes to food formulation and incorporation of food powders into food products, higher solubility is a desirable attribute for most industrial operations since it simplifies production protocols and minimizes preparation time and costs [14]. Significant differences were observed for solubility in water: while CB-SD was highly soluble (almost 80%), CB-FD had poor solubility (approx. 52%; Table 1). The tested solubility of chickpea protein in water was 39.75%. Spray-dried samples are smaller and more uniform particles (Figure 2) than freeze-dried particles, which might play an important role in the observed differences [23]. It was found that chemical interactions between protein and polyphenols may result in protein–polyphenol structures with either higher or lower solubility, depending on the type of protein, polyphenol, and pH [24]. In our study, both CB-SD and CB-FD particles had higher solubility than chickpea protein alone (39.95 ± 0.75%) when evaluated in their natural pH (approx. pH 6.5). Higher protein surface hydrophilicity induced by phenolic binding might play a role in the observed higher solubility of CB-SD and CB-FD compared to non-complexed chickpea protein [25].

In addition, the drying method significantly affected the hygroscopicity of resulting protein–betalain particles ($p < 0.05$; Table 1). The observed difference results lead to distinct definitions according to GEA classification [23], while CB-FD is classified as slightly hygroscopic (results between 10% and 15%), and CB-SD is considered hygroscopic (results between 15% and 20%). A similar trend was observed for moringa leaf extract with maltodextrin and high methoxyl pectin [13] and beetroot juice with pumpkin protein encapsulated by spray drying and freeze drying [20]. Both treatments had higher hygroscopicity when compared to the chickpea protein alone (5.17 ± 0.06%), which indicates that the increased capacity to absorb water from the environment is directly related to the presence of the beetroot extract solids. Our results are higher than spray-dried and freeze-dried protein–polyphenol particles produced with wild blueberry pomace extracts and different protein (soy protein isolate, chickpea, and wheat flour) sources [12] but lower than spray-dried elderberry juice and pomace extract with soy protein isolate [24].

3.1.3. Phytochemical Composition

Betalains were measured using the photometric method. The recorded concentration of red betacyanins pigments is 12.37 ± 0.85 mg BE/g for the beet extract (BEx), 2.59 ± 0.12 mg BE/g for freeze-dried concentrated beet (CB-FD), and 5.48 ± 0.36 mg BE/g for spray-dried concentrated beet (CB-SD). For the yellow-orange pigments called betaxanthins, the measurements were 7.68 ± 0.71 mg VE/g for BEx, 1.84 ± 0.19 mg VE/g for CB-FD, and 3.84 ± 0.29 mg VE/g for CB-SD. In total, the betalain content was 19.85 ± 1.54 mg/g dry weight (dw) for BEx, 4.40 ± 0.92 mg/g dw for CB-FD, and 9.30 ± 0.61 mg/g dw for CB-SD (Table 1). The DPPH antioxidant capacity assay recorded 82.95 ± 10.00 and 64.75 ± 4.20 μMol Trolox equivalent for CB-SD and CB-FD, respectively.

The betalain content in CB-FD was found to be lower than that in CB-SD, as shown in Table 1. This difference can be attributed to the complexation method employed in preparing the CB-FD complex. Specifically, during this process, centrifugation was used, which effectively separated the supernatant rich in excess betalains from the pellet containing the betalain–protein complex. Subsequently, the pellet was freeze-dried to produce the final CB-FD particles [9]. Moreover, spray drying is a drying technique that uses a high temperature-short time principle. Due to the evaporative cooling effect, the particles are exposed to a much lower processing temperature than the inlet temperature, which means that the exposure to high temperatures is minimal. In fact, drying takes place almost instantaneously and an intensive evaporation happens at the surface of each droplet [26,27]. Other studies have shown similar trends. For example, compared to freeze drying, spray drying showed a better or similar performance regarding the retention of betalains and phenolics from red cactus pear [28].

HPLC-DAD detection of major betalains recorded at 480 and 532 nm wavelengths compared to reference standard beet root extract displayed profile similarity. Three betacyanins were identified: betanin, isobetanin, and neobetanin. Vulgaxanthin-I showed up at 480 nm (Figure 3). Betanin is the major component in red beetroot. HPLC peak areas showed that the ratio of betanin to violaxanthin-1 is 1.0:0.2.

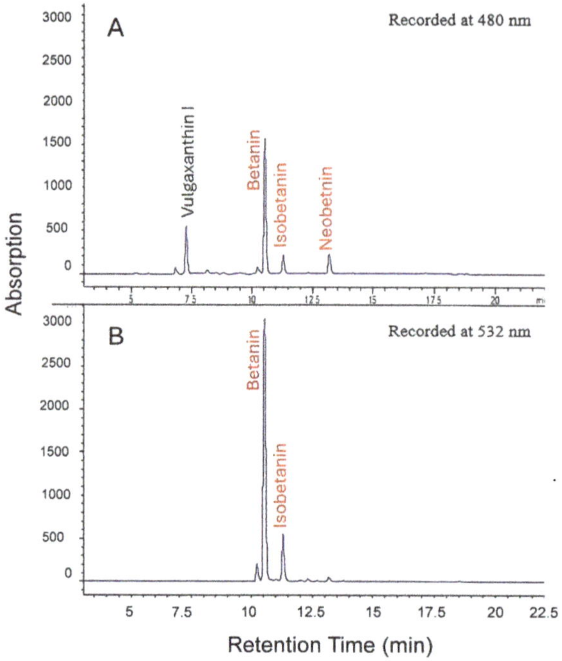

Figure 3. HPLC profiles of betalains recorded at 480 nm showing the major betxanthin (Vulgaxanthin I) and major betacyanins (Betanin, Isobetanin and Neobetanin) (**A**) and betacyanins (Betanin, Isobetanin) recorded at 532 nm (**B**).

LC-MS and MS2 allowed us to predict the molecular formula. With the aid of UV/Vis spectra and retention time, the identification process was accomplished by comparing the experimental MS data to previous studies [29–32]. Twelve betacyanins, including isomers of betanin, betanidin, and 17-decarboxybetanin, and six betaxanthins were identified in CB-FD and CB-SD particles (Table 2).

Table 2. Betalain compounds identified in protein–betalain particles produced by freeze drying and spray drying from beetroot extract and chickpea protein isolate.

No.	Compound	Rt (min)	MS m/z $[M + H]^+$	MS/MS m/z	FD-CB [1]	SD-CB [2]	References
			Betacyanins (Betanins)				
1	Betanin	8.73	551.15	389	+	+	[29]
2	Betanidin	8.78	389.09	345	+	+	[29]
3	17-Decarboxybetanin	8.78	507.16	345	+	+	[29]
4	2,17-Bidecarboxy-neobetanin	8.78	461.15	299	+	+	[29]
5	17-Decaboxyneobetanin	8.80	505.14	343, 297	+	+	[29]
6	2′-*O*-Glucosyl-isobetanin	8.90	713.2	551, 389	+	+	[29]
7	Prebetanin	9.05	631.1	551, 389	+	+	[29]
3′	17-Decarboxyisobetanin	9.78	507.16	345	+	+	[29]
2′	Isobetanidin	9.95	389.09	345	+	+	[29]
1′	Isobetanin	9.96	551.15	389	+	+	[29]
10	Neobetanin	13.47	549.13	387	+	+	[29]
11	2-Decarboxy-neobetanin	14.19	505.14	343, 297	+	+	[29]

Table 2. *Cont.*

No.	Compound	Rt (min)	MS m/z $[M + H]^+$	MS/MS m/z	FD-CB [1]	SD-CB [2]	References
			Betaxanthins (bx)				
12	Phenyl alanine-isobx	2.51	359.12	315	+	+	[29,30]
13	Asparagine-bx (Vulgaxanthin III)	2.59	326.1	280	+	+	[29,30]
14	Glutamine-bx (Vulgaxanthin I)	2.62	340.11	323, 277	+	+	[29,30]
15	Proline-bx (Indicaxanthin)	8.59	309.11	291	+	+	[29,30]
16	Valine-bx/-IsoBx	14.55	311.12	267	+	+	[29]
17	Leucine-bx (Vulgaxanthin IV)	18.72	325.14	281	+	+	[29]

[1] Freeze-dried chickpea protein–betalain; [2] spray-dried chickpea protein–betalain particles.

There is no commercially [33]. However, the profiles of the compounds could change during the purification process and degradation is possible.

Nemzer [29], using their proprietary method of large- available betalain standards for performing qualitative and quantitative analysis of betalains. Semi-quantitative methods relied on measuring HPLC peak area relative to total peak areas [29,32]. This method requires suitable separation chromatography. Some researchers isolated the major betacyanin, betanin, and the betaxanthin, vulgaxanthin I, from beetroot extract and used them for semi-quantification of betacyanins as betanin equivalent and betaxanthins as betaxanthin I equivalent [34]. Others had to isolate major betacyanins and semi-synthesized betanin derivatives, and betaxanthins scale chromatographic purification of red beet root extract, allowed the production of more concentrated betalain formulations. Thirty betacyanins, including ten isomers, and seventeen betaxanthins with seven isomers were tentatively identified using HPLC-MS. Betanin/isobetanin represented 41% of total betacyanins. Others included decarboxylated betanin derivatives and neobetanin.

In another study of thirteen different varieties of beetroot, Sawicki [31] indicated that betanin, isobetanin, and vulgaxanthin I constituted the predominant compounds. Betanin and isobetanin represented 41–64% and 15–20%, respectively, while vulgaxanthin I ranged from 13.3 to 28.3% of total betalain content in the examined varieties. Compared to the previous studies [29,31], the LC-MS profile of our FD and SD particles indicated that the two encapsulation methods captured the main betalain components from beetroot extract.

3.2. Storage Stability

3.2.1. Physicochemical Attributes—A_w and Color

Only freeze-dried particles had a significant A_w increase during 4-week storage at different temperatures (Figure 4A). CB-FD had an initial lower A_w than CB-SD, and therefore, the higher water gradient between the protein–betalain particles and the environment creates favorable conditions for water uptake. This tendency agrees with previous reports assessing food powders kept under similar storage conditions [35]. However, despite the observed variations during storage, all protein–betalain particles had A_w levels within the microbiologically safe range [36].

The International Commission on Illumination (CIE) recommends the following interpretation of results: $\Delta E < 5$: negligible color variation, $5 \leq \Delta E < 12$: detectable color difference, and $\Delta E \geq 12$: remarkable color difference [37]. According to this scale, after 2 weeks of storage, the color differences were negligible, while after 4 weeks, higher values were detected but still under the threshold of $\Delta E < 12$, except for CB-SD. Indeed, after 4 weeks, treatment CB-SD stored at 22 °C reached the highest ΔE, which coincides with higher a* and b* values (Figure 4C,E). Although most groups showed increased ΔE as storage progressed, CB-FD stored at 40 °C had higher ΔE after 2 weeks followed by a decrease after 4 weeks. This atypical behavior is justified by an abruptly higher L* after 2 weeks, followed by a significant decrease after 4 weeks (Figure 4D). Our ΔE results are lower than observed for betalain extracts produced from beetroots and stored at 4, 25, and 40 °C [37].

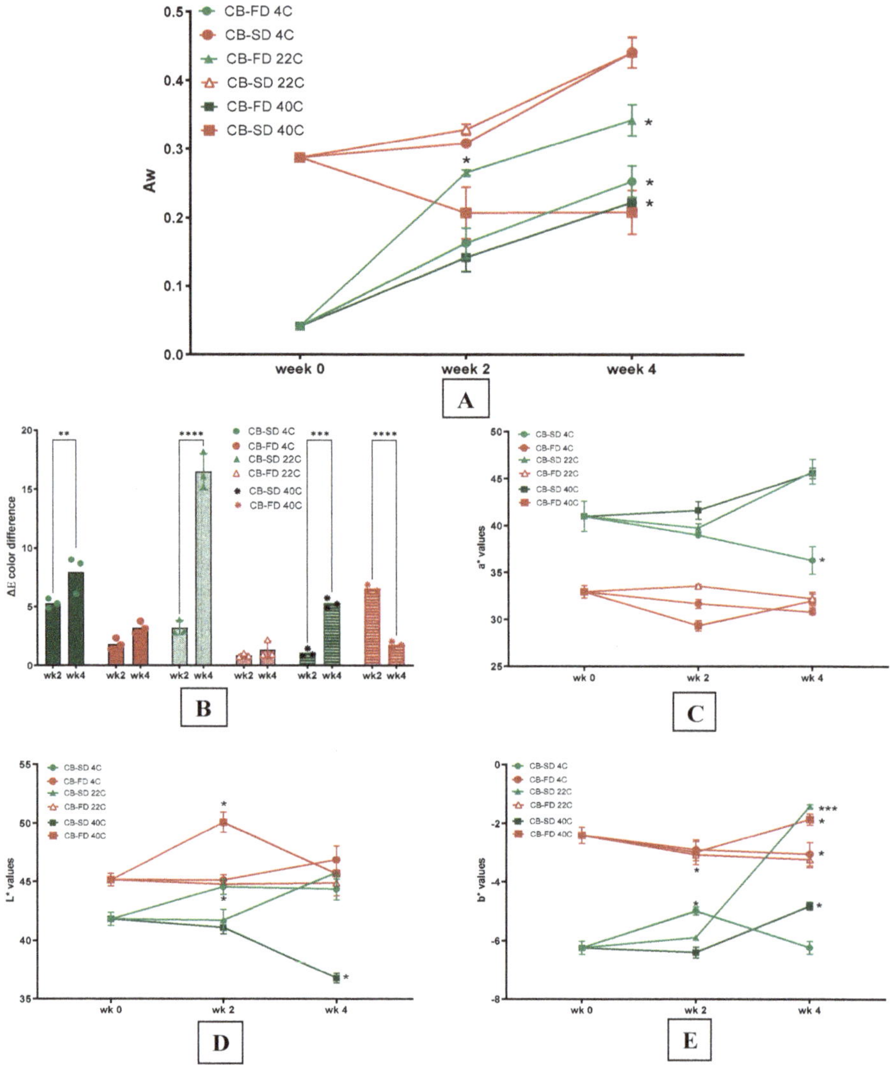

Figure 4. A_w (**A**), ΔE total color difference (**B**), L* (**C**), a* (**D**), and b* (**E**) results of protein–betalain particles during 28-day storage at different temperatures. Legend: CB-FD 4C: freeze-dried chickpea protein–betalain particles stored at 4 °C; CB-FD 22C: freeze-dried chickpea protein–betalain particles stored at 22 °C; CB-FD 40 C: freeze-dried chickpea protein–betalain particles stored at 40 °C; CB-SD 4C: spray-dried chickpea protein–betalain particles stored at 4 °C; CB-SD 22C: spray-dried chickpea protein–betalain particles stored at 22 °C; CB-SD 40 C: spray-dried chickpea protein–betalain particles stored at 40 °C. Asterisks indicate significant statistical differences by ANOVA analysis and Dunnett's test between samples analyzed after 2 or 4 weeks of storage and right after production (week 0). *: $p \leq 0.05$; **: $p \leq 0.01$; ***: $p \leq 0.001$; ****: $p \leq 0.0001$. For clarity, some asterisks are shown on the side of the experimental point.

Moreover, results show a clear color differentiation between treatments. Independent of the temperature, during the 28-day storage, CB-SD particles maintained a more intense red color (higher a* values) and darker tones (lower L*) compared to CB-FD (Figure 4C,D). The a* values of CB-SD are higher than beetroot juice concentrate spray-dried with maltodextrin, gum Arabic, and whey protein concentrate [38], while our L* values are lower than encapsulated beetroot extracts produced by spray and freeze drying [20].

Red is the dominant color in beetroot extracts, and a* values are important color markers for protein–betalain particles. Interestingly, as storage progressed, the red color of CB-SD particles was maintained or intensified, as demonstrated by stable or increased a* values (Figure 4C). This finding is a positive indication of the stabilization of encapsulated natural beetroot pigments. Indeed, it has been shown that the degradation of betacyanins and betaxanthins can be minimized by exposing the extracts to mild temperatures around or below 60 °C [39]. Therefore, besides the effective betalains encapsulation by chickpea protein [38], the low outlet spray drying temperature (kept around 65 °C) and the short spray drying time [40] justify the observed preservation of protein–betalain particle redness.

3.2.2. Thermal Stability of Betalains and Antioxidant Activity

The retention of betacyanins in CB-FD, CB-SD, and the non-encapsulated BEx remained consistent over a six-week period when stored at 4 °C (Figure 5A). In contrast, BEx exhibited significant degradation starting at the four-week mark ($p > 0.5$), with an increased decline at week six under 22 °C storage. The elevated temperature had a greater impact on BEx, maintaining only 73% betacyanin retention at 40 °C. On the other hand, the encapsulated forms, CB-FD and CB-SD, demonstrated superior stability compared to BEx, particularly CB-SD, which maintained its stability without significant changes throughout six weeks of storage at 4, 22, and 40 °C. While CB-FD showed some decrease in stability, with a slight drop in betacyanin content—retaining 89% after six weeks at both 22 °C and 40 °C—CB-SD remained the more stable option. Betaxanthin pigments in CB-FD and CB-SD exhibited remarkable stability, with no significant differences observed over six weeks (Figure 5B).

Despite a few limitations, the DPPH radical scavenging assay remains a popular method for assessing the antioxidant activities found in natural products. This assay is known for its simplicity, affordability, speed, reproducibility, and compatibility with thermally unstable compounds. Additionally, there is a strong correlation between the results and the content of bioactive compounds, with a regression factor exceeding 0.8 [41]. Some precautions were applied, like performing the assay in organic media and avoiding exposure to light. The DPPH radical scavenging activity results indicated a linear correlation coefficient with total betalain concentration ($r = 0.86$), shown in Figure 5C.

At higher temperatures, betalains undergo several degradation processes, including hydrolysis, isomerization, dehydrogenation, deglycosylation, and decarboxylation [42]. Specifically, increased heat prompts the decarboxylation of betanin, resulting in the formation of neobetanin, which alters the color due to creating a less stable aglycone [42]. Additionally, the thermal breakdown of betalains leads to the production of mono-, di-, and tricarboxylic betacyanins [32].

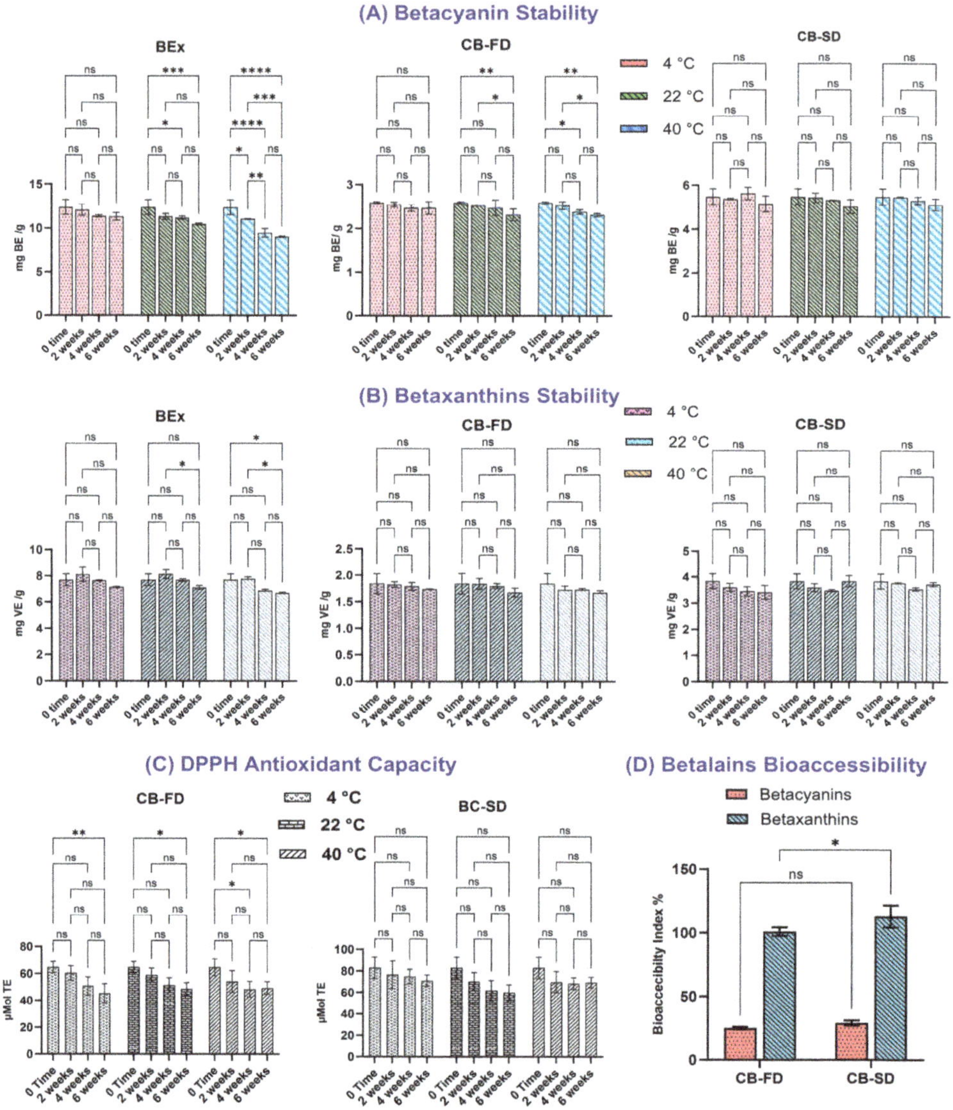

Figure 5. Retention of betacyanins (**A**), betaxanthins (**B**), and DPPH antioxidant capacity (**C**) of beetroot extract (BEx), freeze-dried chickpea protein–betalain particles (BC-FD) and spray-dried chickpea protein–betalain particles (CB-SD) during six weeks of storage at 5 °C, 22 °C, and 40 °C. (**D**) Bioaccessibility index of betacyanins and betaxanthins after intestinal phase digestion. According to ANOVA analysis and Dunnett's test, asterisks indicate significant differences between samples analyzed after 2, 4, or 6 weeks of storage and right after production (week 0). *: $p \leq 0.05$; **: $p \leq 0.01$; ***: $p \leq 0.001$; ****: $p \leq 0.0001$.

Pigment concentration plays a significant role in the thermal degradation of betalains. Increased pigmentation levels enhance betalains' stability [43]. The low total betalain content in CB-FD (4.40 ± 0.29 mg/g), which is less than half the concentration found in CB-SD (9.34 ± 1.0 mg/g), slightly lowers the stability of CB-FD compared to CB-SD particles.

An increase in water activity (A_w) can lead to the degradation of phytochemical compounds throughout storage. Water absorption and the resultant rise in water activity led to greater carrier plasticity and promoted phytochemical degradation, for instance, under identical bay losses in total phenolic content and anthocyanins [44]. Only CB-FD demonstrated a notable A_w increase at 4 and 22 °C by the fourth week (Figure 5A). Initially, CB-FD had a lower A_w than CB-SD, but it increased significantly over time, adversely impacting the stability of CB-FD.

Overall, our findings demonstrate the efficacy of chickpea encapsulation in prolonging the shelf-life of betalains, with spray drying being the most effective encapsulation method for beetroot extracts.

Many proteins have been proven to transport and enhance the stability of bioactives of small molecules. Soy protein encapsulated with beetroot extract by FD increased the thermal retention of betalain from 55.3% to 75.9%, as reported by Hu et al. [45]. In another citation, rice protein and pea protein significantly decreased the thermal degradation rate of betalain from 94% to 56%. They referred that to the hydrophobic forces and hydrogen bonding between protein and betalains [46].

3.2.3. In Vitro Simulated Gastrointestinal Digestion

Bioaccessibility was assessed by measuring the contents of betacyanins and betaxanthins following in vitro simulated gastrointestinal digestion (SGD) of CB-FD and CB-SD particles. Figure 5D illustrates the % bioaccessibility of betacyanins and betaxanthins in the intestinal supernatant after SGD. We observed a loss of 74.9% of total betacyanins for CB-FD and 70.84% for CB-SD, with no significant differences, suggesting that betacyanins were less stable under digestion conditions. Previous reports noted even lower % bioaccessibility levels of betacyanins [47]. In another study, betacyanins was decreased by 22% during the simulated oral digestion, did not change in the gastric digestion stage, whereas a significant decrease was noted after the intestinal digestion step [48]. The reduction in betacyanins during SGD may be attributed to the hydrophobic nature of betanin, the primary component, coupled with digestive factors such as high temperatures, enzymes, and alterations in pH [31,47]. In contrast, betaxanthins displayed greater stability during SGD, achieving recovery rates $\geq$ 100%. This remarkable stability is in line with the previously mentioned storage stability and aligns with the research by Tesoriere [49]. Their findings show that betaxanthins were completely soluble in the aqueous (bioaccessible) fraction after the ultracentrifugation of post-intestinal digesta, while the release of betacyanins from the matrix was not fully achieved. Betacyanins are more sensitive to oxygen and temperature, and they display less stability in low pH environments compared to betaxanthins. In contrast, betaxanthins demonstrate greater stability in acidic conditions and during exposure to hydrolytic enzymes [49].

4. Conclusions

This study explored how encapsulating beetroot extract in chickpea protein isolate by freeze drying and spray drying affects various properties. CB-SD showcased a more vibrant and darker appearance than CB-FD, and different morphology was observed, with CB-SD being smaller, round-shaped particles and CB-FD having irregular fractured structures. CB-SD had higher A_w and water solubility compared to CB-SD. Interestingly, both particles outperformed regular chickpea protein in terms of solubility.

CB-FD experienced a significant increase in A_w at 4, 22, and 40 °C during storage, but results remained within microbiological safety limits. Betalains were more stable when encapsulated in spray-dried particles over a six-week period at 22 and 40 °C. The antioxidant capacity (DPPH) is strongly correlated with the betalain content. When subjected to simulated gastrointestinal digestion, there was a substantial loss in betacyanidin. However, betaxanthins demonstrated remarkable stability and retention following the intestinal phase of digestion. In summary, encapsulating betalain compounds from beetroot extract in chickpea protein isolate via spray drying is a clean, efficient, and cost-effective technology to produce food ingredients with preserved phytochemical content and extended shelf life. In future research, it will be important to explore additional components in beetroot, including phenolics, vitamins, and sugars, while assessing their bioaccessibility data. Furthermore, contrasting the results of this study with in vivo bioaccessibility data could yield significant insights.

Supplementary Materials: The following supporting information can be downloaded at: https://www.mdpi.com/article/10.3390/foods14020281/s1, Table S1. Betalain content in chickpea protein–beetroot extract (CP-BRE) mixture before and after spray drying at three protein concentrations: 6, 8, and 10%. Table S2. Betalain content in chickpea protein–beetroot extract (CP-BRE) after freeze drying at three protein concentration: 6, 8, and 10%.

Author Contributions: Conceptualization, M.A.L., M.H.G. and R.T.H.; methodology, V.I.C., M.H.G., R.T.H. and M.A.; software, M.H.G. and R.T.H.; validation, M.H.G., R.T.H. and V.I.C.; formal analysis, M.H.G., R.T.H., V.I.C. and M.A.; investigations, R.T.H., M.H.G. and M.A. resources, M.A.L.; data curation, V.I.C., R.T.H. and M.H.G.; writing—original draft preparation, R.T.H. and M.H.G., writing—review and editing, M.A.L. and R.T.H. All authors have read and agreed to the published version of the manuscript.

Funding: Morphological analyses were performed at the Analytical Instrumentation Facility (AIF) at North Carolina State University, which is supported by the State of North Carolina and the National Science Foundation (award number ECCS-2025064).

Institutional Review Board Statement: Not applicable.

Informed Consent Statement: Not applicable.

Data Availability Statement: The original contributions presented in this study are included in the article/Supplementary Materials. Further inquiries can be directed to the corresponding author.

Acknowledgments: Appreciation is expressed to Fulbright Bulgaria for the Fulbright Scholarship provided to V.I.C. and fruitful hospitality of M.A.L.

Conflicts of Interest: The authors declare no conflicts of interest.

References

1. Akan, S.; Tuna Gunes, N.; Erkan, M. Red Beetroot: Health Benefits, Production Techniques, and Quality Maintaining for Food Industry. *J. Food Process. Preserv.* **2021**, *45*, e15781. [CrossRef]
2. Fernández-García, E.; Carvajal-Lérida, I.; Pérez-Gálvez, A. In Vitro Bioaccessibility Assessment as a Prediction Tool of Nutritional Efficiency. *Nutr. Res.* **2009**, *29*, 751–760. [CrossRef] [PubMed]
3. Gaona-Ruiz, M.; Vallejo-García, J.L.; Arnaiz, A.; Sedano-Labrador, C.; Trigo-López, M.; Rodríguez, A.; Carrillo, C.; Vallejos, S. Smart Polymers and Smartphones for Betalain Measurement in Cooked Beetroots. *Food Chem.* **2024**, *459*, 140358. [CrossRef] [PubMed]
4. Adjei, M.L.; Boakye, A.; Deku, G.; Pepra-Ameyaw, N.B.; Jnr, A.S.A.; Oduro, I.N.; Ellis, W.O. Development of Yoghurt Incorporated with Beetroot Puree and Its Effect on the Physicochemical Properties and Consumer Acceptance. *Heliyon* **2024**, *10*, e25492. [CrossRef] [PubMed]
5. Jafari, S.; Jafari, S.M.; Ebrahimi, M.; Kijpatanasilp, I.; Assatarakul, K. A Decade Overview and Prospect of Spray Drying Encapsulation of Bioactives from Fruit Products: Characterization, Food Application and in Vitro Gastrointestinal Digestion. *Food Hydrocoll.* **2023**, *134*, 108068. [CrossRef]
6. Samborska, K.; Barańska, A.; Boostani, S.; Riazi, M.; Jafari, S.M. Introduction to the Spray Drying Process. In *Spray Drying for the Food Industry*; Elsevier: Amsterdam, The Netherlands, 2024; pp. 3–28. ISBN 978-0-12-819799-8.
7. Woo, M.W.; Bhandari, B. Spray Drying for Food Powder Production. In *Handbook of Food Powders*; Elsevier: Amsterdam, The Netherlands, 2024; pp. 19–36. ISBN 978-0-323-98820-9.

8. Ratti, C. Freeze Drying for Food Powder Production. In *Handbook of Food Powders*; Elsevier: Amsterdam, The Netherlands, 2024; pp. 37–56. ISBN 978-0-323-98820-9.
9. Grace, M.H.; Hoskin, R.T.; Hayes, M.; Iorizzo, M.; Kay, C.; Ferruzzi, M.G.; Lila, M.A. Spray-Dried and Freeze-Dried Protein-Spinach Particles; Effect of Drying Technique and Protein Type on the Bioaccessibility of Carotenoids, Chlorophylls, and Phenolics. *Food Chem.* **2022**, *388*, 133017. [CrossRef]
10. Hoskin, R.T.; Grace, M.H.; Xiong, J.; Lila, M.A. Spray-drying Microencapsulation of Blackcurrant and Cocoa Polyphenols Using Underexplored Plant-based Protein Sources. *J. Food Sci.* **2023**, *88*, 2665–2678. [CrossRef]
11. Lin, Y.; Cheng, N.; Jiang, Y.; Grace, M.H.; Lila, M.A.; Hoskin, R.T.; Zheng, H. Colloidal and Interfacial Properties of Spray Dried Pulse Protein-Blueberry Polyphenol Particles in Model Dispersion Systems. *Food Chem.* **2024**, *457*, 140073. [CrossRef]
12. Correia, R.; Grace, M.H.; Esposito, D.; Lila, M.A. Wild Blueberry Polyphenol-Protein Food Ingredients Produced by Three Drying Methods: Comparative Physico-Chemical Properties, Phytochemical Content, and Stability during Storage. *Food Chem.* **2017**, *235*, 76–85. [CrossRef]
13. Dadi, D.W.; Emire, S.A.; Hagos, A.D.; Eun, J.-B. Physical and Functional Properties, Digestibility, and Storage Stability of Spray- and Freeze-Dried Microencapsulated Bioactive Products from Moringa Stenopetala Leaves Extract. *Ind. Crop. Prod.* **2020**, *156*, 112891. [CrossRef]
14. Tumbas Šaponjac, V.; Čanadanović-Brunet, J.; Ćetković, G.; Jakišić, M.; Vulić, J.; Stajčić, S.; Šeregelj, V. Optimisation of Beetroot Juice Encapsulation by Freeze-Drying. *Pol. J. Food Nutr. Sci.* **2020**, *70*, 25–34. [CrossRef]
15. Bobo-García, G.; Davidov-Pardo, G.; Arroqui, C.; Vírseda, P.; Marín-Arroyo, M.R.; Navarro, M. Intra-Laboratory Validation of Microplate Methods for Total Phenolic Content and Antioxidant Activity on Polyphenolic Extracts, and Comparison with Conventional Spectrophotometric Methods: Comparison of Microplate and Conventional Methods for Folin-Ciocalteu and DPPH. *J. Sci. Food Agric.* **2015**, *95*, 204–209. [CrossRef] [PubMed]
16. Zhang, Z.-H.; Li, X.; Ma, A.; Gao, X.; Zhu, S.; Li, B. Characteristics of Pomegranate (*Punica Granatum* L.) Peel Polyphenols Encapsulated with Whey Protein Isolate and β-Cyclodextrin by Spray-Drying. *Int. J. Biol. Macromol.* **2024**, *278*, 135279. [CrossRef] [PubMed]
17. Minekus, M.; Alminger, M.; Alvito, P.; Ballance, S.; Bohn, T.; Bourlieu, C.; Carrière, F.; Boutrou, R.; Corredig, M.; Dupont, D.; et al. A Standardised Static *in Vitro* Digestion Method Suitable for Food—An International Consensus. *Food Funct.* **2014**, *5*, 1113–1124. [CrossRef] [PubMed]
18. Ledari, S.A.; Milani, J.M.; Shahidi, S.-A.; Golkar, A. Comparative Analysis of Freeze Drying and Spray Drying Methods for Encapsulation of Chlorophyll with Maltodextrin and Whey Protein Isolate. *Food Chem. X* **2024**, *21*, 101156. [CrossRef]
19. Franceschinis, L.; Salvatori, D.M.; Sosa, N.; Schebor, C. Physical and Functional Properties of Blackberry Freeze- and Spray-Dried Powders. *Dry. Technol.* **2014**, *32*, 197–207. [CrossRef]
20. Čakarević, J.; Šeregelj, V.; Tumbas Šaponjac, V.; Ćetković, G.; Čanadanović Brunet, J.; Popović, S.; Kostić, M.H.; Popović, L. Encapsulation of Beetroot Juice: A Study on the Application of Pumpkin Oil Cake Protein as New Carrier Agent. *J. Microencapsul.* **2020**, *37*, 121–133. [CrossRef]
21. Laokuldilok, T.; Kanha, N. Effects of Processing Conditions on Powder Properties of Black Glutinous Rice (*Oryza Sativa* L.) Bran Anthocyanins Produced by Spray Drying and Freeze Drying. *LWT-Food Sci. Technol.* **2015**, *64*, 405–411. [CrossRef]
22. Laureanti, E.J.G.; Paiva, T.S.; De Matos Jorge, L.M.; Jorge, R.M.M. Microencapsulation of Bioactive Compound Extracts Using Maltodextrin and Gum Arabic by Spray and Freeze-Drying Techniques. *Int. J. Biol. Macromol.* **2023**, *253*, 126969. [CrossRef]
23. GEA—Engineering for a Better World. Available online: https://www.gea.com/ (accessed on 22 November 2024).
24. Ravichandran, K.S.; Silva, E.S.; Moncada, M.; Perkins-Veazie, P.; Lila, M.A.; Greenlief, C.M.; Thomas, A.L.; Hoskin, R.T.; Krishnaswamy, K. Spray Drying to Produce Novel Phytochemical-Rich Ingredients from Juice and Pomace of American Elderberry. *Food Biosci.* **2023**, *55*, 102981. [CrossRef]
25. Yan, X. A Review of the Structure, Function, and Application of plant-Based Protein–Phenolic Conjugates and Complexes. *Compr. Rev. Food Sci. Food Saf.* **2023**, *22*, 1312–1336. [CrossRef] [PubMed]
26. Bhandari, B.R.; Datta, N.; Howes, T. Problems Associated with Spray Drying of Sugar-Rich Foods. *Dry. Technol.* **1997**, *15*, 671–684. [CrossRef]
27. Gharsallaoui, A.; Roudaut, G.; Chambin, O.; Voilley, A.; Saurel, R. Applications of Spray-Drying in Microencapsulation of Food Ingredients: An Overview. *Food Res. Int.* **2007**, *40*, 1107–1121. [CrossRef]
28. Morales, N.X.C.; Gómez, K.Y.V.; Schweiggert, R.M.; Delgado, G.T.C. Stabilisation of Betalains and Phenolic Compounds Extracted from Red Cactus Pear (*Opuntia Ficus-Indica*) by Spray and Freeze-Drying Using Oca (*Oxalis Tuberosa*) Starch as Drying Aid. *Food Sci. Technol. Int.* **2021**, *27*, 456–469. [CrossRef]
29. Nemzer, B.; Pietrzkowski, Z.; Spórna, A.; Stalica, P.; Thresher, W.; Michałowski, T.; Wybraniec, S. Betalainic and Nutritional Profiles of Pigment-Enriched Red Beet Root (*Beta Vulgaris* L.) Dried Extracts. *Food Chem.* **2011**, *127*, 42–53. [CrossRef]
30. Kugler, F.; Graneis, S.; Stintzing, F.C.; Carle, R. Studies on Betaxanthin Profiles of Vegetables and Fruits from the Chenopodiaceae and Cactaceae. *Z. Naturforschung C* **2007**, *62*, 311–318. [CrossRef]

31. Sawicki, T.; Martinez-Villaluenga, C.; Frias, J.; Wiczkowski, W.; Peñas, E.; Bączek, N.; Zieliński, H. The Effect of Processing and in Vitro Digestion on the Betalain Profile and ACE Inhibition Activity of Red Beetroot Products. *J. Funct. Foods* **2019**, *55*, 229–237. [CrossRef]
32. Wybraniec, S. Formation of Decarboxylated Betacyanins in Heated Purified Betacyanin Fractions from Red Beet Root (*Beta Vulgaris* L.) Monitored by LC-MS/MS. *J. Agric. Food Chem.* **2005**, *53*, 3483–3487. [CrossRef]
33. Xie, G.-R.; Chen, H.-J. Comprehensive Betalain Profiling of Djulis (*Chenopodium Formosanum*) Cultivars Using HPLC-Q-Orbitrap High-Resolution Mass Spectrometry. *J. Agric. Food Chem.* **2021**, *69*, 15699–15715. [CrossRef]
34. Spórna-Kucab, A.; Tekieli, A.; Grzegorczyk, A.; Świątek, Ł.; Rajtar, B.; Skalicka-Woźniak, K.; Starzak, K.; Nemzer, B.; Pietrzkowski, Z.; Wybraniec, S. Metabolite Profiling Analysis and the Correlation with Biological Activity of Betalain-Rich Portulaca Grandiflora Hook. Extracts. *Antioxidants* **2022**, *11*, 1654. [CrossRef]
35. Ramírez, M.J.; Giraldo, G.I.; Orrego, C.E. Modeling and Stability of Polyphenol in Spray-Dried and Freeze-Dried Fruit Encapsulates. *Powder Technol.* **2015**, *277*, 89–96. [CrossRef]
36. Looi, Y.F.; Ong, S.P.; Julkifle, A.; Alias, M.S. Effects of Pretreatment and Spray Drying on the Physicochemical Properties and Probiotics Viability of Moringa (*Moringa Oleifera* Lam) Leaf Juice Powder. *J. Food Process. Preserv.* **2019**, *43*, e13915. [CrossRef]
37. Lombardelli, C.; Benucci, I.; Mazzocchi, C.; Esti, M. Betalain Extracts from Beetroot as Food Colorants: Effect of Temperature and UV-Light on Storability. *Plant Foods Hum. Nutr.* **2021**, *76*, 347–353. [CrossRef]
38. Bazaria, B.; Kumar, P. Comparative Analysis of Bio-Polymers Addition on Structural and Physical Properties of Spray Dried Beetroot Juice Concentrate. *J. Food Process. Preserv.* **2017**, *41*, e13232. [CrossRef]
39. Chew, Y.M.; Hung, C.-H.; King, V.A.-E. Accelerated Storage Test of Betalains Extracted from the Peel of Pitaya (*Hylocereus Cacti*) Fruit. *J. Food Sci. Technol.* **2019**, *56*, 1595–1600. [CrossRef]
40. Shofinita, D.; Fawwaz, M.; Achmadi, A.B. Betalain Extracts: Drying Techniques, Encapsulation, and Application in Food Industry. *Food Front.* **2023**, *4*, 576–623. [CrossRef]
41. Nabeelah Bibi Sadeer, N.B. The Versatility of Antioxidant Assays in Food Science and Safety—Chemistry, Applications, Strengths, and Limitations. *Antioxidants* **2020**, *9*, 709. [CrossRef]
42. Herbach, K.M.; Stintzing, F.C.; Carle, R. Betalain Stability and Degradation—Structural and Chromatic Aspects. *J. Food Sci.* **2006**, *71*, R41–R50. [CrossRef]
43. Merin, U.; Gagel, S.; Popel, G.; Bernstein, S.; Rosenthal, I. Thermal Degradation Kinetics of Prickly-Pear-Fruit Red Pigment. *J. Food Sci.* **1987**, *52*, 485–486. [CrossRef]
44. Feng, Z. Effect of Spray Drying and Storage on the Stability of Bayberry Polyphenols. *Food Chem.* **2011**, *129*, 1139–1147. [CrossRef]
45. Hu, T.; Dai, T.; He, X.; Deng, L.; Li, T.; Sun, J.; Liu, C.; Chen, J. Non-Covalent Interaction of Complex Plant Protein and Betanin: Mechanism of Improving Thermal Stability of Betanin. *Food Hydrocoll.* **2023**, *138*, 108456. [CrossRef]
46. Hu, T.; Chen, J.; He, X.; Tang, Y.; Sun, J.; Liu, C.; Dai, T. Complex Plant Protein Prepared from Rice Protein and Pea Protein: Improve the Thermal Stability of Betanin. *Food Res. Int.* **2023**, *164*, 112341. [CrossRef] [PubMed]
47. Igual, M.; Fernandes, Â.; Dias, M.I.; Pinela, J.; García-Segovia, P.; Martínez-Monzó, J.; Barros, L. The In Vitro Simulated Gastrointestinal Digestion Affects the Bioaccessibility and Bioactivity of Beta Vulgaris Constituents. *Foods* **2023**, *12*, 338. [CrossRef] [PubMed]
48. Trych, U.; Buniowska-Olejnik, M.; Marszałek, K. Bioaccessibility of Betalains in Beetroot (*Beta Vulgaris* L.) Juice under Different High-Pressure Techniques. *Molecules* **2022**, *27*, 7093. [CrossRef] [PubMed]
49. Tesoriere, L. In Vitro Digestion of Betalainic Foods. Stability and Bioaccessibility of Betaxanthins and Betacyanins and Antioxidative Potential of Food Digesta. *J. Agric. Food Chem.* **2008**, *56*, 10487–10492. [CrossRef]

Disclaimer/Publisher's Note: The statements, opinions and data contained in all publications are solely those of the individual author(s) and contributor(s) and not of MDPI and/or the editor(s). MDPI and/or the editor(s) disclaim responsibility for any injury to people or property resulting from any ideas, methods, instructions or products referred to in the content.

foods MDPI

Article

Effect of Pea Protein Isolate–Soybean Meal Ratio on Fiber Structure and Texture Properties of High-Moisture Meat Analogs

Zhongjiang Wang, Yachao Tian, Fangxiao Lou and Zengwang Guo *

College of Food Science, Northeast Agricultural University, Harbin 150030, China
* Correspondence: gzwname@163.com

Abstract: Inadequate fibrous attributes and prohibitive pricing are pivotal barriers to the broader market penetration of meat analogs (MAs). This research endeavors to address these impediments by formulating a blend of cost-effective soybean meal (SM) and pea protein isolate (PPI) across a spectrum of ratios (PPI:SM = 1:0, 8:2, 6:4, 4:6, 2:8, and 0:1). The analysis of textural properties elucidated that the integration of SM markedly diminished the textural rigidity and mastication resistance of MAs. Employing scanning electron microscopy (SEM) and fibrillation degree metrics, it was ascertained that the most favorable fibrous architecture of MAs was attained at a PPI to SM ratio of 6:4. Further experimental evidence underscored that the synergistic interaction between SM and PPI catalyzed the conversion of free sulfhydryl groups into disulfide linkages, a pivotal mechanism in the augmentation of MAs' fibrous matrices. The conclusions drawn from this study provide substantive contributions to the formulation of superior-quality, economically viable MAs, and could potentially accelerate their market acceptance.

Keywords: pea protein isolate; soybean meal; meat analogs; textural properties; scanning electron microscopy

1. Introduction

With the growth of the global population, the demand for meat as a food source continued to rise [1,2]. However, traditional methods of animal meat production not only consumed a large amount of resources, but also had a serious impact on the environment [3]. In addition, the high cholesterol and fat content of animal meat, along with potential food safety issues, attracted widespread attention [4]. Therefore, developing healthier, environmentally friendly, and sustainable meat substitutes became the focus of current research. As a new meat substitute, with vegetable protein serving as the main raw material, plant-based meat analogs (MAs) could imitate the fiber structure of meat [5]. Extrusion was the most effective feasible technology for processing plant-based MAs [6]. Products produced through low-moisture extrusion typically required rehydration before consumption and further processing, and their texture tended to be relatively rough [7]. High-moisture extrusion technology could significantly enhance the chewing sensation and elasticity of MAs, aiding in retaining nutrients within MAs and promoting a uniform distribution of nutrients. This improved the overall nutritional value of MAs, making them closer to the eating experience of traditional animal meat [8].

The plant proteins utilized in the production of MAs have been primarily sourced from soybeans [9,10], peas [11,12], and wheat [12,13]. Pea protein isolate (PPI) garnered increasing attention due to its advantages, such as its nutritional value, digestibility, non-toxicity, and non-transgenic origins [14,15]. However, it was found that the texture of MAs prepared by PPI through high-moisture extrusion was poor, and the fiber structure was not similar to that of animal meat [12]. It was found that MAs produced with protein from two different sources under high-moisture conditions exhibited a superior fiber structure.

For instance, the combination of soybean protein isolate (SPI) and wheat gluten protein (WG) could effectively enhance the structural and textural properties of MAs [16]. Based on this observation, we believed that the fiber structure and quality of MAs could be effectively improved through high-moisture extrusion when PPI was compounded with other types of protein. In fact, the high cost was an important factor limiting the promotion of MAs. This was primarily due to the relatively high prices of SPI, PPI, and WG used in the manufacturing of MAs. Soybean meal (SM), another prevalent source of plant protein, is abundant in amino acids and nutrients, and its price is significantly lower than that of SPI, PPI, and WG due to the simplicity of its process. Soybean meal (SM) represents the most basic form of soybean protein [17]. The allergenicity of PPI is low, and its allergens primarily originate from two types of globulins, Pis s1 and Pis s2 [18]. SM is the main byproduct of soybean processing, and its allergenic proteins primarily include β-conglycinin and glycinin [19]. Although PPI and SM contain some allergenic substances, the research shows that high-temperature and high-pressure treatment in the process of high-moisture extrusion can effectively reduce allergenicity [20]. Through previous experiments, we found that MAs prepared by SM alone have almost no fiber structure, which obviously cannot meet the requirements of the market and consumers for MA quality. The combination of PPI and SM can not only reduce the cost, but also further improve the texture and fiber structure of MAs through their complementary effects. However, plant proteins may have different physical and chemical properties and interaction mechanisms. Therefore, it is necessary to systematically study the influence of the ratio of PPI to SM on the texture and fiber structure of MAs during high-moisture extrusion.

In this study, PPI and SM were mixed in different proportions (PPI:SM = 1:0, 8:2, 6:4, 4:6, 2:8, and 0:1). MAs were produced through a high-moisture extrusion process. The texture properties of MAs were measured by a texture analyzer. The changes in the fiber structure of MA samples was observed by a scanning electron microscope (SEM). In order to deeply understand the interaction between PPI and SM, the secondary structure of MA samples was analyzed. We assessed the textural characteristics of MAs using a texture analyzer. SEM imaging was employed to observe alterations in the fiber structure of MA samples. In addition, in order to deeply understand the interaction between PPI and SM, we detected the number of sulfhydryl groups and disulfide bonds in MAs. The research results can be applied to the development of new MA foods and health products.

2. Materials and Methods

2.1. Materials

The PPI and SM samples were sourced from Yuwang Group (Yucheng City, Shandong Province, China). The preparation method of PPI was as follows: Soak peas in water for about 20 h, and then grind them with a grinding wheel. Bean dregs were repeatedly ground 3 times, and the residue was removed. Add acid to the supernatant to adjust the pH to about 4.5 to precipitate protein, add alkali to adjust the pH of protein after acid precipitation to about 7.2 to dissolve protein, and then homogenize and spray dry to obtain PPI. PPI comprises approximately 85% protein content, with 12.5% carbohydrates, 1% fat, and a moisture level of 3%. The preparation method of SM was as follows: clean soybean, soak it in water, and cook to inactivate lipase. Then, the crude soybean meal was obtained by low-temperature pressing to remove oil and part of the water. Then, the squeezed SM was dried at a low temperature to further reduce the moisture content and obtain SM. SM has a composition of roughly 57% protein, 20.5% carbohydrates, 2.5% fat, and 7% moisture. Additional chemicals and reagents were supplied by Yuanye Biotechnology Co., Ltd. (Shanghai, China).

2.2. MA Preparation

MAs were prepared using an HT-36 extruder (Shandong Arrow Machinery Co., Ltd., Jinan, China). PPI and SM were fully mixed in a WLDH-500 mixer (Jiangsu Hongda Powder Equipment Co., Ltd., Jiangyin, China) to prepare the feed mixture (PPI:SM = 1:0,

8:2, 6:4, 4:6, 2:8, and 0:1). The extruder barrel comprised ten sections. The first section included the gate for solid feed, while the tenth section featured individual temperature controls. The extruder parameters were set as follows: the feeding speed of the PPI and SM mixture was 10.0 kg/h, the water flow rate was 15 L/h, and the moisture content was 60% (*w*/*w*). The screw speed was set to 280 rpm. Temperature profiles were established at 50, 70, 110, 140, 145, 150, 150, 140, and 120 °C for sections 2–10, respectively. During the process, the cooling zone was maintained at a temperature below 50 °C via the circulation of water. Subsequently, MA samples were sliced into strips measuring 15 cm in width and 25 cm in length. These MA strips were then preserved in a −20 °C refrigerator for future analytical purposes. A schematic diagram of the extruder is shown in Figure 1.

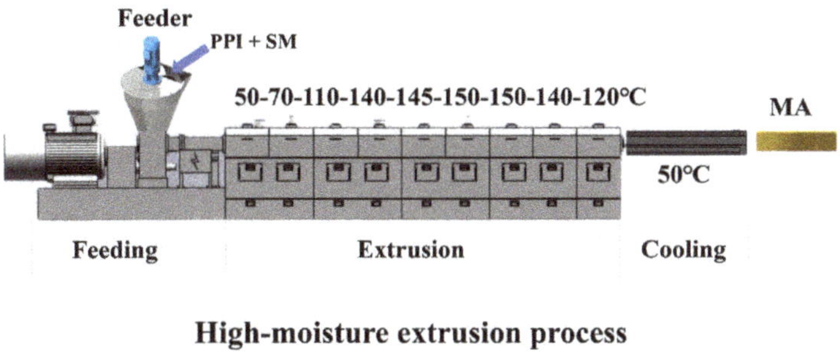

Figure 1. Schematic diagram of the extruder.

2.3. Texture Properties' Measurement

The texture distribution of the MA samples was detected using a Brookfield CT3 texture analyzer (Brookfield, WI, USA). The MA samples were cut into rectangular specimens measuring 15 mm in length, 15 mm in width, and 10 mm in height. These specimens were then doubly compressed to 50% of the original height under the texture profile analysis (TPA) mode. The descending and ascending speeds of the probe were set at 2 mm/s, and the testing speed at 0.5 mm/s. Furthermore, the MAs were cut into rectangular samples. The samples were then reduced to 75% of their original thickness at a speed of 1.5 mm/s, both vertically and parallel to the extrusion direction. The lengthwise shear force (FL) and crosswise shear force (FC) were defined as the maximum forces required to cut the sample in the vertical and parallel directions, respectively. The fiber degree (FD) was shown as the ratio of FL to FC.

2.4. Color Measurement

An NR60CP+ handheld colorimeter, manufactured by Shenzhen Sanenshi Technology Co., Ltd. in Shenzhen, China, was utilized to monitor the color alterations of the MAs. The device was used to record L* (brightness), a* (redness), and b* (yellowness) values. Subsequently, these values were contrasted with the initial readings to ascertain the magnitude of color alteration. The lightness (L_0*), redness (a_0*), and yellowness (b_0*) of the standard white plate were 97.85, −0.01, and 1.43, respectively. ΔE is given by the following formula:

$$\Delta E = \sqrt{\Delta L^2 + \Delta a^2 + \Delta b^2} \quad (1)$$

where ΔL, Δa, and Δb are the differences between samples and the standard white plate in terms of L_0*, a_0*, and b_0* values, respectively.

2.5. SEM Observation

Based on our previous method [21], the microstructure of the sample was observed by SEM (Hitachi Manufacturing Co., Ltd. in Tokyo, Japan). MA samples were freeze-dried and ground into powder using a mortar. Before SEM observation, all samples were sprayed with gold. The magnification was 300 and 1500 times.

2.6. Rheological Properties' Measurement

We refined our existing methodology to determine the rheological properties of the samples [22]. Firstly, the MA freeze-dried sample was ground into powder, and deionized water was added to prepare a 15% (w/v) solution. After magnetic stirring for 2 h, it was stored at 4 °C overnight. Then, an MCR 101 rheometer (Anton Paar, Graz, Austria) was used to measure it. The linear viscoelastic region was determined by stress scanning prior to measurement. As well as the shear stress and shear rate, apparent viscosity was also measured.

2.7. DSC Measurement

The thermal properties of the MA samples were recorded using a DSC-25 differential scanning calorimeter (TA Instrument, New Castle, DE, USA), following methods employed in previous studies [23]. Specifically, 5 mg of the MA sample was accurately weighed, placed into a crucible, and lightly pressed with a mold to ensure that the sample was sealed in the plate. Nitrogen was selected as the protective gas, and the experiment was conducted with a starting temperature of 20 °C, an ending temperature of 170 °C, and a heating rate of 15 °C/min.

2.8. Low-Field Nuclear Magnetic Resonance (LF-NMR) Measurement

The moisture distribution of MA samples was analyzed using an NMI20-030V-I LF-NMR analyzer (Niumag Inc., Shanghai, China). The equipment was fitted with a permanent magnet (the sample cavity was placed vertically), the magnetic field intensity was 0.5 ± 0.05 T, the RF pulse frequency range was 2~30 MHz, the RF frequency control accuracy was ≤0.1Hz, the RF pulse accuracy was ≤100 ns, and the maximum sampling bandwidth was ≥2000 KHz. The MA sample was placed at the bottom of the special glass tube for nuclear magnetic resonance, and then inserted into the center of the LF-NMR analyzer, and the spin–spin relaxation time of fresh extrudates was studied. The waiting time was 4000 ms, the echo time was 0.10 ms, and scanning was conducted 6 times. The measured relaxation time was converted into the corresponding relaxation signal components.

2.9. Free Sulfhydryl (FS) and Disulfide Bonds (DBs)

An enhanced version of our prior technique was adopted to assess the rheological properties [24]. The DTNB method was utilized to measure the FS quantity of the MA sample. The MA sample was mixed with dH2O, diluted with buffer solution (Tris-Gly Urea), and then DTNB was added. Here is the FS calculation formula:

$$FS\left(\frac{\mu\text{mol}}{\text{g}}\right) = \frac{73.53 \times D \times A_{412}}{C} \quad (2)$$

where 73.53, D, A_{412}, and C respectively represent the DTNB extinction coefficient, dilution coefficient, sample absorbency, and MA concentration (mg/mL).

Tris-Gly urea buffer (15 mL) was added to the MA sample, and it was mixed with β-mercaptoethanol (0.3 mL) and reacted at room temperature for 1 h. Meanwhile, trichloroacetic acid (TCA) was added. After centrifugal precipitation, the precipitate was dissolved in Tris-Gly urea buffer (20 mL), and DTNB (0.2 mL) was added to react for 20 min. The formulas for calculating the total sulfhydryl (TS) content and DB content are as follows:

$$TS\left(\frac{\mu\text{mol}}{\text{g}}\right) = \frac{73.53 \times A_{412}}{C} \quad (3)$$

$$DB\left(\frac{\mu\text{mol}}{\text{g}}\right) = \frac{TS - FS}{2} \quad (4)$$

where 73.53, A_{412}, and C respectively represent the $DTNB$ extinction coefficient, sample absorbency, and MA concentration (mg/mL).

2.10. Statistical Analysis

The samples underwent three replicate tests, and the resulting data were summarized using mean values accompanied by standard deviations. Statistical analysis of the experimental data was conducted using ANOVA and the Tukey test in SPSS 21 software (Chicago, IL, USA), with statistical significance set at $p < 0.05$.

3. Results and Discussion

3.1. Texture Properties' Analysis

The texture properties were one of the most important indices used to evaluate the similarity between MAs and animal meat [25]. By testing MA texture properties, such as hardness, springiness, chewiness, and fibrosity, the overall quality of the product could be assessed [26]. This would provide a scientific basis for product development and production. The results of texture data are shown in Table 1. The results show that when the ratio of PPI to SM changed from 1:0 to 0:1, the hardness, springiness, and chewiness of MAs increased first and then decreased. When the ratio of PPI to SM was 6:4, the hardness (62.63 ± 2.77 N), springiness (1.01 ± 0.03), and chewiness (4897 ± 206 g) were the highest. By increasing the proportion of SM, the hardness, springiness, and chewiness of MAs decreased gradually. This could be because there may be an interaction between SM and PPI. In a certain range, the addition of SM may help to enhance the structural stability of MAs, thus improving their texture properties. However, when the content of SM is too high, it may destroy the stable structure originally formed by PPI, resulting in a decline in texture characteristics. When an MA consists of SM only, the hardness, elasticity, and chewiness are the lowest. The taste may be too soft, inelastic, or chewy, which may affect consumers' eating experience and satisfaction. A previous study found that the hardness of beef is about 76 N, and that of pork is about 55 N [27]. When the ratio of PPI to SM is 6:4, its hardness is closest to that of beef and pork, and this is the best preparation technology of MAs.

Table 1. Effect of the ratio of PPI to MS on the texture properties of MAs.

Samples	Hardness (N)	Springiness (mm)	Chewiness (g)	Degree of Fibrosity
PPI:SM = 1:0	44.54 ± 1.69 ^{b}	0.80 ± 0.02 ^{c}	3173 ± 78 ^{c}	0.92 ± 0.03 ^{c}
PPI:SM = 8:2	46.71 ± 4.65 ^{b}	0.96 ± 0.05 ^{a}	3721 ± 146 ^{b}	1.15 ± 0.06 ^{b}
PPI:SM = 6:4	62.63 ± 2.77 ^{a}	1.01 ± 0.03 ^{a}	4897 ± 206 ^{a}	1.28 ± 0.05 ^{a}
PPI:SM = 4:6	38.36 ± 3.07 ^{c}	0.85 ± 0.02 ^{b}	2728 ± 186 ^{d}	0.93 ± 0.05 ^{c}
PPI:SM = 2:8	36.03 ± 4.80 ^{c}	0.87 ± 0.03 ^{b}	2457 ± 353 ^{d}	0.85 ± 0.04 ^{c}
PPI:SM = 0:1	11.27 ± 1.40 ^{d}	0.73 ± 0.02 ^{d}	573 ± 90 ^{e}	0.74 ± 0.02 ^{d}

Note: The results are presented as the mean value along with its standard deviation. Significant differences ($p < 0.05$) are indicated by distinct lowercase letters within the same column.

The degree of fibrosity is one of the most important indexes to evaluate the quality of plant meat, which is determined according to the ratio of crosswise shear force to lengthwise shear force [28]. Generally speaking, the higher the degree of fibrosity, the better the fiber structure of MAs, which means the higher the similarity between MAs and animal meat [12]. As shown in Table 1, when the ratio of PPI to SM is 6:4, the degree of fibrosity reaches its maximum (1.28 ± 0.05). This shows that adding the proper amount of SM in the extrusion process can lead to the formation of longer MA fiber structures along the extrusion direction. Previous studies have also found that compounding two different plant proteins can effectively improve the degree of fibrosity of MAs [29]. When MAs are only composed of SM, the degree of fibrosity is the lowest (0.74 ± 0.02). This may be because too much SM is not conducive to the orientation of protein molecules in the extrusion process, thus reducing the degree of fibrosity. A previous study found that MAs prepared by adding 30% WG to PPI had the highest degree of fibrosity (2.510 + 0.036) [12]. Based on the importance of the degree of fibrosity in MAs, they think that adding 30%

WG is the best process to prepare MAs. Similarly, in our research, PPI:SM = 6:4 is the best preparation technology for MAs.

3.2. Color Analysis

Color was an important property of MAs, which affected consumers' acceptance of the products [28]. The color of MAs primarily depended on the reaction that occurred during extrusion [30]. The influence of the PPI and SM ratio on MA color is shown in Table 2. L* represents brightness (black and white), and the higher the L* value, the brighter the color appeared. If the L* value was positive, it indicated that the sample was brighter than the standard plate; if it was negative, it indicated that the sample was darker [31]. After adding SM, the L* value increased slightly, but only the PPI:SM = 6:4 sample was significantly different. The a* represented red–green chromaticity, where a positive value indicated that the sample was redder than the standard, and a negative value signified that it was greener than the standard [32]. Table 2 shows that a* is positive, indicating that MAs are redder than the standard. After adding SM, the value of a* increased slightly, but there was a significant difference only when PPI:SM = 4:6. In Table 2, b* represents the degree of yellow–blue, and a positive value of b* indicates that the color of the sample was yellower compared to the standard, while a negative value signifies that it was bluer. Table 2 shows that after adding SM to PPI, the value of b* increases significantly. This may be because SM is yellower than PPI. After adding SM to PPI, the ΔE value increased significantly ($p < 0.05$). However, there was no significant difference in ΔE between MA samples containing different proportions of SM ($p > 0.05$). The enhanced color properties might be leveraged in food products where a richer appearance is desirable, potentially increasing consumer appeal.

Table 2. Effect of the ratio of PPI to MS on the color values of MAs.

Samples	L*	a*	b*	ΔE
PPI:SM = 1:0	61.76 ± 2.03 ^{b}	6.18 ± 1.04 ^{b}	22.94 ± 1.00 ^{d}	9.53 ± 0.15 ^{b}
PPI:SM = 8:2	64.92 ± 4.96 ^{ab}	6.61 ± 1.48 ^{ab}	27.01 ± 2.27 ^{c}	9.92 ± 0.13 ^{a}
PPI:SM = 6:4	68.08 ± 2.33 ^{a}	6.38 ± 0.98 ^{ab}	28.91 ± 2.54 ^{bc}	10.16 ± 0.21 ^{a}
PPI:SM = 4:6	66.92 ± 6.22 ^{ab}	7.93 ± 1.69 ^{a}	33.60 ± 4.26 ^{a}	10.40 ± 0.31 ^{a}
PPI:SM = 2:8	65.60 ± 2.05 ^{ab}	7.46 ± 0.41 ^{ab}	32.43 ± 2.29 ^{ab}	10.27 ± 0.24 ^{a}
PPI:SM = 0:1	65.54 ± 2.08 ^{ab}	6.81 ± 1.05 ^{ab}	32.79 ± 3.45 ^{a}	10.25 ± 0.27 ^{a}

Note: The results are presented as the mean value along with its standard deviation. Significant differences ($p < 0.05$) are indicated by distinct lowercase letters within the same column.

3.3. SEM Analysis

Microstructure observation has played a key role in the research on MAs in the past, and the fibrotic structure in MAs can be directly observed via SEM [28]. The microstructures of MAs with different PPI and SM contents were observed under SEM 300 times and 1500 times, as shown in Figure 2. When the ratio of PPI to SM is 1:0, the MA has a few fibrous knots. With the addition of SM, the orientation structure of the MA is more orderly, and the fiber structure is more obvious and detailed along the extrusion direction. When the ratio of PPI to SM is 6:4, the fiber structure in the MA is the most abundant and evenly distributed. This may be because SM and PPI formed two incompatible and independent phases, which prevented protein aggregation and promoted the formation of the fiber structure [33]. With the increase in SM content, the fiber property of the MA decreases. When the ratio of PPI to SM is 0:1, a high level of protein aggregation and dispersion is observed, and the fiber structure of the MA is almost completely destroyed. This may be because excessive SM (PPI:SM ratio is 2:8 and 0:1) interferes with protein recombination, thus damaging the fiber structure of the MA. These changes in microstructure are also reflected in the surface diagram of MA (Figure S1). Previous studies have also found that too much wheat protein will obviously weaken the chewiness of the extrudate, which will hinder the arrangement of protein fibers in the extrusion direction [12]. Although it is not exactly the same as the raw materials used in previous studies, the mechanism of MA fiber structure change may be the same. Generally speaking, when the ratio of PPI to SM is 6:4, MAs contain an obvious fibrous structure, and a small amount of SM can promote the fibrous structure of the MA, while excessive SM will destroy the fibrous structure. This may also be one of the reasons for the decrease in the chewiness and hardness of the MA. The findings suggest that controlling the ratio of PPI to SM is crucial for optimizing the fibrous structure of MAs, which is important for achieving the desirable texture and chewiness in the final product. For industrial applications, this means manufacturers can fine-tune ingredient ratios to produce MAs that closely mimic the texture of real meat, potentially increasing consumer acceptance. Additionally, understanding the interaction between PPI and SM can help in designing efficient extrusion processes, leading to the cost-effective production of high-quality meat alternatives.

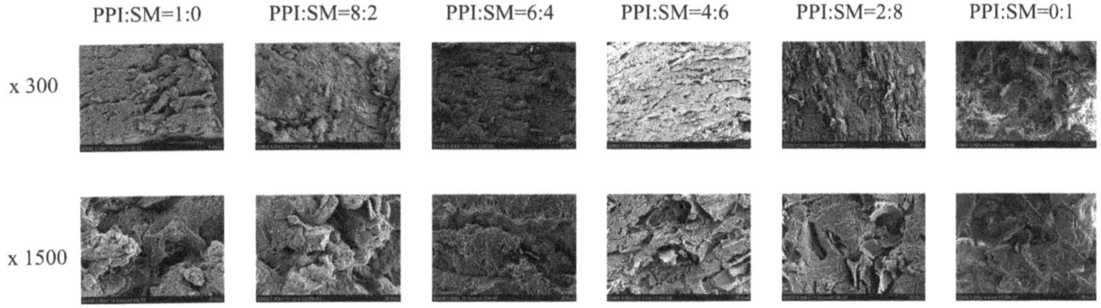

Figure 2. Influence of PPI:SM ratio on MA microscopic image.

3.4. Rheological Properties' Analysis

Rheological characteristics can reflect the fluidity and viscosity of MAs, as well as the interaction and protein aggregation between different components [28]. The G′ and G″ of MA powder are shown in Figure 3. With the increase in shear frequency, both the G′ and G″ of MA samples increase, and the value of G′ is greater than that of G″. This shows that the elastic properties of MA samples are dominant, indicating that the change in frequency in the linear viscoelastic range will not destroy the structure of an MA. At the same time, it can also be found that with the increase in the SM ratio, both the G′ and G″ of an MA decrease significantly. This may be related to the fact that SM contains more fat, which plays a role in lubrication. This finding is consistent with previous studies, which found that the storage modulus of pea-protein-based MAs decreased significantly with the increase in fat content [34]. Another study on plant-based foods also found that an increase in fat content leads to a decrease in storage modulus and viscosity [35]. Furthermore, the addition of SM may also form protein aggregation pellets with a lubrication function, which can be verified in the SEM images of MAs with a high proportion of SM. Previous studies have also found that adding yeast protein to PPI forms protein aggregate pellets with a lubricating effect, which leads to the decrease in the G′ and G″ of an MA [28].

Figure 3C illustrates the variation in the apparent viscosity of the MA samples as a function of shear rate. As the shear rate rises, the viscosity of the MA sample decreases, demonstrating the sample's shear-thinning behavior. This is related to the destruction of protein interaction during shearing [26]. In addition, it was found that the viscosity of MAs decreases rapidly at a low-speed shear rate, while it decreases slowly at a high-speed shear rate. This may be due to the destruction of protein clusters in MAs, especially at a high-speed shear rate. At the same shear rate, the viscosity of MAs decreases with the increase in SM content. This may be due to the fact there is more fat in SM, and the lubrication function of protein aggregate pellets. The SEM images of MA samples support this observation.

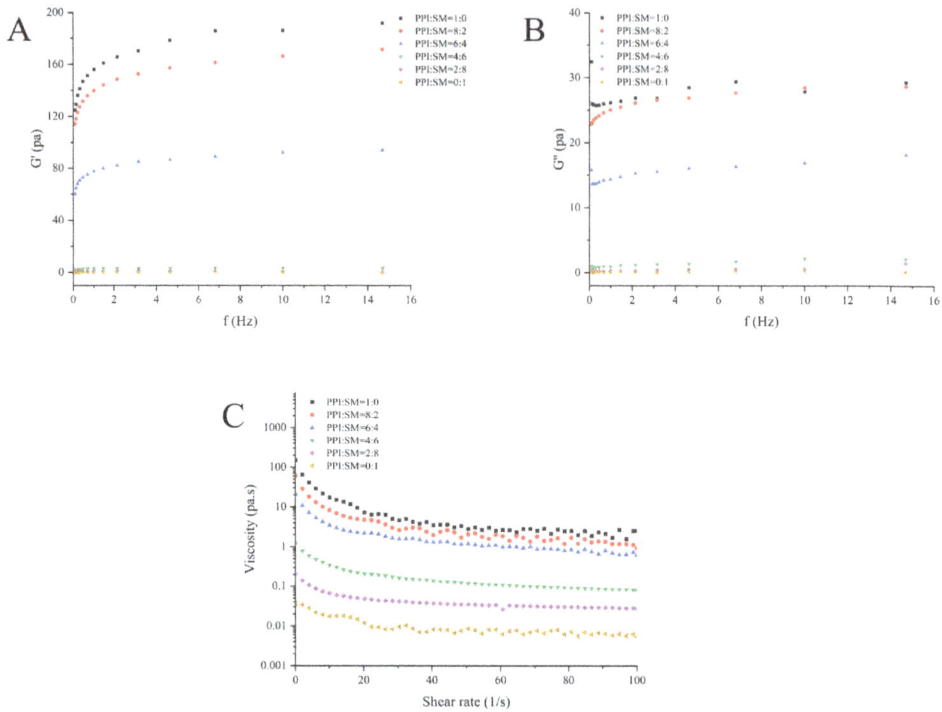

Figure 3. Influence of the PPI:SM ratio on MA rheological properties. (**A–C**) respectively represent G′, G″, and the viscosity diagrams of the MA.

3.5. DSC Analysis

DSC is a common analytical technique used to measure the degree of protein denaturation [28]. The effects of different proportions of PPI and SM on the thermal stability of MAs are shown in Figure 4. As can be seen from Figure 4, only one endothermic peak was observed on the DSC curve of the MA samples, and the temperature range of thermal denaturation was 51.43–55.81 °C. With the increase in the proportion of SM, the denaturation temperature gradually decreased from 55.81 °C to 51.43 °C. This decrease in denaturation temperature could be due to the interaction between SM and the protein structure of the MA samples, leading to a destabilization of intermolecular bonds. Additionally, the presence of SM might interfere with the protein network, reducing the energy required for denaturation. These results show that SM reduces the thermal stability of MAs, indicating that the molecular structure of MAs with SM can be expanded by lower energy [36]. This discovery shows that adjusting the ratio of PPI to SM can be used to fine-tune the thermal performance of MAs to adapt to specific applications.

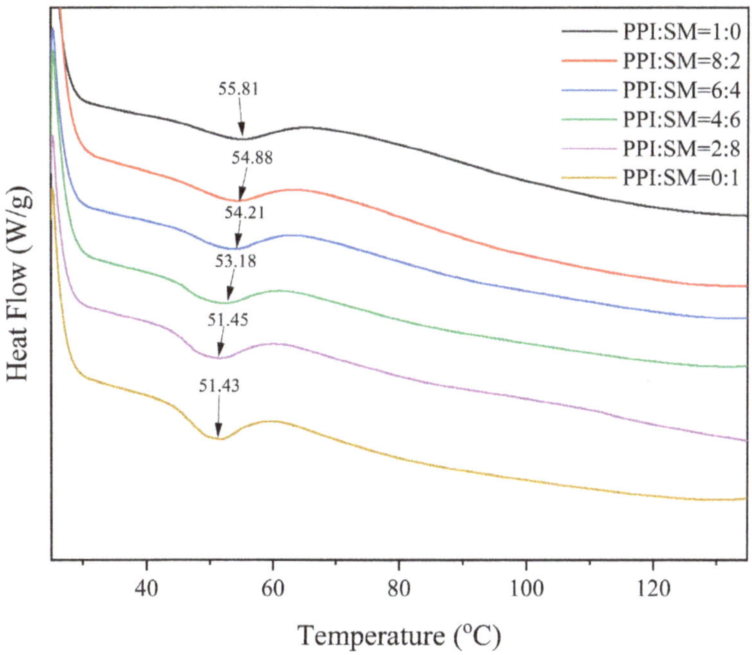

Figure 4. Influence of PPI:SM ratio on DSC thermal profiles.

3.6. LF-NMR Analysis

The distribution and migration of water in MAs is an important index, which can be characterized by LF-NMR [28]. For example, the relaxation time distribution curve of the PPI to SM ratio versus MAs is shown in Figure 5. Examining the figure, it becomes evident that as the degrees of freedom increase, the MA peaks can be differentiated into distinct categories: T2b and T21, which represent bound water, T22, corresponding to immobilized water, and T23, signifying free water. Table 3 shows the proportion of T2 distribution, with P22 accounting for the highest proportion, so water molecules mainly exist in the form of fixed water in MAs. With the increase in the proportion of SM, the proportion of P21 in the MA samples showed an upward trend, while P22 and P23 showed a downward trend. The increase in the proportion of bound water may be due to the fact that SM contains more polysaccharides with stronger binding ability to water. Understanding the T2 distribution in MAs is very important for optimizing their structure and hydration properties. By analyzing different forms of water in the sample, researchers can adjust the ratio of PPI to SM to achieve the required moisture content and stability.

Table 3. Influence of the PPI:SM ratio on MA T2 relaxation time.

Samples	P_{21} (%)	P_{22} (%)	P_{23} (%)
PPI:SM = 1:0	5.945 ± 0.43 [d]	93.457 ± 0.87 [a]	0.598 ± 0.02 [a]
PPI:SM = 8:2	5.252 ± 0.35 [d]	94.37 ± 1.21 [a]	0.378 ± 0.03 [b]
PPI:SM = 6:4	6.51 ± 0.23 [c]	93.231 ± 1.11 [a]	0.259 ± 0.04 [c]
PPI:SM = 4:6	9.725 ± 0.11 [a]	90.021 ± 1.15 [b]	0.254 ± 0.06 [c]
PPI:SM = 2:8	9.088 ± 0.13 [b]	90.855 ± 0.73 [b]	0.057 ± 0.01 [d]
PPI:SM = 0:1	9.151 ± 0.09 [b]	90.825 ± 0.45 [b]	0.024 ± 0.01 [e]

Note: The results are presented as the mean value along with its standard deviation. Significant differences ($p < 0.05$) are indicated by distinct lowercase letters within the same column.

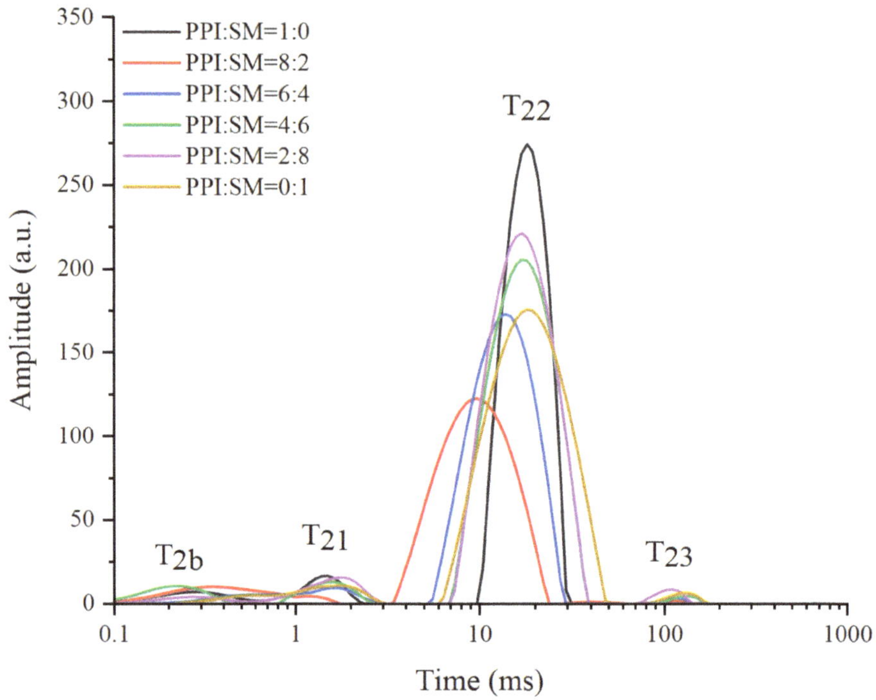

Figure 5. Influence of the PPI:SM ratio on MA T2 relaxation time.

3.7. FS and DB Analysis

FS and DB are the key substances used to maintain protein spatial conformation and endow proteins with specific functions. The effects of different proportions of PPI and SM on the contents of FS and DB in MAs are shown in Table 4. When the ratio of PPI to SM changed from 1:0 to 0:1, with the increase in SM content, FS content first decreased and then increased. When the ratio of PPI to SM is 6:4, the FS content is the lowest (0.015 0.004 μ mol/g). However, the DB content first increased and then decreased with the increase in SM content. When the ratio of PPI to SM is 6:4, the DB content is the highest (0.582 0.004 μ mol/g). The experimental results show that a certain proportion of SM can promote the formation of DB in MAs. This phenomenon is consistent with the degree of fibrosis and SEM results. This shows that an increase in DB content may be one of the reasons for the strengthening of MAs' fiber structure.

Table 4. Effect of the PPI and SM compound ratio on the free sulfhydryl and disulfide bonds of MAs.

Sample	Free Sulfhydryl (μmol/g)	Total Sulfhydryl (μmol/g)	Disulfide Bonds (μmol/g)
PPI:SM = 1:0	0.032 ± 0.004 ^{a}	1.079 ± 0.005 ^{c}	0.533 ± 0.004 ^{d}
PPI:SM = 8:2	0.024 ± 0.005 ^{ab}	1.146 ± 0.04 ^{a}	0.556 ± 0.008 ^{c}
PPI:SM = 6:4	0.015 ± 0.004 ^{b}	1.155 ± 0.007 ^{a}	0.582 ± 0.004 ^{a}
PPI:SM = 4:6	0.025 ± 0.007 ^{ab}	1.123 ± 0.003 ^{b}	0.549 ± 0.007 ^{c}
PPI:SM = 2:8	0.035 ± 0.005 ^{a}	1.129 ± 0.007 ^{b}	0.569 ± 0.002 ^{b}
PPI:SM = 0:1	0.035 ± 0.004 ^{a}	1.077 ± 0.008 ^{c}	0.511 ± 0.005 ^{e}

Note: The results are presented as the mean value along with its standard deviation. Significant differences ($p < 0.05$) are indicated by distinct lowercase letters within the same column.

4. Conclusions

As an important source of protein, SM is rich in protein, fat, dietary fiber, vitamins, and minerals. MAs produced by a PPI–SM mixture are helpful to improve the dietary structures and health levels of consumers. In summary, the blending of PPI and SM has been demonstrated to be an effective approach for enhancing the textural properties and fibrous structures of MAs. SEM results indicate that the optimal fibrous structure of MAs is achieved at a PPI to SM ratio of 6:4. Rheological data suggest that the incorporation of SM reduces the viscosity and storage modulus of MAs, which may have a positive impact on reducing energy input during processing. Experimental results on water distribution reveal that water molecules in MAs predominantly exist in a bound state. The addition of SM leads to an increase in the content of bound water in MAs, which is advantageous for improving their storage and processing characteristics. The inclusion of SM not only ameliorates the fibrous structure of MAs but also significantly reduces production costs. However, anti-nutritional compounds, such as trypsin inhibitor and phytic acid in SM, may affect the digestibility and quality of MAs. In the future, it is necessary to further study the effects of anti-nutritional compounds and allergenic elements in SM on the taste and quality of MAs.

Supplementary Materials: The following supporting information can be downloaded at: https://www.mdpi.com/article/10.3390/foods13233818/s1, Figure S1: The influence of PPI:SM ratio on the surface diagram of MA.

Author Contributions: Z.W.: conceptualization, methodology, software, writing, and original draft; Y.T.: data curation and visualization; F.L.: investigation, writing; and Z.G.: supervision, funding acquisition, and project administration. All authors have read and agreed to the published version of the manuscript.

Funding: This research was funded by the "HeiLongjiang Natural Science Fund" (YQ2022C021), the Shandong Provincial Key R&D Plan (2024CXPT023), and the Talent Cultivation Project funded by the Central Government to Support the Reform and Development of Local Universities.

Institutional Review Board Statement: Not applicable.

Informed Consent Statement: Not applicable.

Data Availability Statement: The original contributions presented in the study are included in the article/Supplementary Material, further inquiries can be directed to the corresponding author.

Conflicts of Interest: The authors declare no conflicts of interest.

References

1. Henchion, M.; Hayes, M.; Mullen, A.M.; Fenelon, M.; Tiwari, B. Future protein supply and demand: Strategies and factors influencing a sustainable equilibrium. *Foods* **2017**, *6*, 53. [CrossRef]
2. Aiking, H. Future protein supply. *Trends Food Sci. Technol.* **2011**, *22*, 112–120. [CrossRef]
3. Nijdam, D.; Rood, T.; Westhoek, H. The price of protein: Review of land use and carbon footprints from life cycle assessments of animal food products and their substitutes. *Food Policy* **2012**, *37*, 760–770. [CrossRef]
4. Pogorzelska-Nowicka, E.; Atanasov, A.G.; Horbańczuk, J.; Wierzbicka, A. Bioactive compounds in functional meat products. *Molecules* **2018**, *23*, 307. [CrossRef] [PubMed]
5. Mazumder, M.A.R.; Panpipat, W.; Chaijan, M.; Shetty, K.; Rawdkuen, S. Role of plant protein on the quality and structure of meat analogs: A new perspective for vegetarian foods. *Future Foods* **2023**, *8*, 100280. [CrossRef]
6. Zhang, X.; Shen, A.; Zhang, Z.; Zhang, T.; Jiang, L.; Zhou, W.; Zhang, Y.; Sui, X. Advancing molecular understanding in high moisture extrusion for plant-based meat analogs: Challenges and perspectives. *Food Chem.* **2024**, *460*, 140458. [CrossRef]
7. Offiah, V.; Kontogiorgos, V.; Falade, K.O. Extrusion processing of raw food materials and by-products: A review. *Crit. Rev. Food Sci. Nutr.* **2019**, *59*, 2979–2998. [CrossRef]
8. Bryant, C.J. Plant-based animal product alternatives are healthier and more environmentally sustainable than animal products. *Future Foods* **2022**, *6*, 100174. [CrossRef]

9. Zahari, I.; Ferawati, F.; Helstad, A.; Ahlström, C.; Östbring, K.; Rayner, M.; Purhagen, J.K. Development of high-moisture meat analogues with hemp and soy protein using extrusion cooking. *Foods* **2020**, *9*, 772. [CrossRef]
10. Lyu, B.; Li, J.; Meng, X.; Fu, H.; Wang, W.; Ji, L.; Wang, Y.; Guo, Z.; Yu, H. The protein composition changed the quality characteristics of plant-based meat analogues produced by a single-screw extruder: Four main soybean varieties in china as representatives. *Foods* **2022**, *11*, 1112. [CrossRef]
11. Zhu, H.-G.; Tang, H.-Q.; Cheng, Y.-Q.; Li, Z.-G.; Tong, L.-T. Potential of preparing meat analogue by functional dry and wet pea (*Pisum sativum*) protein isolate. *LWT* **2021**, *148*, 111702. [CrossRef]
12. Zhao, Y.-R.; Peng, N.; Li, Y.-Q.; Liang, Y.; Guo, Z.-W.; Wang, C.-Y.; Wang, Z.-Y.; Wang, C.; Ren, X. Physicochemical properties, structural characteristics and protein digestibility of pea protein-wheat gluten composited meat analogues prepared via high-moisture extrusion. *Food Hydrocoll.* **2024**, *156*, 110283. [CrossRef]
13. Zhang, X.; Zhao, Y.; Zhang, T.; Zhang, Y.; Jiang, L.; Sui, X. Potential of hydrolyzed wheat protein in soy-based meat analogues: Rheological, textural and functional properties. *Food Chem. X* **2023**, *20*, 100921. [CrossRef] [PubMed]
14. Wang, J.; Kadyan, S.; Ukhanov, V.; Cheng, J.; Nagpal, R.; Cui, L. Recent advances in the health benefits of pea protein (*Pisum sativum*): Bioactive peptides and the interaction with the gut microbiome. *Curr. Opin. Food Sci.* **2022**, *48*, 100944. [CrossRef]
15. Guo, Y.; Gong, Q.; Sun, F.; Cheng, T.; Fan, Z.; Huang, Z.; Liu, J.; Guo, Z.; Wang, Z. Interaction mechanism of pea proteins with selected pyrazine flavors: Differences in alkyl numbers and flavor concentration. *Food Hydrocoll.* **2024**, *147*, 109314. [CrossRef]
16. Chiang, J.H.; Loveday, S.M.; Hardacre, A.K.; Parker, M.E. Effects of soy protein to wheat gluten ratio on the physicochemical properties of extruded meat analogues. *Food Struct.* **2019**, *19*, 100102. [CrossRef]
17. Nahashon, S.N.; Kilonzo-Nthenge, A.K. Advances in soybean and soybean by-products in monogastric nutrition and health. In *Soybean and Nutrition*; InTech: Rijeka, Croatia, 2011; pp. 125–156.
18. Taylor, S.L.; Marsh, J.T.; Koppelman, S.J.; Kabourek, J.L.; Johnson, P.E.; Baumert, J.L. A perspective on pea allergy and pea allergens. *Trends Food Sci. Technol.* **2021**, *116*, 186–198. [CrossRef]
19. Zhang, T.; Li, W.; Li, H.; Liu, X. Soybean protein and soybean peptides: Biological activity, processing technology, and application prospects. *Food Sci. Technol. Res.* **2023**, *29*, 277–288. [CrossRef]
20. Islam, M.; Huang, Y.; Islam, M.S.; Lei, N.; Shan, L.; Fan, B.; Tong, L.; Wang, F. Effect of high-moisture extrusion on soy meat analog: Study on its morphological and physiochemical properties. *Ital. J. Food Sci.* **2022**, *34*, 9–20. [CrossRef]
21. Tian, Y.; Yuan, C.; Cui, B.; Lu, L.; Zhao, M.; Liu, P.; Wu, Z.; Li, J. Pickering emulsions stabilized by β-cyclodextrin and cinnamaldehyde essential oil/β-cyclodextrin composite: A comparison study. *Food Chem.* **2022**, *377*, 131995. [CrossRef]
22. Sun, F.; Xu, J.; Wang, Z.; Cheng, T.; Wang, D.; Liu, J.; Guo, Z.; Wang, Z. Effect of glycosylation on soy protein isolate–sodium carboxymethyl cellulose conjugates heat-induced gels and their applications as carriers of riboflavin. *Food Hydrocoll.* **2024**, *153*, 110072. [CrossRef]
23. Jiang, W.; Feng, J.; Yang, X.; Li, L. Structure of pea protein-based complexes on high-moisture extrusion: Raw materials and extrusion zones. *LWT* **2024**, *194*, 115823. [CrossRef]
24. Li, C.; Tian, Y.; Liu, C.; Dou, Z.; Diao, J. Effects of heat treatment on the structural and functional properties of *Phaseolus vulgaris* L. Protein. *Foods* **2023**, *12*, 2869. [CrossRef] [PubMed]
25. Samard, S.; Ryu, G.H. A comparison of physicochemical characteristics, texture, and structure of meat analogue and meats. *J. Sci. Food Agric.* **2019**, *99*, 2708–2715. [CrossRef]
26. McClements, D.J.; Weiss, J.; Kinchla, A.J.; Nolden, A.A.; Grossmann, L. Methods for testing the quality attributes of plant-based foods: Meat-and processed-meat analogs. *Foods* **2021**, *10*, 260. [CrossRef]
27. Bakhsh, A.; Lee, S.-J.; Lee, E.-Y.; Hwang, Y.-H.; Joo, S.-T. Evaluation of rheological and sensory characteristics of plant-based meat analog with comparison to beef and pork. *Food Sci. Anim. Resour.* **2021**, *41*, 983. [CrossRef]
28. Xia, S.; Shen, S.; Ma, C.; Li, K.; Xue, C.; Jiang, X.; Xue, Y. High-moisture extrusion of yeast-pea protein: Effects of different formulations on the fibrous structure formation. *Food Res. Int.* **2023**, *163*, 112132. [CrossRef]
29. Li, F.; Zhang, Y.; Li, M.; Don, C.; Zhang, B.; Guo, B.; Wei, Y. The impact of extrusion parameters on the glutenin macropolymer content of flour-water dough. *J. Cereal Sci.* **2019**, *90*, 102849. [CrossRef]
30. Žilić, S.; Mogol, B.A.; Akıllıoğlu, G.; Serpen, A.; Delić, N.; Gökmen, V. Effects of extrusion, infrared and microwave processing on Maillard reaction products and phenolic compounds in soybean. *J. Sci. Food Agric.* **2014**, *94*, 45–51. [CrossRef]
31. Ketnawa, S.; Chaijan, M.; Grossmann, L.; Rawdkuen, S. High-moisture soy protein-mushroom-based meat analogue: Physicochemical, structural properties and its application. *Int. J. Food Sci. Technol.* **2024**, *59*, 596–614. [CrossRef]
32. Marcus, R.T. The measurement of color. In *AZimuth*; Elsevier: Amsterdam, The Netherlands, 1998; Volume 1, pp. 31–96.
33. Wang, H.; van den Berg, F.W.; Zhang, W.; Czaja, T.P.; Zhang, L.; Jespersen, B.M.; Lametsch, R. Differences in physicochemical properties of high-moisture extrudates prepared from soy and pea protein isolates. *Food Hydrocoll.* **2022**, *128*, 107540. [CrossRef]
34. Saavedra Isusi, G.I.; Pietsch, V.; Beutler, P.; Hoehne, S.; Leister, N. Influence of rapeseed oil on extruded plant-based meat analogues: Assessing mechanical and rheological properties. *Processes* **2023**, *11*, 1871. [CrossRef]

35. McClements, D.J. Modeling the rheological properties of plant-based foods: Soft matter physics principles. *Sustain. Food Proteins* **2023**, *1*, 101–132.
36. Peng, H.; Zhang, J.; Wang, S.; Qi, M.; Yue, M.; Zhang, S.; Song, J.; Wang, C.; Zhang, D.; Wang, X.; et al. High moisture extrusion of pea protein: Effect of l-cysteine on product properties and the process forming a fibrous structure. *Food Hydrocoll.* **2022**, *129*, 107633. [CrossRef]

Disclaimer/Publisher's Note: The statements, opinions and data contained in all publications are solely those of the individual author(s) and contributor(s) and not of MDPI and/or the editor(s). MDPI and/or the editor(s) disclaim responsibility for any injury to people or property resulting from any ideas, methods, instructions or products referred to in the content.

foods

MDPI

Review

Peach Palm (*Bactris gasipaes*) as a Sustainable Source of Plant Proteins, Dietary Fiber and Other Functional Ingredients: Recovery Techniques and Functional Food Applications

Kartik Sharma [1], Nattaya Konsue [1], Samart Sai-Ut [2], Ekasit Onsaard [3], Wanli Zhang [4], Shusong Wu [5], Jia-Qiang Huang [6], Young Hoon Jung [7] and Saroat Rawdkuen [1,*]

1 Unit of Innovative Food Packaging and Biomaterials, School of Agro-Industry, Mae Fah Luang University, Chiang Rai 57100, Thailand; kartik.coa@gmail.com (K.S.); nattaya.kon@mfu.ac.th (N.K.)
2 Department of Food Science, Faculty of Science, Burapha University, Chonburi 20131, Thailand; samarts@go.buu.ac.th
3 Department of Agro-Industry, Faculty of Agriculture, Ubon Ratchathani University, Ubon Ratchathani 34190, Thailand; ekasit.o@ubu.ac.th
4 School of Food Science and Engineering, Hainan University, Haikou 570228, China; zwl@hainanu.edu.cn
5 College of Animal Science and Technology, Hunan Agricultural University, Changsha 410128, China; wush688@hunau.edu.cn
6 Department of Nutrition and Health, China Agricultural University, Beijing 100083, China; jqhuang@cau.edu.cn
7 School of Food Science and Biotechnology, Kyungpook National University, Daegu 41566, Republic of Korea; younghoonjung@knu.ac.kr
* Correspondence: saroat@mfu.ac.th

Abstract

The current rise in global population and the subsequent demand for food supply to meet the current population has directed the attention of researchers towards sustainable, plant-based sources, particularly underutilized crops. *Bactris gasipaes* is one such underutilized crop with significant functional food value. During processing, 84% of the total weight of the palm is discarded in the form of waste, or so-called by-products, which are a rich source of bioactive compounds. These compounds can be effectively recovered through modern extraction and valorization techniques. This review critically examines the extraction methods, nutritional profiles, and valorization opportunities of peach palm, highlighting both traditional uses and innovative processing strategies. Recent advances enable the targeted recovery of multiple peach palm fractions, e.g., proteins are commonly extracted using alkaline methods, lipids and carotenoids via green solvents or supercritical CO_2, and starch and dietary fiber through hydrothermal or downstream separation processes. Key nutritional findings demonstrate that peach palm fractions offer significant protein content (with isolates reaching 40 to 60%), a favorable starch profile (up to 79%), and abundant unsaturated lipids and carotenoids, making them suitable for gluten-free, protein-enriched, and functional ingredient applications. Previous studies have focused mainly on the edible pulp of peach palm for protein, lipid, and carotenoid extraction, whereas other fractions such as peel, seed, and processing residues remain comparatively underexplored due to technological and safety constraints. This review provides a consolidated and critical overview of recent advances in fractionation and green extraction strategies for multiple value-added streams (proteins, lipids, carotenoids, starch, and dietary fiber), highlighting knowledge gaps and opportunities for sustainable food ingredient development.

Keywords: peach palm; plant-based foods; protein isolate; bioactive compounds; waste valorization

1. Introduction

The alarming rise of global population, which is projected to reach up to 10 billion within the next 25 years, has emerged as an urgent challenge to ensure adequate and sustainable nutrition for all. This escalating figure demands an estimated 70% increase in food production to meet future needs [1]. The growing demand for food is not only driven by population expansion but also by increasing consumer awareness towards sustainable and plant-based foods. While animal-derived foods are nutritionally rich, their production is associated with considerable environmental burdens. This includes high land and water usage with elevated greenhouse gas emissions [2]. As a result, global priorities are shifting more towards sustainable, plant-based foods rather than animal-based sources. This trend has directed researchers' interest towards underutilized tropical crops that possess higher nutritional and functional properties [3]. Figure 1 depicts the sustainability shift from animal-based to plant-based food systems, emphasizing environmental impacts and the circular bioeconomy benefits associated with plant-based diets.

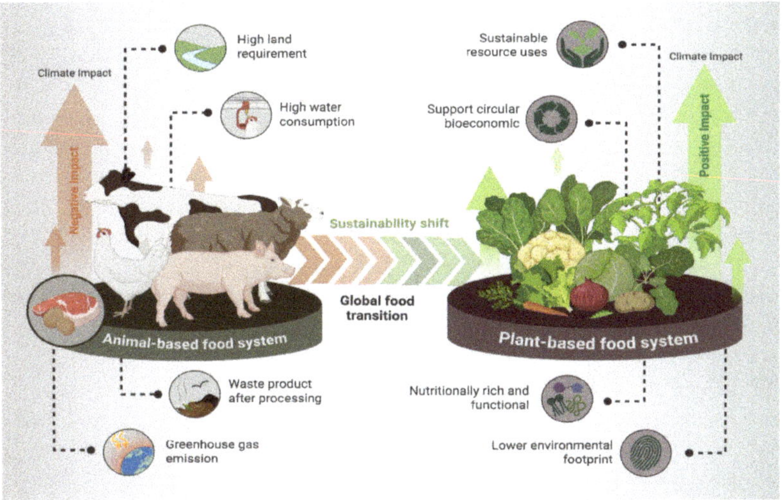

Figure 1. Sustainability-driven transition in global food systems from animal-based to plant-based production. https://biorender.com/2btt43z. Accessed on 5 February 2026.

Bactris gasipaes, commonly known as *Pupunha*, belonging to the *Arecaceae* family, is one such underutilized crop with significant potential for valorization in future food systems [4]. Its leaves, fruit, seeds, roots, and wood have been used traditionally for various purposes. For instance, the fruit is used to cure body aches, headaches, anti-inflammatory for gallbladder and eyes; leaves are employed for curing epilepsy; seeds are used in stomach aches; roots are used during urinary or menstrual problems, uterine infections, and many more [5]. Nutritionally, the fruit is a rich source of dietary fiber, carbohydrates, fats and bioactive compounds. The fruit possesses less protein but contains all essential amino acids [5]. It is also rich in essential minerals, with potassium, selenium and chromium being the dominating ones.

The fruits and by-products, which are dumped directly into the environment after processing, are rich in nutrients and bioactive compounds. These compounds can be effectively recovered through modern extraction and valorization techniques. Such strategies mirror

https://doi.org/10.3390/foods15040736

the broader shift of the food industry toward circular bioeconomy principles, emphasizing clean label formulation and the expansion of plant-based protein alternatives. Notably, the valorization of residual fractions, including basal segments and both inner and outer sheaths, offers pathways for zero-waste processing, functional ingredient development, and integration within biorefinery systems [6].

Recent progress in technologies and advancements in extraction technologies has opened numerous ways to fully harness the potential of peach palm, including its by-products. Following protein fractionation, functional optimization can enable its incorporation into meat analog formulations, while starch fractions can be extracted and modified for texturizing or prebiotic purposes. Likewise, lipid and carotenoid isolates serve as natural colorants and antioxidant-rich ingredients in clean label food systems. However, several technical challenges remain, including the need to improve extraction efficiency, address safety risks linked to calcium oxalate crystals, and enhance the techno-functional performance of recovered compounds to satisfy industrial processing standards.

While our previous review 'Unlocking the potential of peach palm for plant-based foods' comprehensively summarized the nutritional composition, processing effects, and general food applications of peach palm, the present review is intentionally focused on extraction, fractionation, and valorization strategies for generating ingredient-grade fractions. Emphasis is placed on processing pathways, green extraction technologies, and the functional potential of isolated proteins, lipids/carotenoids, starch, and dietary fiber for sustainable food systems.

Therefore, the aim of the current review is to critically examine the current knowledge on the extraction, processing and valorization of peach palm and to explore its potential in future food applications. The discussion focuses on strategies for recovering high-value compounds from both fruits and processing by-products, evaluates their applicability in plant-based food systems, and identifies persistent research gaps and technological challenges. By integrating technological, nutritional, and sustainability perspectives, the review positions peach palm as a promising circular raw material capable of contributing to the evolving sustainable food industry.

2. Processing Potential for Plant-Based Food Applications

Peach palm has been used traditionally for various applications, making it a promising species for diversifying and modernizing plant-based food systems. The application of modern technologies can facilitate its scalability and sustainable processing. Historically, peach palm was consumed after cooking in the form of cooked fruit, slowly fermented silage and fermented chichi. This process inactivates the antinutritional factors, including phytates, trypsin inhibitors, tannins and majorly calcium oxalate crystals [5]. As the effects of traditional processing on nutritional quality and antinutritional factors have been comprehensively reviewed in our previous review, the current discussion is limited to aspects relevant to downstream ingredient processing and functionality. Cooking peach palm softens the texture of the fruit, thereby making the pulp suitable for direct consumption or for preparing flour. The flour thus obtained has attracted attention as a gluten-free ingredient for use in bakery applications, broadening the dietary choices for individuals with celiac disease or gluten intolerance. Functionally, peach palm flour shows high water- and oil-holding capacity, strong emulsifying ability, and good stability in complex food matrices. Nevertheless, conventional processing treatments can influence lipid and fiber content, promote Maillard reaction browning, and alter the sensory characteristics of the product. Such changes ultimately affect the nutritional quality and techno-functional performance of the resulting flours or starches [7]. Figure 2 illustrates the conventional processing pathway

https://doi.org/10.3390/foods15040736

of peach palm fruit, highlighting each stage, from harvesting and boiling to drying and milling, along with its main purpose and associated drawbacks.

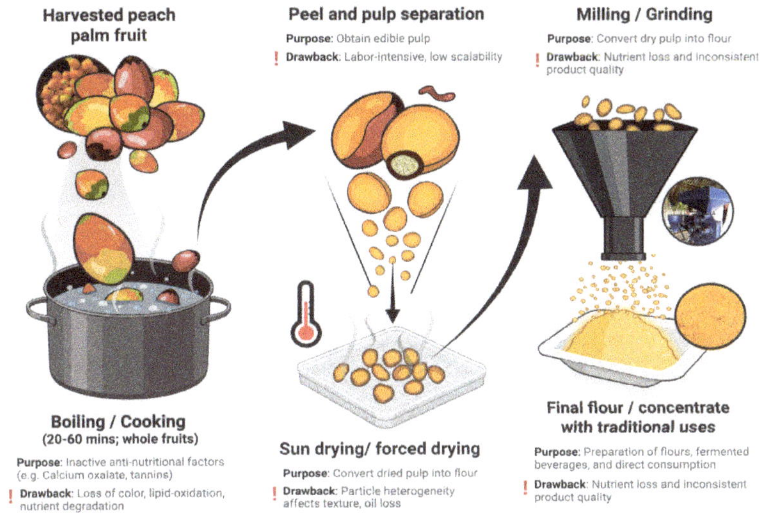

Figure 2. Traditional processing of peach palm fruit: main steps, purposes, and drawbacks. https://biorender.com/3axwi6v. Accessed on 5 February 2026.

Different types of peach palm vary in terms of varieties, ecotypes, and physical and biochemical composition. Therefore, there is a need for industrial-scale utilization and product development, which will offer consistency in the product. According to González-Jaramillo et al. [5], adapting processing methods to the intrinsic properties of peach palm varieties, including starch and carotenoid content, is key to consistent and innovative product development. In addition, it was also observed that the optimization of hydrothermal and drying parameters enhances product safety and functionality during industrial production or large-scale manufacturing operations [5,7].

The processing strategies summarized in Table 1 illustrate the transition from conventional thermal treatments toward greener and more selective technologies for peach palm valorization. Although traditional processing ensures detoxification and basic functionality, the integration of green technologies is required to enhance both functional and economic value. Microwave- and ultrasound-assisted extractions have been reported to improve carotenoid and phenolic recovery from pulp and peel while reducing solvent consumption, thereby offering advantages in efficiency and sustainability [8,9]. Ionic liquid systems further demonstrate high selectivity and bioactive yields suitable for functional ingredient development, although their broader application requires careful solvent management [10]. In parallel, supercritical CO_2 extraction combined with targeted physicochemical starch modification enables the design of ingredients with tailored techno-functional properties, such as low-digestibility starches for glycemic control and carotenoid-rich oils for natural coloration and antioxidant applications [11,12]. Collectively, these technological developments support the production of functional foods that align with the increasing consumer demand for health-oriented and sustainable products [5].

https://doi.org/10.3390/foods15040736

Table 1. Processing methods for peach palm and their plant-based food applications.

Methods	Parameters	Targeted Component	Key Quality/Yield (Basis as Reported)	Food Applications	Advantages
Traditional Boiling/Cooking	Boil whole fruit (20–60 min), peel, manual pulp separation	Pulp, cooked fruit	Reduced oxalate, increased starch	Direct eating, flour, puree	Inactivates antinutrients, ready for gluten-free use
Sun or Forced-Air Drying	Dry cooked pulp or flour at 50–60 °C	Flour, pulp	Lower water activity, stable	Baking, bread, extrudates	Extended shelf life, stable color/nutrients
Milling/Grinding	Mechanical grinding, post-cooking and drying, sieving (particle size control)	Flour (pulp/peel), starch	Gluten-free, fiber-rich flour	Gluten-free bakery, cookies, extruded snacks	Different mesh size for texture/functionality
Traditional/Conventional Oil Extraction	Solvent-based (hexane, ether) or mechanical pressing	Oil (pulp/seed), carotenoids	7–21% *w*/*w* (oil yield, basis as reported)	Spreads, enriched foods, functional oils	High vitamin A, tocopherols, phytosterols
Ultrasound-Assisted Extraction (UAE)	Ethanol as solvent, 30 min, 20 kHz, 50 °C, variety-dependent	Lipids, carotenoids, polyunsaturates	2–8% (variety), enriches carotene	Nutraceutical oils, colorants, PUFA for health	"Green", less solvent, higher antioxidant retention
Supercritical CO_2 Extraction	Supercritical CO_2 at 300 bar, 40 °C, pulped fruit or peel	Carotenoids, oils	Max carotenoid, low solvent residue	Functional/therapeutic food colors/ingredients	Clean, scalable, preserves labile compounds
Ionic Liquid Extraction	Imidazolium-based IL, recyclable solvent, 30–60 min, controlled recycle	Carotenoids, phenolics	172 µg/g, 94% IL recovery	Emulsified food, supplements	High yield, greener, repeated use
Starch Extraction and Fractionation	Water or acid extraction, sieving, separation of granules	Amylose, amylopectin, resistant starch	55–72%, 14–20% amylose	Low glycemic breads, porridge, gluten-free flour	Produces slow-digesting starch for metabolic health
Enzymatic/Physical Protein Extraction	Solubilization and precipitation post-cooking, mechanical pressing	Protein concentrate/isolate	Up to 4–7% protein in flour	Fortified flours, alternative protein foods	Functionality as foaming/emulsifying agent
Modern Gluten-Free Flour Production	Combined cooking, drying (forced air or freeze-drying), fine milling	Whole flour, composite ingredients	High water- and oil-binding capacities	Gluten-free cakes, bread, pasta, batters	Provides structure, color, fiber, minerals
Extrusion Cooking	High-temp, short-time, peach palm/corn blends	Texturized flour, enriched extrudate	Maintains pigments, texture control	Cereal, breakfast, snack extrudates	Enhances carotenoid retention, fiber enrichment
Microwave-Assisted Extraction	Use with solvent (e.g., ethanol), short bursts for extraction of bioactives	Carotenoids, phenolics	Higher yields, darker color oil	Functional oils, provitamin A concentrates	Efficient, less time, gentle for sensitive compounds
Enzyme Production (from Waste/Peel)	Solid-state fermentation with Trichoderma or Pleurotus spp., supplement N-source	Amylase, hydrolytic enzymes	29–53 U/g (amylase, optimized mix)	Starch hydrolysis in baking/foods, fermentation	Waste reduction and creates value for by-products
Color and Phytochemical Extraction	Sequential solvent extraction, use of green solvents	Flavonoids, tocopherols, carotenoids	357 mg/kg oil for total carotenoids	Natural colorants, antioxidant-rich foods	"Green" chemistry, for pigment/nutrient-rich foods
Formulation in Plant-Based Foods	Blending with other cereal or legume flours, texturization, color stabilization	Final formulated ingredient	High sensory acceptance (>70%)	Cookies, breads, vegan spreads, health snacks	Increases micronutrient density, functional fibers

Note: Percentage values refer to yields as reported in the original studies (e.g., based on fresh material, dry material, or oil fraction). Source: [4,6,7,13–15].

2.1. *Protein Extraction and Fractionation Potential*

(a) Theoretical protein concentrate/isolate development

Peach palm has the potential of being a promising source of plant-based proteins. Peach palm's several anatomical parts, most notably the fruit pulp, derived flour, and palm heart, contain a substantial number of proteins with balanced essential amino acid profiles. The protein contents in peach palm pulp ranges from 1.8 to 4.6 g/100 g on a dry basis, with higher values reported after dehydration and milling into flour, as reported in various

 https://doi.org/10.3390/foods15040736

studies. Evidence from Soares et al. [13] indicates that the protein content in peach palm flour can reach up to 20% on a dry matter basis. The high protein content in peach palm in comparison with other tropical fruits and starchy staples provides a strong basis for the production of protein concentrates and enrichment using peach palm. The development of protein concentrates and isolates typically entails mechanical disruption (grinding or milling), followed by aqueous or alkaline extraction and isoelectric precipitation or ultrafiltration to recover discrete protein fractions [16,17]. Some of the existing literature highlights that the nutritional value of flour derived from the fruit pulp and peel of peach palm is not only relatively rich in protein but also contains all essential amino acids in amounts that meet or exceed FAO reference patterns [18,19]. This unique nutritional profile of peach palm makes it suitable not just as a nutritional fortifier but also as a base for functional isolates with balanced amino acid profiles.

(b) Functional protein modification opportunities

Functional protein modification aims to improve solubility, water-/oil-binding capacity, emulsification and foaming properties to expand their value in plant-based food systems [4,7,13]. Recent studies emphasize several innovative methods, including enzymatic hydrolysis, physical processes and thermal treatments to improve the techno-functional profile of peach palm proteins [13,15]. Enzymatic hydrolysis is one of the approaches to produce bioactive peptides with various functional and biological properties [20,21]. This has been well documented in various legumes and oil seeds, where enzymatic hydrolysis significantly enhanced the digestibility and solubility of the resulted peptides along with augmented antioxidant and antihypertensive activities [22,23]. Enzymatic hydrolysis is responsible for altering the size of the peptides and this in turn improves their emulsification properties. Consequently, such processes provide more ways for their incorporation into vegan dairy products, baking formulations, and beverage systems [5,13]. Physical processes like ultrasonication and high-pressure processing (HPP) and thermal treatments such as hydrothermal treatment also provide opportunities for improving functional properties by altering the shape of proteins [24]. Justino et al. [25] demonstrated that the use of ultrasound technology results in an effective breakdown of protein aggregates and a reduction in particle size. This results in enhanced dissolution and greater surface coverage. These properties make proteins suitable as natural stabilizers for emulsified and aerated systems [25]. Meanwhile, HPP causes compact protein structures (globular shape) to unfold, thereby exposing the hydrophobic domains that increase both foaming and emulsifying capacities [26]. Overall, enzymatic treatments mainly improve digestibility and bioactivity, physical processes (ultrasound and high-pressure processing) enhance solubility and interfacial properties, controlled chemical reactions influence flavor and thermal stability, and fermentation-based approaches reduce antinutritional factors while generating novel functional attributes.

Chemical modification offers another approach for altering protein functionality when conducted under controlled food-processing conditions. Non-enzymatic glycation (Maillard-type reactions) may occur between amino groups of proteins and carbonyl groups of reducing sugars during thermal treatment, influencing both functional properties and flavor development [21,27]. However, such reactions are not considered deliberate protein modification methodologies, as uncontrolled Maillard reactions can lead to nutritional losses and the formation of advanced glycation end products. Previous studies have shown that mild and controlled glycation can improve solubility and thermal stability in proteins from tropical and Amazonian crops [28,29]. In this context, carefully regulated glycation reactions could potentially be explored for peach palm proteins to modulate functional performance in plant-based formulations, while avoiding excessive thermal damage.

https://doi.org/10.3390/foods15040736

Overall, these diverse protein enhancement techniques, including enzymatic treatment, physical processing, chemical modification and fermentation-based methods, allow for the precise improvement of peach palm proteins. Each approach offers a way to adjust their structures or functions to suit modern food production needs. As a result, these proteins are finding a wider use in sustainable, plant-based foods that demand both performance and quality. Table 2 summarizes the recent advances in protein extraction and functional modification techniques from peach palm, whereas Figure 3 summarizes major physical, chemical, enzymatic and biotechnological modification strategies used to enhance the functional properties of plant-derived proteins, including solubility, emulsification and bioactivity.

Table 2. Recent advances in protein extraction, fractionation, and functional modification from peach palm and plant sources.

Extraction Method	Source	Protein Content (%, Basis as Reported)	Modification Steps	Functional Traits Developed	Food Applications
Hydrothermal extraction/milling	Cooked pulp, flour	1.8–4.6	Cooking, milling, sieving	Increased solubility, partial denaturation	Gluten-free bakery, breads, snack powders
Dry fractionation	Dried flour	Up to 17–20	Fine grinding, dry sieving	Improved water/oil holding, emulsification	Cakes, biscuits, extruded cereals
Alkaline/aqueous protein extraction	Pulp, flour	8–25 (concentrate)	Homogenization, aqueous or alkali leaching, centrifuge	Higher protein purity, reduced antinutrients	Protein concentrates, supplement blends
Enzymatic assisted modification	Pulp or flour	Variable	Protease-assisted extraction, hydrolysis	Enhanced solubility, antioxidant peptides	Functional protein isolate, protein beverages
Ultrasound-assisted extraction	Pulp, flour	Up to 25	Acoustic-assisted aqueous extraction	Improved yield and foaming	Protein enrichment in bakery, snacks
Isoelectric precipitation	Aqueous extract	40–60 (isolate)	pH shift, protein separation, drying	Concentrated fractions, tailored gelling	Vegan cheese, meat analogues
Functional protein modification	Protein flour/extract	Application-specific (variable)	Enzymatic or heat-induced changes, blending	Water and oil binding, improved emulsification	Texturizers, beverage stabilizers

Note: Protein contents are reported as provided in the cited studies and may be expressed on a fresh-weight (fw), dry-weight (dw), or isolate basis. Source: [4,5,7,13–16,30–34].

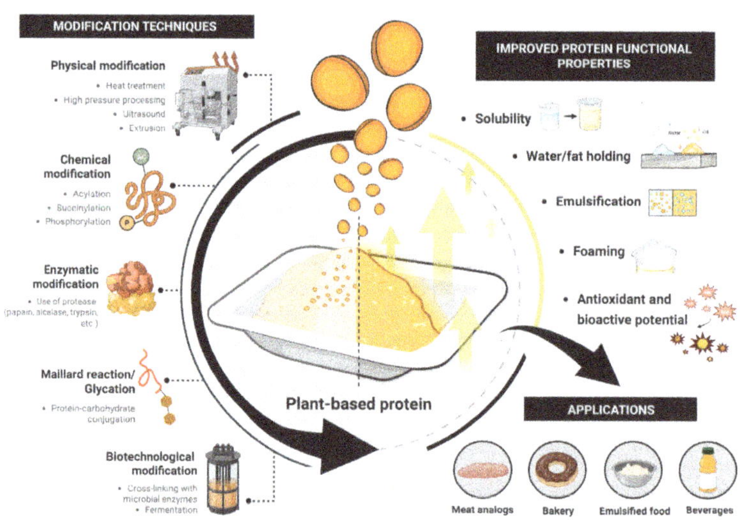

Figure 3. Functional modification strategies to improve techno-functional properties of plant proteins. https://biorender.com/p7zdxpo. Accessed on 5 February 2026.

https://doi.org/10.3390/foods15040736

2.2. Starch and Carbohydrate Utilization

Peach palm possesses high starch concentrations. Rosário et al. [35] reported that peach palm contains about 79% total starch, with approximately 12.4% amylose and 66.6% amylopectin in the fruit matrix. The low amylose content in peach palm aids in the production of starch with desirable properties. Therefore, such flours can be used for various food and industrial purposes where variable mechanical or functional properties of starch is required for product formulations [35]. The native starches from peach palm show varied granule shapes and sizes; however, this variation is strongly influenced by the maturation stage of the fruit. As such, starches from unripe fruits do not show distinct morphologies [4,35,36].

(a) Starch isolation and modification

Peach palm starch is obtained from the peach palm pulp. It contains residual levels of ash, proteins, lipids and fibers. Starch extraction involves isolating starch granules from plant tissues through processes such as grinding, steeping, sieving, and centrifugation. The separated starch is then washed and dried to achieve high purity and yield [36]. Extraction methods like wet milling and alkaline treatment can influence starch quality, purity, and functional properties by affecting its molecular structure and amylose-to-amylopectin ratio. Extraction techniques, whether mechanical, chemical, or physical, are selected based on the raw material and the desired characteristics of the final product. Each method presents a trade-off between yield, purity, and functional properties [37,38].

The processing methods entail the extraction and alteration of starch, as the native form of starch is not suitable for direct use in industrial applications due to its unfavorable characteristics such as propensity towards retrogradation, high viscosity even at low concentrations, handling problems, poor freeze–thaw stability, low process tolerance, and gel opacity, which prevent its use in food processing [39]. To enhance starch's physicochemical and functional properties to meet industrial standards, it must undergo various modifications, such as physical, chemical, and enzymatic modifications. Chemical modification is highly effective, but it does have disadvantages, such as the high cost of chemical residues, which makes it environmentally unfriendly. On the other hand, enzymatic methods are more intricate and time-consuming [40]. Conversely, physical modification techniques lack chemicals and do not produce wastewater with toxic residues. Furthermore, the wastewater treatment required after modification is minimal. Modern physical modification techniques such as cold plasma processing, irradiation, and hydrothermal treatment are used in the starch industry to produce starch with direct commercial applications that negatively impact the environment and consumer consumption. The experimental studies demonstrate that treatments like annealing, hydrothermal treatment (HMT), autoclaving, and cross-linking with citric acid can significantly improve starch characteristics such as crystallinity, thermal stability, and resistance to retrogradation. Thus, with the aid of these techniques, the starch of peach palm can be used in starch-based food systems and for developing biodegradable materials. For example, Soares et al. [4] reported the production and characterization of a peach palm starch-based biodegradable thermoplastic. This thermoplastic showed high tensile strength along with high thermal degradation ability. Consequently, it can be concluded that peach palm starch may serve as an important environmentally friendly alternative in the production of this type of material, which is of great interest to the packaging industry [40–42].

(b) Fiber extraction and functionality

Peach palm pulp is classified as a fiber-rich food [4]. Dietary fiber plays an important role in food systems by enhancing food texture through water retention and supporting digestive health, functional properties, and nutritional quality. The key dietary fiber

https://doi.org/10.3390/foods15040736

components, viz. cellulose, hemicellulose, and lignin, have been successfully obtained from the processing discards or by-products. Dietary fibers are not digested and absorbed in the human small intestine. They are threatened by complete or partial fermentation in the large intestine. Dietary fibers have beneficial physiological functions, including laxation and improving bowel health by stimulating the growth of beneficial gut micro-flora, lowering blood cholesterol and glucose levels, preventing obesity, coronary heart diseases, diabetes, blood pressure, and lowering energy intake [15]. Insoluble dietary fibers remain intact during digestion and are involved in reducing the risk of coronary heart disease and type 2 diabetes. Giombelli et al. [42] recovered dietary fiber concentrates from peach palm waste using subcritical water and low-pressure aqueous systems as extraction techniques. The fibers obtained using these methods possess high water- and oil-holding capacities. Under subcritical extraction conditions, the fibers become more fragmented and exhibit higher porosity. The substantial cellulose composition, combined with altered conformational attributes detected via FT-IR analysis, leads to improved hydration and emulsification capabilities, thereby expanding the potential applications of peach palm fibers in functional food systems. Furthermore, Giombelli et al. [42] reported that scanning electron microscopy revealed a more open, porous microstructure in starch after subcritical water treatment, which enhanced its functional properties, including oil-binding capacity and swelling index.

(c) Prebiotic and functional carbohydrate applications

The carbohydrates in peach palm, mainly its soluble dietary fiber and processed starches, exhibit prebiotic potential. Šárka et al. [43] observed that pectic polysaccharides extracted from peach palm stimulate beneficial gut microbial fermentation in vitro. Their resistance to enzymatic degradation further enhances their prebiotic activity. Likewise, [44] also demonstrated that resistant starch fractions produced through controlled modification digest slowly and promote favorable fecal fermentation profiles. In addition to these components, xylooligosaccharides (XOS), composed of 2–20 D-xylose units linked by β $(1 \rightarrow 4)$ glycosidic bonds, can also contribute to prebiotic functionality. XOS are typically derived from the hydrolysis of xylan present in lignocellulosic biomass, and, because humans lack the enzymes to hydrolyze β-linkages, these oligosaccharides reach the large intestine intact, where they serve as substrates for beneficial microbes such as Bifidobacterium and Lactobacillus [15]. Their fermentation not only supports gut microbial balance but also provides secondary health benefits, including improved calcium absorption, better lipid metabolism, and reduced risk of chronic metabolic disorders. Collectively, these dietary components play vital roles in blood sugar regulation, colon health, and in sustaining probiotic bacteria that help maintain a balanced intestinal microbiome [40].

2.3. Lipid and Bioactive Compound Extraction

(a) Oil extraction for food applications

Considering that the peach palm fruit generally has a high lipid content, it is important to consider its lipid profile, particularly the quantity and quality of the fatty acids present in the pulp. The fruit of peach palm is naturally rich in oils, making it a promising source for the production of edible oil. It is recognized for its high yield and impressive nutritional quality. The oil extracted from peach palm pulp can serve as an intermediate source of omega-6 fatty acids (4.9–8.6%). In addition, it also contains substantial proportions of oleic, linoleic, and linolenic acids, which are known for their cardiovascular benefits and oxidative stability. Refs. [4,15] show a clear move away from old solvent-based extraction methods toward cleaner, safer technologies for obtaining peach palm oil, and [45] also supported the transition to green extraction processes that meet the growing demand for clean label, sustainable ingredients while maintaining oil quality. Among these processes, ultrasound-assisted extraction combined with ethanol achieved oil yields of up to 8.9%

https://doi.org/10.3390/foods15040736

while preserving key bioactive constituents like carotenoids and unsaturated fatty acids, particularly in red and yellow peach palm fruits. However, peach palm oil has been reported to be relatively susceptible to oxidation, likely due to its high omega-9 (oleic acid) content. Fatty acids are inherently prone to oxidative degradation because of the double bonds in their molecular structure, leading to free radical chain reactions that can compromise both the nutritional and the sensory quality of the oil. In parallel, supercritical CO_2 extraction enables the recovery of high-purity oils with negligible solvent residues, making them ideal for health-oriented applications. These lipids, naturally rich in oleic, linoleic, and linolenic acids, are prized for their heart health benefits and resistance to oxidation, fitting neatly with the growing demand for nutritious, sustainable, plant-based foods [4,45]. Despite this, Soares et al. [4] also documented the antimicrobial characteristic of oil extracted using hexane from peach palm bark against strains of *Staphylococcus aureus* 24 h after the addition of 10 μL oil. Given the growing interest in peach palm as a potential sustainable lipid source for food applications, the fatty acid profile of peach palm pulp lipid extracts is summarized in Table 3. This profile highlights the relative proportions of saturated and unsaturated fatty acids and provides compositional information relevant to nutritional quality and oxidative stability.

Table 3. Fatty acid composition of peach palm pulp lipid extract obtained by ultrasound-assisted extraction.

Fatty Acid (% Total FA)	Red	Yellow	Green	White
Lauric (C12:0)	0.01 ± 0.00	0.02 ± 0.00	0.01 ± 0.00	0.01 ± 0.00
Myristic (C14:0)	0.08 ± 0.00	0.15 ± 0.00	0.08 ± 0.00	0.10 ± 0.00
Palmitic (C16:0)	23.77 ± 0.15	28.96 ± 0.23	33.86 ± 0.34	42.62 ± 0.43
Stearic (C18:0)	Nd	0.70 ± 0.02	Nd	1.87 ± 0.23
Arachidic (C20:0)	0.14 ± 0.02	0.10 ± 0.00	0.12 ± 0.03	0.19 ± 0.00
Palmitoleic (C16:1)	9.89 ± 0.34	13.23 ± 0.12	3.98 ± 0.91	4.99 ± 0.03
Oleic (C18:1n−9)	60.20 ± 0.50	44.85 ± 0.41	57.62 ± 0.14	40.73 ± 0.54
Linoleic (C18:2n−6)	4.04 ± 0.61	8.05 ± 0.91	2.03 ± 0.14	6.95 ± 0.17
α-Linolenic (C18:3n−3)	1.48 ± 0.24	2.50 ± 0.07	0.54 ± 0.23	2.14 ± 0.34

Note: Values are expressed as percentage of total fatty acids as reported in the original study. Nd= Not detected. Source: [45].

(b) Carotenoid concentration for natural coloring

Among the fat-soluble constituents of peach palm, carotenoids are also considered important bioactive nutrients and are present in high concentrations in the fruit. The intensely colored carotenoid compounds from peach palm constitute valuable natural colorants and bioactive substances. Various studies have found that yellow, orange, and red peach palms are rich in carotenoids, and the stage of fruit maturation strongly influences their total carotenoid content. Moreover, fresh fruits of the orange peach palm variety exhibit the highest total carotenoid content, followed by the red and yellow cultivars, demonstrating a clear relationship between color intensity and carotenoid concentration.

Spacki et al. [15] demonstrated that carotenoid-rich fractions from peach palm can be effectively extracted using ultrasound-assisted extraction (UAE), ionic liquid systems, and supercritical CO_2 methods with recyclable, low-toxicity solvents. In addition, ref. [4,46] reported that the red and yellow cultivars possess the highest carotenoid concentrations, underscoring their potential in clean label product development and functional food formulation.

Carotenoids are subject to instability, which can be influenced by their chemical composition (carotene or xanthophyll), molecular structural configuration (cis or trans), esterification, and the cellular matrix, as well as by processing and storage conditions. Notably, higher contents of total carotenoids after cooking have been reported in peach

 https://doi.org/10.3390/foods15040736

palm fruit compared with fresh samples, possibly due to the release of carotenoids from cell walls. This thermal effect may also lead to the formation of isomers such as Z-β-carotene, Z-γ-carotene, and Z-lycopene, which have been identified in peach palm fruit and may contribute to variations in bioactivity. Furthermore, the bioaccessibility of carotenoids from peach palm can be enhanced by incorporating them into lipid-based matrices. For instance, carotenoids extracted from peach palm pulp by ultrasound and added to a mayonnaise emulsion were found to be 11 times more bioaccessible after in vitro digestion compared with those in freeze-dried fruit. The utilization of peel fractions also supports natural pigment recovery and promotes sustainability in ingredient procurement [4,15,45].

(c) Antioxidant activity of peach palm extracts

Lipid and carotenoid-rich extracts from peach palm exhibit notable antioxidant activity resulting from their polyphenol, tocopherol, and phytosterol contents. Various extraction methods such as ultrasound and microwave assistance have been demonstrated to improve the extraction efficiency of phenols and antioxidant potential, exceeding the performance of traditional methods. These bioactive extracts contribute to functional food and nutraceutical development, corresponding with present-day demands for health-beneficial, sustainable ingredient sources [4,15]. An overview of advanced extraction technologies available for lipid and bioactive compound recovery from peach palm, including their process parameters, compositional highlights, and functional applications, is presented in Table 4. Unless otherwise stated, quantitative values are reported on the basis used in the original studies (fresh weight, dry weight, or extract basis), which is explicitly indicated to avoid misinterpretation.

Current advancements in extraction technology enable zero-waste processing approaches and circular ingredient development for the sustainable food industry, exploiting lipids and carotenoids from complete peach palm biomass fractions for innovative plant-based food applications.

Figure 4 shows an integrated schematic of modern extraction technologies for peach palm, outlining how different green methods yield functional ingredients for food, nutraceuticals and biomaterial applications. Current advancements in extraction technology enable zero-waste processing approaches and circular ingredient development for the sustainable food industry, exploiting lipids and carotenoids from complete peach palm biomass fractions for innovative plant-based food applications.

(d) Bioactive compounds identified in peach palm and their biological activities

Several studies have identified diverse classes of bioactive compounds in peach palm, including carotenoids, phenolic compounds, tocopherols, phytosterols, and organic acids. These compounds contribute to antioxidant, anti-inflammatory, antimicrobial, and cardioprotective activities, supporting the functional food potential of peach palm. Table 5 summarizes the main bioactive compounds reported in different peach palm fractions, together with their biological activities and supporting references.

 https://doi.org/10.3390/foods15040736

Table 4. Overview of advanced extraction technologies for lipid and bioactive compound recovery from peach palm.

Extraction Method	Solvent Type	Main Compound	Extraction Yield (%, as Reported in Original Studies)	Highlights	Functional Properties	Applications	Disadvantages	Reference
Ultrasound-Assisted Extraction	Ethanol	Carotenoids, Lipids	Up to 8.9 (red var.)	High beta-carotene, unsat. FA	Preserves thermolabile pigments	Clean label oil, natural pigment	Limited scalability; possible oxidation if ultrasound intensity is not controlled; higher equipment cost than conventional extraction	[8,45]
Supercritical CO_2 Extraction	CO_2	Carotenoids, Lipids	6.1–8.2	Strong antioxidant retention	Solvent-free, scalable	Functional oil, nutraceuticals	High capital and operating cost; requires high pressure; low efficiency for polar compounds without co-solvent	[11]
Ionic Liquid Extraction	Ionic liquids	Carotenoids, Phenolics	Up to 172 µg/g extract	Enables selectivity	High recyclability, green tech	High-value pigment fractions	Solvent recovery required; regulatory acceptance for food use is limited; potential toxicity depending on ionic liquid type	[10]
Microwave-Assisted Extraction	Ethanol/Water	Lipids, Bioactives	6–10	Moderate yield, rapid process	Retains bioactivity	Pigment/antioxidant extracts	Risk of uneven heating; possible degradation of heat-sensitive compounds; scale-up challenges	[15]
Enzymatic-Assisted Extraction	Enzyme-buffer	Phenolics, Lipids	5–9	Increases extractability	Mild conditions, low residue	Antioxidant/fat ingredient	High enzyme cost; long processing time; sensitive to pH and temperature variations	[6,9]
Conventional Solvent	Hexane/Ether	Lipids	7–21	High yield, less selectivity	Food-grade restrictions	Bulk oil ingredient	Use of toxic/flammable solvents; environmental burden; solvent residues; low selectivity	[4,18]
[18] Mechanical Pressing	—	Lipids	3–8	Preserves natural composition	Low yield, safe process	Whole oil, unrefined ingredient	Low extraction efficiency; requires pretreatment; not suitable for bound lipids	[18]
Solid-State Fermentation	—	Phenolics, Bioactives	Variable	Generates novel bioactives	Nutritional and functional	Enriched flours, extracts	Long processing time; contamination risk; batch-to-batch variability; difficult process control	[6]

Note: Data are reported as provided in the original studies. Percentage values refer to extraction yield relative to the initial material or oil fraction, as specified in the cited sources.

https://doi.org/10.3390/foods15040736

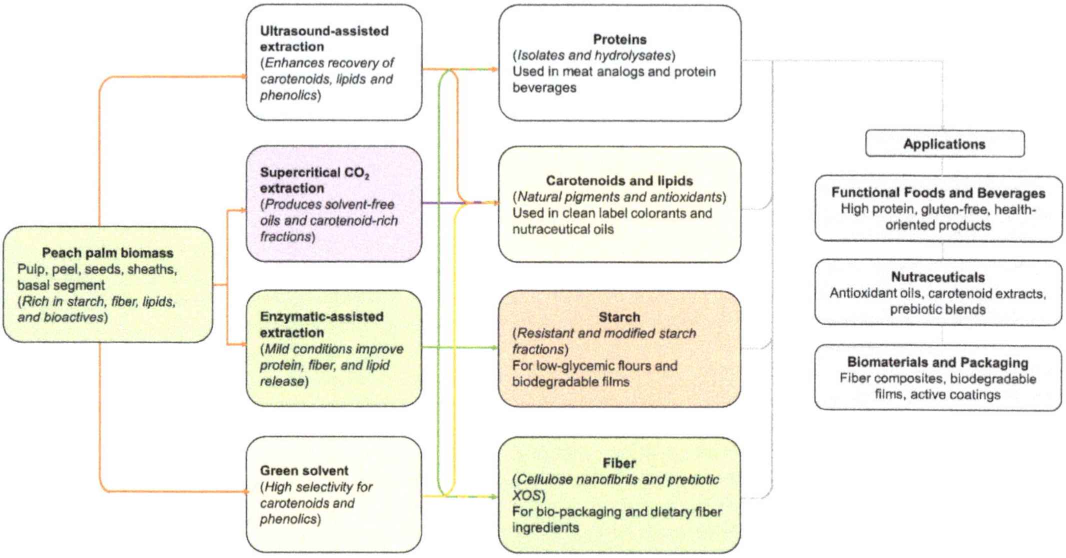

Figure 4. Modern extraction technologies for peach palm; from biomass to functional ingredients.

Table 5. Bioactive compounds identified in peach palm and reported biological activities.

Compound Class	Identified Compounds	Plant Fraction	Reported Biological Activity
Carotenoids	β-carotene, α-carotene, lutein, zeaxanthin, lycopene	Fruit pulp, peel	Antioxidant, provitamin A activity, immune modulation
Phenolic compounds	Gallic acid, ferulic acid, caffeic acid, chlorogenic acid	Peel, pulp, sheath	Antioxidant, anti-inflammatory
Tocopherols	α-tocopherol, γ-tocopherol	Fruit oil	Lipid oxidation inhibition, cardiovascular protection
Phytosterols	β-sitosterol, campesterol, stigmasterol	Fruit oil, seed	Cholesterol-lowering, anti-inflammatory
Organic acids	Citric acid, malic acid, succinic acid	Pulp, peel	Antimicrobial, pH regulation, antioxidant synergy
Polysaccharides	Pectin, hemicellulose, resistant starch	Pulp, by-products	Prebiotic, gut microbiota modulation
Myo-inositol	Myo-inositol	Sheath, basal portion	Metabolic regulation, prebiotic effect

Source: [4,7,9–11,15,42–45].

2.4. Integrated Processing Approaches

(a) Zero-waste processing concepts

Although the concept of "zero-waste" is frequently used to describe fully integrated biorefinery and valorization strategies, the complete elimination of waste is rarely achievable in practice due to unavoidable material losses, process residues, and energy demands. In the case of peach palm, a near zero-waste approach can nevertheless be pursued through the sequential utilization of different plant fractions. Edible pulp can be directed to food applications, while peels, sheaths, and basal portions can be converted into dietary fiber concentrates, carotenoid-rich extracts, fermentable substrates, or biopolymer precursors. Residual solids may further be valorized through fermentation or energy recovery processes. This cascading use of biomass reduces disposal streams and improves resource efficiency, even if true zero-waste conditions remain theoretical rather than absolute [6,15].

https://doi.org/10.3390/foods15040736

(b) By-product valorization strategies

A wide range of peach palm by-products, including peels, inner sheaths, and stem segments, are currently acknowledged as abundant sources of dietary fiber, xylans (for prebiotic xylooligosaccharide production), cellulose (for nanofibril applications), phenolic compounds, and carotenoids [42]. Principal valorization approaches include the production of high-fiber flours for the application as functional food ingredients and emulsifying agents; extraction of natural pigments and antioxidant compounds, supporting food and nutraceutical sectors; alkaline or enzymatic hydrolysis of xylans for xylooligosaccharide (prebiotic) generation, producing antioxidant-rich fractions; and extraction with nanoprocessing of cellulose for bio-based packaging materials or rheological modifiers. Through proper process integration and sequencing, these methodologies reduce disposal expenses and establish new revenue opportunities for processing facilities and rural farming communities [15].

(c) Multi-component extraction systems

Recent progress in multi-component extraction have made it possible to isolate lipids, carotenoids, dietary fibers, starch, and phenolic compounds, either sequentially or simultaneously, from both edible and inedible parts of peach palm. Using techniques such as ultrasound-assisted extraction, supercritical CO_2, and ionic liquid-based extraction enables the targeted recovery of specific compounds without significant loss in biological activity. These compounds can be used in food, cosmetics, pharmaceutical, and biomaterial applications [4,14,15]. When combined with enzymatic or fermentation-based processes, for instance, fermentation using Trichoderma or Lentinula, these systems generate additional value through enzyme production (amylases, xylanases) and prebiotic oligosaccharide synthesis. [15].

For instance, Lima et al. [47] studied the performance of *B. gasipaes* residues as a substrate for the growth of a mycelium-based composite on *Lentinula edodes*. The composite formed displayed close values compared to other mycelium-based composites on compressive strength and elastic modulus. The authors concluded that pupunha residues are a potential alternative for mycelium-based composites. The production and commercialization of mushrooms (healthy functional foods) using peach palm residues could result in socio-environmental benefits by increasing the income of the involved individuals and by reducing environmental liability. Therefore, developing sustainable, integrated biorefinery models is essential for the full exploitation of peach palm resources. Such approaches not only minimize environmental impact but also enhance economic potential, positioning peach palm residues as key contributors to future zero-waste and circular bioeconomy systems. Table 6 summarizes the specialized valorization strategies for various peach palm by-products, highlighting the functional materials generated and their role in the circular bioeconomy. Sustainable, integrated biorefinery strategies are therefore critical for the comprehensive exploitation of peach palm resources, corresponding with global movements towards zero-waste systems and circular bioeconomy advancement. Such frameworks address environmental liabilities while enhancing economic value, establishing peach palm residues as essential resources in future food and bioproduct supply chains.

https://doi.org/10.3390/foods15040736

Table 6. Specialized valorization strategies for peach palm by-products; functional materials and circular bioeconomy innovations.

By-Product Fraction	Main Component	Processing Technology	Function	Application	Innovation	Limitations	Reference
External Sheath	Dietary fiber (cellulose, hemicellulose), low protein	Alkaline/enzymatic extraction, hydrothermal treatment	XOS (xylo-oligosaccharides), cellulose nanofibrils	Prebiotic blends, biopolymer composite materials	Edible packaging, XOS-based gut health supplement	Requires chemical pretreatment; generation of alkaline wastewater; variability in fiber composition depending on maturity	[14,48,49]
Internal Sheath	Complex fiber, pectin, trace minerals	Enzyme refinement, composting, microbial fermentation	Natural pectin fraction, biofertilizer	Functional hydrocolloid, soil amendment	Microbial valorization for organic farming	Low pectin yield compared with citrus sources; microbial processes require strict control; slow processing rate	[14,42,50]
Basal Segment	Lignocellulose, residual carbohydrate	Saccharification, anaerobic digestion, drying	Biogas, resistant starch flour	Renewable energy, specialty feeds	Circular plant energy, starch for clean label baking	High lignin content limits enzymatic hydrolysis; requires energy-intensive pretreatment; limited food-grade applications	[6,35,48]
Fruit Residue	Pulp fiber, polyphenols, organic acids	Aqueous extraction, enzyme hydrolysis, drying	Polyphenol-rich colorant, citric extracts	Functional colorant, natural acidulant	Active packaging ingredient for shelf-life extension	Polyphenols are sensitive to heat and oxidation; aqueous extracts are dilute and require concentration; seasonal variability	[7,51,52]
Seed/Kernel Waste	Oil, lignin, bioavailable micronutrients	Cold pressing, pyrolysis, ultrafiltration	Bio-oil, micronutrient concentrate	Biolubricant/green solvent, fortificant	Micronutrient delivery for new nutraceuticals	Low oil yield by pressing; pyrolysis products not suitable for food use; additional refining required for edible applications	[18,53]

https://doi.org/10.3390/foods15040736

3. Valorization of By-Products and Circular Uses

The valorization potential of different anatomical parts of peach palm, detailing their composition, principal uses, and value-added outputs within a circular bioeconomy framework, is shown in Figure 5. The production and consumption of peach palm generates a huge volume of by-products [48]. The utilization of processing by-products represents a key strategy for achieving the circular biomass management of peach palm within sustainable food systems. During peach palm extraction, it is estimated that approximately 84% of the total weight of the palm is wasted. Nevertheless, chemical analysis of these fractions reveals high dietary fiber contents ranging from 59% to 68%, alongside protein levels of 8–12%, confirming their potential as sustainable raw materials for high-fiber or protein-enriched functional ingredients [6,15].

Figure 5. Valorization potential of peach palm parts and derived outputs. https://BioRender.com/v81vyht. Accessed on 5 February 2026.

In addition to the macronutrient composition of these by products, these so-called wastes (by-products) are also rich in nutraceutical compounds such as myo-inositol—a bioactive compound with documented metabolic and prebiotic functions—and a spectrum of organic acids that contribute to potential antioxidant and antimicrobial activities. Recent progress in extraction and processing has made it possible to transform the sheath and basal fractions into marketable products, including high-fiber flours and dietary supplements, xylooligosaccharide-based prebiotic formulations, and cellulose nanofibrils suitable for use in packaging materials or as food texture modifiers [6,15].

A number of valorization pathways have been established within circular economy frameworks, encompassing enzymatic hydrolysis for xylooligosaccharide production, solid-state fermentation for edible mushroom cultivation or enzyme biosynthesis, and nanoprocessing techniques designed to improve cellulose recovery from processing residues [54]. Implementing these strategies not only reduces environmental burden but also opens new avenues for ingredient innovation and economic benefits for processors and local producers [6,15].

In conclusion, the external and internal sheaths together with the basal segment of peach palm should no longer be considered waste but should be recognized as valuable

https://doi.org/10.3390/foods15040736

co-products with transformative potential for the development of fiber- and protein-rich foods, nutraceutical formulations, and bio-based materials within modern sustainable food systems. A comparative overview of the valorization pathways for major peach palm by-product is provided in Table 7.

3.1. Potential Uses of Peach Palm By-Products

(a) Dietary fiber ingredients

Peach palm residues, particularly those derived from the median sheaths and stem portions, are rich in non-starch polysaccharides and can be processed into fiber-enriched flours suitable for improving the nutritional profile of cereals, baked products, and similar food matrices [4,13]. The fractions of these flours predominantly consist of insoluble dietary fibers, known to support intestinal function, aid in cholesterol regulation, and contribute to glycemic control when consumed regularly. In addition, the flour obtained from pulp and peel retains substantial amounts of total dietary fiber and bioactive compounds. This extends the potential of peach palm as functional ingredients across a range of modern food formulations [15].

(b) Functional food additives

The by-products of peach palm after processing can be used as food additive materials. The one such example includes the derived flour from peach palm. The flour exhibits remarkable water and oil absorption capacities, emulsifying ability, and structural stability properties. Due to these unique properties of flour, it can be employed as an effective natural emulsifier or textural modifier. It can also be used as the matrix for encapsulating bioactive molecules in baked goods, dairy formulations, and meat product applications [52]. In addition, the antioxidant properties and carotenoid concentrations in these by-products can often be used as natural coloring agents and provitamin A reservoirs. This contributes to nutritional value augmentation and conformity with clean label consumer preferences [4,13,15,45].

(c) Mushroom substrates

Lignocellulosic materials such as leaf sheath and middle sheath components derived from peach palm act as suitable substrates for solid-state fermentation and edible or medicinal mushroom production systems [15]. Camilleri et al. [55] reported that peach palm residues create an optimal environment for fungal growth, particularly for *Lentinula edodes* and *Pleurotus ostreatus*, resulting in mycelium-based composites with notable structural strength and nutritional value. In parallel, findings from [13,15] highlighted the potential of fermentation-derived by-products as functional feed components that influence glycemic and lipid metabolism, as well as promising precursors for developing sustainable mycelium-based biomaterials [13,15]. As per [50], others evaluated *P. ostreatus* yield in peach palm leaves supplemented with rice bran and reached yields between 20.6 and 42.3% (*w*/*w*). They recorded the protein content of 24.1 g/100 g in *P. ostreatus* produced in peach palm leaves. The protein content is one of the desirable and most important parameters in mushrooms, especially when used in plant-based diets, due to the restrictions in the consumption of sources of proteins from animal origin [50].

https://doi.org/10.3390/foods15040736

Table 7. Comparative overview of valorization pathways for major peach palm by-product fractions; composition, key nutraceuticals, functional applications and recent trends.

By-Product	Fraction of Biomass	Fiber (%)	Protein (%)	Key Nutraceuticals	Advanced Valorization Products	Applications	Recent Trends	Disadvantages	Reference
External sheath	~83.6	59–68	8–12	Myo-inositol, organic acids, polyphenols	Fibrous flour, cellulose nanofibrils, polyphenol-rich extracts	Dietary fiber supplements, biodegradable packaging materials	Global growth in bio-based packaging; fiber used in active packaging	High fiber content can reduce palatability; requires particle-size control and pretreatment; possible contamination if poorly handled	[15,42,49]
Internal sheath	~83.6	59–68	8–12	Xylooligosaccharides, myo-inositol	XOS, bioactive-rich flour, bioplastic precursors	Prebiotics, functional food fortification, bioplastics	XOS gaining market share as functional ingredient	XOS production requires controlled hydrolysis; possible bitter taste at high concentrations; processing cost	[15,48,49]
Basal portion	~83.6	59–68	8–12	N-acetyl-D-glucosamine, amino acids	Substrate for fermentation, food hydrocolloid source	Feedstocks, enzyme production, food thickeners	New bioprocessing for cell-cultured meat media	Composition varies with plant maturity; fermentation efficiency depends on pretreatment; limited direct food applications	[6,15,50]

https://doi.org/10.3390/foods15040736

(d) Biomaterials

By-products from peach palm subjected to upcycling processes are gaining recognition for their applicability in biomaterial development and environmentally sustainable chemistry. Recent developments indicate that the peach palm's lignocellulosic wastes hold great potential for being upcycled into valuable biotechnological products such as prebiotics, enzymes, cellulose and high fiber flours. Combination of chemical and mechanical treatments of peach palm sheaths led to the production of cellulose nanofibrils. Martins et al. [49] used these cellulose nanofibrils to improve the characteristics of cassava starch films. It was observed that physical reinforcement was the main effect observed in cassava starch films containing cellulose nanofibrils according to the analysis of mechanical strength and permeability. The spectroscopic data further revealed a possible formation of crosslinking between starch and cellulose nanofibrils, which can positively influence the tensile strength of such films. Additionally, starch extracted from non-traditional sources, including white peach palm varieties, displays distinctive functional characteristics and pasting behavior profiles, thereby broadening utilization prospects in biodegradable film technologies, pharmaceutical delivery systems, and as gel-forming or binding agent alternatives [4,13,15].

(e) Biorefinery resources

Peach palm by-products are rich in lignocellulosic materials, mainly cellulose, hemicellulose, and lignin, which makes them suitable feedstocks for biorefinery-based processing systems. In a biorefinery context, biomass is fractionated into multiple value-added streams rather than treated as waste, allowing the sequential conversion of structural polysaccharides and associated compounds into food ingredients, chemicals, and energy carriers.

The effective application of this concept requires consideration of the macromolecular organization of peach palm tissues, in which cellulose and hemicellulose are embedded within lignin-rich matrices that limit enzymatic accessibility. Accordingly, fungal pretreatment has been reported to enhance enzymatic hydrolysis efficiency by partially disrupting these structural associations, thereby enabling the release of fermentable sugars for the production of second-generation bioethanol, organic acids, and other renewable chemicals [56].

Spacki et al. [6,15] and Soares et al. [13] further demonstrated that xylan derivatives obtained from the inner sheath can be converted into prebiotic xylooligosaccharides. Following carbohydrate recovery, the remaining fiber-rich residues may be directed toward energy recovery processes or used as feedstock for bio-based materials. Such a cascading utilization of peach palm fractions illustrates how understanding macromolecular connectivity can guide the selection of appropriate processing routes within a biorefinery framework. These integrated pathways support circular bioeconomy strategies by linking agricultural residues with value-added production systems. Table 8 provides a comparative overview of commercial and research-stage peach palm products applied in plant-based foods.

3.2. *Commercial Perspectives of Peach Palm Valorization*

Currently, the commercial exploitation of peach palm is mainly focused on edible fruits and heart-of-palm products, while most bioactive and functional ingredients derived from by-products remain at the research or pilot scale. Dietary fiber concentrates, starch-based ingredients, and carotenoid-rich extracts represent the most promising product categories for food and nutraceutical applications. Among the processing technologies discussed, conventional solvent extraction and mechanical pressing remain the most widely used methods at the industrial scale due to their operational simplicity, lower capital cost, and established regulatory acceptance. Emerging techniques such as ultrasound-assisted, enzymatic, and supercritical carbon dioxide extraction have demonstrated superior selectivity and product quality; however, their commercial adoption is still limited by equipment costs, scale-up constraints, and energy requirements. Nevertheless, increasing demand for clean

https://doi.org/10.3390/foods15040736

label, sustainable, and plant-based ingredients is driving interest in greener extraction and fractionation technologies, suggesting that peach palm by-products could progressively transition from laboratory-scale valorization to industrial biorefinery-based applications.

Table 8. Comparison of major peach palm-based products for plant-based food applications.

Product Type	Nutritional Profile	Key Functional Properties	Example Foods	Consumer/Market Relevance	Sustainability/Valorization Aspects
Peel Flour	13–14% lipids, ~6% protein, ~62% carbohydrates, high fiber	Oil/water binding, texture, fiber	Gluten-free bread, cakes, snacks	Celiac/health segment	Produced from waste streams, valorizes by-products
Protein Isolate	40–60% protein, all essential AAs	Foaming, emulsifying, texture	Alt meat, dairy analogs, protein bars	Premium/high-protein seekers	Low amylose/high amylopectin aids processing, sustainable source
Lipid Extracts	High USFA (up to 70%), ω-3, ω-6, β-carotene up to 748 μg/100 g extract	Antioxidant, cardiovascular function	Margarine, supplements, enriched drinks	Heart health, wellness focus	Green extraction, edible oil from by-products
Colorant Extract	Carotenoids up to 172 μg/g (peels), up to 748 μg/100 g in oil	Natural pigment, antioxidant	Coconut drinks, baked goods	Clean label, fortification	Utilization of epicarp waste, green solvent extraction
Starch Flour	68–79% starch, low protein, gluten-free	Binder, texture, gelation	Cookies, cakes, meat product fillers	Processed food innovation	Seasonal valorization, replaces wheat for GF options

Source: [4,10,45,52,57,58].

4. Challenges and Research Needed

4.1. Toxicity Issues

The raw consumption of peach palm fruit is limited by the presence of calcium oxalate crystals, which induce irritation and necessitate prior processing. The sharp, needle-like crystals are concentrated in the pulp as well in the peel of peach palm fruit. When consumed raw, these crystals cause an immediate burning sensation, irritation, and pain in mouth or throat [15]. Soares et al. [4] documented that even small amounts of uncooked fruit can trigger airway swelling and, in severe cases, respiratory distress. Moreover, long-term or high-level exposure to these crystals may also affect kidney and liver health, underscoring the importance of proper processing before consumption [5].

Heat treatment, either cooking or boiling, serves as an effective way to make peach palm edible without any harmful effect [13]. Thermal processing not only solubilizes these crystals but also deactivates antinutritional and irritants present in the fruit. Santos et al. [59] claimed that cooking the fruit at 105 °C for 20 min eliminates these oxalate crystals from the peel. In addition, cooking also inhibits antinutritional factors such as trypsin inhibitors, inactivating peroxidase enzymes present in the pulp. These components can otherwise irritate the throat mucosa. Overall, it improves the safety and flavor profile of the fruit. In Amazonian cuisines, the fruits are often boiled in salted water. This enhances the taste while eliminating toxic elements, thereby providing complementary benefits. This essential step preserves the fruit's valuable nutrients, including fiber, vitamins, and bioactive phytochemicals [4,5,13,15].

Despite the proven effectiveness of thermal processing in enhancing the safety of peach palm fruit, important research gaps persist. Optimized thermal parameters for cooking are still required to achieve complete detoxification without compromising nutritional or

 https://doi.org/10.3390/foods15040736

sensorial properties. As peach palm has emerged beyond its native regions as a plant-based food ingredient, there is a need to ensure safety and consistency under industrial scale processing conditions. Moreover, improved methods for detecting and quantifying calcium oxalate residues are necessary to support both regulation and quality control. Addressing these gaps will enhance consumer safety and help establish peach palm as a reliable, sustainable ingredient in future food innovations [15].

4.2. Nutritional Variability

There is variability in the nutritional values that can be explained by the different *B. gasipaes* varieties, size, and starch oil composition of the fruit [4,5]. The fruit also varies genetically and from region to region. The major macronutrients vary significantly among accessions [13]. Detailed discussions on nutritional composition, fatty acid profiles, and bioactive compounds have been comprehensively addressed in our previous review ('Unlocking the potential of peach palm for plant-based foods') and are therefore only briefly summarized here. For instance, peach palm is known for its high starch, fiber and carotenoid content with modest protein levels when compared with most cereal grains. In certain processed forms, such as flours from albino varieties or certain industrial by-products, protein content can increase to 16–20 g per 100 g (dry basis), though such cases are uncommon.

The variations in nutritional profiles among the varieties, particularly pigmented and albino varieties, present novel opportunities for selective breeding among the varieties. This will result in enhanced protein quality and overall nutritional value for specific food applications or population needs. Unlocking the full potential of peach palm will depend on detailed genetic studies, improved protein isolation methods, and thoughtful formulation strategies to achieve optimal amino acid balance in next-generation plant-based food systems [4,5].

4.3. Functional Limitation

The functional properties of peach palm are limited, majorly in terms of gelling and foaming ability. These properties restrict its direct use in products that rely on viscosity control, emulsification, or stable foam structures. The high gelatinization temperature, low oil absorption and emulsifying capacity of peach palm requires blending with certain other flours or use of certain modification methods (as detailed above) to improve texture and performance in bakery or processed food applications [4,5].

4.4. Scalability Barriers

The large-scale commercial utilization of peach palm as a food component is limited due to a number of factors, such as processing difficulties, sensory challenges, evolving regulations for novel plant proteins, etc. As stated earlier, optimized and effective processing is still required to reduce toxicity without compromising the nutritional and sensorial properties. Because peach palm is considered an emerging protein source, manufacturers may also face regulatory hurdles that demand thorough safety assessments, toxicological studies, and detailed compliance documentation before products can enter the market [15,60,61]. Table 9 summarizes the key scalability barriers and mitigation strategies that currently limit the large-scale valorization of peach palm in sustainable food systems.

https://doi.org/10.3390/foods15040736

Table 9. Key scalability barriers and mitigation strategies for peach palm valorization in sustainable food systems.

Barrier Type	Description/Example	Impact on Scale-Up	Potential Solutions/Strategies
Genetic and Raw Material Variability	Limited development of high-yielding, high-quality varieties; genetic erosion; fragmented wild populations	Inconsistent product quality	Breeding programs, in vitro culture protocols, conservation
Agro-Industrial Waste Management	80–90% of the palm mass is by-product; waste disposal challenges impede processing scale-up	Increased costs and complexity	Circular economy valorization, upcycling, fiber bioproducts
Processing Technology Limitations	Slow adoption of green extraction, enzymatic modification, and waste bioproduct tech; high energy/water use	Low efficiency, high costs	Biotechnological pretreatment, solid-state fermentation
Market and Supply Chain Fragmentation	Multi-stakeholder chains, price fluctuations, long farm-to-market chains	Low profitability for smallholders	Producer associations, direct marketing, fair trade models
Regulatory and Safety Hurdles	"Novel food" status, need for toxicological and compositional data for non-traditional parts	Slow approvals, compliance risk	Prepare safety dossiers, harmonize with international regs.
Socio-Cultural Acceptance and Skills	Consumer unfamiliarity outside the Amazon, culinary barriers, limited scale of traditional market	Low consumer uptake	Targeted food innovation, education, tailored product dev.
Infrastructure and Value Chain Gaps	Lack of logistics, post-harvest infrastructure, and continuous supply outside local regions	High costs, product loss	Supply chain investment, regional processing hubs

Source: [4,6,13,15,36,62,63].

4.5. In Silico Approaches for Bioactivity Prediction

In silico techniques, including molecular docking, quantitative structure–activity relationship (QSAR) modeling, and bioactivity prediction platforms, are increasingly used to elucidate the mechanistic interactions between food-derived bioactive compounds and biological targets. For peach palm, such computational tools could be applied to predict the binding affinity of carotenoids, phenolic acids, and phytosterols to key molecular targets involved in oxidative stress, inflammation, and lipid metabolism. These approaches may assist in prioritizing compounds for further in vitro and in vivo validation, reducing experimental cost and time while improving mechanistic understanding. Integration of computational modeling with experimental extraction and characterization data could therefore accelerate the development of peach palm-based functional ingredients and nutraceutical formulations.

5. Conclusions

Peach palm is emerging as a nutritionally rich crop with strong adaptability. Some research done on it states that it has adapted well to modern extraction and processing methods. Its extremely rich and diverse composition enables to produce the starches with desirable functional properties, i.e., lipid extracts rich in bioactive compounds and high-quality protein isolates. Such versatility positions peach palm as a promising raw material for the development of next-generation plant-based foods. Valorization of the entire plant, including the peel, pulp, and processing residues, not only enhances resource efficiency but also aligns with circular bioeconomy principles aimed at achieving zero-waste production. With recent advancements in sustainable extraction and green processing methods, the quality and commercial potential of peach palm-derived ingredients has improved considerably.

https://doi.org/10.3390/foods15040736

Continued interdisciplinary research focusing on process optimization, functionality, and product innovation will further establish peach palm as a sustainable, high-value crop. Its integration into food, nutraceutical, and biomaterial applications reflects a step toward a more resource-efficient and environmentally responsible food system.

Author Contributions: K.S.: conceptualization, methodology, investigation, writing—original draft preparation. N.K., S.S.-U., E.O.; conceptualization, W.Z., S.W., J.-Q.H., Y.H.J.: writing—review and editing, S.R.: conceptualization, resources, visualization, writing—review and editing, supervision, project administration. All authors have read and agreed to the published version of the manuscript.

Funding: This research was funded by Mae Fah Luang University, grant number 09/2025 under the Postdoctoral Fellowship to Kartik Sharma. The authors also gratefully acknowledge the additional financial support from Mae Fah Luang University, Chiang Rai, Thailand via the Fundamental Fund: Basic Research [FRB690059/0187] and the Reinventing University Program Fund [F01-683R-04-045], The Office of the Permanent Secretary of the Ministry of Higher Education, Science, Research and Innovation.

Institutional Review Board Statement: Not applicable.

Informed Consent Statement: Not applicable.

Data Availability Statement: No new data were created or analyzed in this study.

Acknowledgments: The authors would like to thank Mae Fah Luang University, Chiang Rai, Thailand for facilities and supports.

Conflicts of Interest: The authors declare no conflicts of interest.

Abbreviations

The following abbreviations are used in this manuscript:

UAE	Ultrasound-Assisted Extraction
HPP	High-Pressure Processing
HMT	Hydrothermal Treatment
XOS	Xylooligosaccharides
QSAR	Quantitative Structure–Activity Relationship

References

1. Choręziak, A.; Rosiejka, D.; Michałowska, J.; Bogdański, P. Nutritional Quality, Safety and Environmental Benefits of Alternative Protein Sources—An Overview. *Nutrients* **2025**, *17*, 1148. [CrossRef] [PubMed]
2. Hertzler, S.R.; Lieblein-Boff, J.C.; Weiler, M.; Allgeier, C. Plant Proteins: Assessing Their Nutritional Quality and Effects on Health and Physical Function. *Nutrients* **2020**, *12*, 3704. [CrossRef]
3. Qin, P.; Wang, T.; Luo, Y. A Review on Plant-Based Proteins from Soybean: Health Benefits and Soy Product Development. *J. Agric. Food Res.* **2022**, *7*, 100265. [CrossRef]
4. Soares, S.D.; Santos, O.V.D.; Nascimento, F.D.C.A.D.; Pena, R.D.S. A Review of the Nutritional Properties of Different Varieties and Byproducts of Peach Palm (*Bactris gasipaes*) and Their Potential as Functional Foods. *Int. J. Food Prop.* **2022**, *25*, 2146–2165. [CrossRef]
5. González-Jaramillo, N.; Bailon-Moscoso, N.; Duarte-Casar, R.; Romero-Benavides, J.C. Peach Palm (*Bactris gasipaes* Kunth.): Ancestral Tropical Staple with Future Potential. *Plants* **2022**, *11*, 3134. [CrossRef]
6. Spacki, K.; Novi, D.M.P.; de Oliveira-Junior, V.A.; Durigon, D.C.; Fraga, F.C.; dos Santos, L.F.O.; Helm, C.V.; de Lima, E.A.; Peralta, R.A.; de Fátima Peralta Muniz Moreira, R.; et al. Improving Enzymatic Saccharification of Peach Palm (*Bactris gasipaes*) Wastes via Biological Pretreatment with *Pleurotus ostreatus*. *Plants* **2023**, *12*, 2824. [CrossRef]
7. Amorim, I.S.; Amorim, D.S.; Godoy, H.T.; Mariutti, L.R.B.; Chisté, R.C.; da Silva Pena, R.; Bogusz Junior, S.; Chim, J.F. Amazonian Palm Tree Fruits: From Nutritional Value to Diversity of New Food Products. *Heliyon* **2024**, *10*, e24054. [CrossRef]
8. Menezes Silva, J.V.; Silva Santos, A.; Araujo Pereira, G.; Campos Chisté, R. Ultrasound-Assisted Extraction Using Ethanol Efficiently Extracted Carotenoids from Peels of Peach Palm Fruits (*Bactris gasipaes* Kunth) without Altering Qualitative Carotenoid Profile. *Heliyon* **2023**, *9*, e14933. [CrossRef]

 https://doi.org/10.3390/foods15040736

9. Han, K.N.; Meral, H.; Demirdöven, A. Recovery of Carotenoids as Bioactive Compounds from Peach Pomace by an Eco-Friendly Ultrasound-Assisted Enzymatic Extraction. *J. Food Sci. Technol.* **2024**, *61*, 2354–2366. [CrossRef]
10. Santamarina, A.B.; de Souza Mesquita, L.M.; Casagrande, B.P.; Sertorio, M.N.; Vitor de Souza, D.; Mennitti, L.V.; Ribeiro, D.A.; Estadella, D.; Ventura, S.P.M.; de Rosso, V.V.; et al. Supplementation of Carotenoids from Peach Palm Waste (*Bactris gasipaes*) Obtained with an Ionic Liquid Mediated Process Displays Kidney Anti-Inflammatory and Antioxidant Outcomes. *Food Chem. X* **2022**, *13*, 100245. [CrossRef]
11. Espinosa-Pardo, F.A.; Martinez, J.; Martinez-Correa, H.A. Extraction of Bioactive Compounds from Peach Palm Pulp (*Bactris gasipaes*) Using Supercritical CO_2. *J. Supercrit. Fluids* **2014**, *93*, 2–6. [CrossRef]
12. Doan, H.T.T.; Kim, T.; Cha, M.; Kim, S.-J. Synthesis and Potential Application of Slowly Digestible Starch. *J. Funct. Foods* **2025**, *131*, 106955. [CrossRef]
13. Soares, S.D.; dos Santos, O.V.; da Conceição, L.R.V.; Costi, H.T.; Silva Júnior, J.O.C.; Nascimento, F.D.C.A.D.; Pena, R.D.S. Nutritional and Technological Properties of Albino Peach Palm (*Bactris gasipaes*) from the Amazon: Influence of Cooking and Drying. *Foods* **2023**, *12*, 4344. [CrossRef] [PubMed]
14. Gillet, S.A.; Silva, M.N.; Carvalho, E.A.; da Silva, E.G.P.; dos Santos Moreau, P.; Uetanabaro, A.P.T.; da Costa, A.M. Utilization of Peach-Palm Waste for Cost-Effective Amylase Production by Trichoderma Stromaticum: Stability and Industrial Potential. *Mycology* **2024**, *15*, 690–701. [CrossRef]
15. Spacki, K.; Corrêa, R.C.G.; Uber, T.M.; Barros, L.; Ferreira, I.C.F.R.; Peralta, R.A.; de Fátima Peralta Muniz Moreira, R.; Helm, C.V.; de Lima, E.A.; Bracht, A.; et al. Full Exploitation of Peach Palm (*Bactris gasipaes* Kunth): State of the Art and Perspectives. *Plants* **2022**, *11*, 3175. [CrossRef]
16. Akyüz, A.; Tekin, İ.; Aksoy, Z.; Ersus, S. Plant Protein Resources, Novel Extraction and Precipitation Methods: A Review. *J. Food Process Eng.* **2024**, *47*, e14758. [CrossRef]
17. Hadidi, M.; Aghababaei, F.; McClements, D.J. Enhanced Alkaline Extraction Techniques for Isolating and Modifying Plant-Based Proteins. *Food Hydrocoll.* **2023**, *145*, 109132. [CrossRef]
18. Pinheiro, R.C.; Ballesteros, L.F.; Cerqueira, M.A.; Rodrigues, A.M.C.; Teixeira, J.A.; Silva, L.H.M. Peach Palm (*Bactris gasipaes* Kunth) and Mammee Apple (*Mammea americana* L.) Seeds: Properties and Potential of Application in Industry. *LWT* **2022**, *170*, 114089. [CrossRef]
19. Llatas, A.Y.; Guzmán, H.; Tello, F.; Ruiz, R.; Vásquez, J.; Chiroque, G.; Mayta-Hancco, J.; Cruzado-Bravo, M.L.M.; Arteaga, H.; Saldaña, E.; et al. Exploring Pijuayo (*Bactris gasipaes*) Pulp and Peel Flours as Fat Replacers in Burgers: A Multivariate Study on Physicochemical and Sensory Traits. *Foods* **2024**, *13*, 1619. [CrossRef]
20. Sharma, K.; Nilsuwan, K.; Zhang, B.; Hong, H.; Benjakul, S. Protein Hydrolysate from Salmon Frame Debittered by Plastein Reaction: Amino Acid Composition, Characteristics and Antioxidant Activities. *Int. J. Food Sci. Technol.* **2022**, *58*, 154–166. [CrossRef]
21. Sharma, K.; Sukkapat, P.; Saetang, J.; Singh, P.; Ma, L.; Benjakul, S. Maillard Reaction Products Derived from Salmon Frame Protein Hydrolysate: Antioxidant Activities in Different Food Model Systems, Cytotoxicity and Bioavailability. *Int. Aquat. Res.* **2024**, *16*, 245–256. [CrossRef]
22. Ashraf, Z.U.; Shah, A.; Gani, A.; Gani, A. Effect of Enzymatic Hydrolysis of Pulse Protein Macromolecules to Tailor Structure for Enhanced Nutraceutical Properties. *LWT* **2024**, *205*, 116502. [CrossRef]
23. Tawalbeh, D.; Al-U'datt, M.H.; Wan Ahmad, W.A.N.; Ahmad, F.; Sarbon, N.M. Recent Advances in In Vitro and In Vivo Studies of Antioxidant, ACE-Inhibitory and Anti-Inflammatory Peptides from Legume Protein Hydrolysates. *Molecules* **2023**, *28*, 2423. [CrossRef] [PubMed]
24. Kong, D.; Liu, Q.; Chen, Q.; Zhang, C.; Liu, H.; Kong, B. A Comprehensive Review on Physical Modifications of Plant Proteins: Mechanism, Influencing Factors, Structural and Functional Properties. *Compr. Rev. Food Sci. Food Saf.* **2025**, *24*, e70273. [CrossRef]
25. Justino, H.D.F.M.; dos Santos, I.F.; de Souza, R.C.N.; Sanches, E.A.; Bezerra, J.D.A.; Lamarão, C.V.; Pires, A.C.D.S.; Campelo, P.H. Exploring Ultrasound-assisted Technique for Enhancing Techno-functional Properties of Plant Proteins: A Comprehensive Review. *Int. J. Food Sci. Technol.* **2024**, *59*, 498–511. [CrossRef]
26. Baldelli, A.; Shi, J.; Singh, A.; Guo, Y.; Fathordoobady, F.; Amiri, A.; Pratap-Singh, A. Effect of High-Pressure on Protein Structure, Refolding, and Crystallization. *Food Chem. Adv.* **2024**, *5*, 100741. [CrossRef]
27. Sharma, K.; Nilsuwan, K.; Hong, H.; Fan, X.; Benjakul, S. Debittering of Salmon Frame Protein Hydrolysate and Plastein Using Maillard Reaction as Affected by Types of Sugar. *Int. J. Food Sci. Technol.* **2024**, *59*, 1560–1571. [CrossRef]
28. Schneider, A.A.; Bu, F.; Ismail, B.P. Enhancement of Pea Protein Solubility and Thermal Stability for Acidic Beverage Applications via Endogenous Maillard-Induced Glycation and Chromatography Purification. *Curr. Res. Food Sci.* **2023**, *6*, 100452. [CrossRef]
29. Khan, H.; Mudgil, P.; Alkaabi, S.A.S.; AlRashdi, Y.H.S.; Maqsood, S. Maillard Reaction-Based Conjugation of Pea Protein and Prebiotic (Polydextrose): Optimization, Characterization, and Functional Properties Enhancement. *Front. Sustain. Food Syst.* **2024**, *8*, 1463058. [CrossRef]

30. Abbaspour, N. Fermentation's Pivotal Role in Shaping the Future of Plant-Based Foods: An Integrative Review of Fermentation Processes and Their Impact on Sensory and Health Benefits. *Appl. Food Res.* **2024**, *4*, 100468. [CrossRef]
31. Fan, M.; He, X.; Cao, Y.; Woldemariam, K.Y.; Cai, M.; Wang, Z.; Jiao, Y.; Tang, W.; Wei, X.; Liu, Y.; et al. Sustainable Microbial Fermentation of Plant Proteins: Potential, Biological Resources, Fermentation Mechanisms, Applications and Challenges in Food Industry. *Food Biosci.* **2025**, *68*, 106727. [CrossRef]
32. Chandran, A.S.; Suri, S.; Choudhary, P. Sustainable Plant Protein: An up-to-Date Overview of Sources, Extraction Techniques and Utilization. *Sustain. Food Technol.* **2023**, *1*, 466–483. [CrossRef]
33. Ceyhan, T.; Tomar, G.S.; Can Karaca, A. Recent Advances in Modification of Plant-Based Proteins for Improved Encapsulation Performance. *Colloids Surf. B Biointerfaces* **2025**, *253*, 114691. [CrossRef]
34. Tang, J.; Yao, D.; Xia, S.; Cheong, L.; Tu, M. Recent Progress in Plant-Based Proteins: From Extraction and Modification Methods to Applications in the Food Industry. *Food Chem. X* **2024**, *23*, 101540. [CrossRef] [PubMed]
35. Rosário, R.C.; Pires, M.B.; do Nascimento, F.D.C.A.; Ribeiro Costa, R.M.; Faial, K.D.C.F.; da Costa Nunes, E.; Teixeira-Costa, B.E.; dos Santos, O.V. Novel Low-Amylose Starch from White Variety of Peach Palm (*Bactris gasipaes* Kunth) Fruit– Nutritional, Functional and Pasting Properties. *LWT* **2025**, *218*, 117531. [CrossRef]
36. Kembabazi, S.; Mutambuka, M.; Shukri, R.; Anwar, F.; Zawawi, N. Unlocking the Potential of Resistant Starches from Underutilized Tropical Fruits as Substrates for Fermentation into Short-Chain Fatty Acids. *J. Funct. Foods* **2025**, *124*, 106630. [CrossRef]
37. Neeraj; Siddiqui, S.; Dalal, N.; Srivastva, A.; Pathera, A.K. Physicochemical, Morphological, Functional, and Pasting Properties of Potato Starch as a Function of Extraction Methods. *J. Food Meas. Charact.* **2021**, *15*, 2805–2820. [CrossRef]
38. Punia Bangar, S.; Ashogbon, A.O.; Singh, A.; Chaudhary, V.; Whiteside, W.S. Enzymatic Modification of Starch: A Green Approach for Starch Applications. *Carbohydr. Polym.* **2022**, *287*, 119265. [CrossRef]
39. He, R.; Li, S.; Zhao, G.; Zhai, L.; Qin, P.; Yang, L. Starch Modification with Molecular Transformation, Physicochemical Characteristics, and Industrial Usability: A State-of-the-Art Review. *Polymers* **2023**, *15*, 2935. [CrossRef]
40. Supreethee, S.; Vashishth, R. Unlocking the Potential of Palm Tree Starch: A Review on Modification Techniques and Its Application. *Int. J. Food Sci. Technol.* **2025**, *60*, vvaf009. [CrossRef]
41. Costa, R.; Hickmann Flôres, S.; Brandelli, A.; Galarza Vargas, C.; Carolina Ritter, A.; Manoel da Cruz Rodrigues, A.; Helena Meller da Silva, L. Development and Properties of Biodegradable Film from Peach Palm (*Bactris gasipaes*). *Food Res. Int.* **2023**, *173*, 113172. [CrossRef] [PubMed]
42. Giombelli, C.; Raspe, D.; Donadone, D.; da Silva, C.; Barros, B. Chemical Composition and Functional Properties of Dietary Fiber Concentrates Obtained from Peach Palm By-Product. *J. Braz. Chem. Soc.* **2023**, *34*, 927–936. [CrossRef]
43. Šárka, E.; Smrčková, P.; Sluková, M. Crystallinity of Starch, Food Composition, and Digestibility of Starch. *Czech J. Food Sci.* **2025**, *43*, 90–104. [CrossRef]
44. Wang, S.; Tian, H.; Du, Y.; Li, X.; Guo, L.; Gao, W. Type 5 Resistant Starch: Structure, Gut Microbiota Modulation, and Nutritional Applications. *Food Hydrocoll.* **2025**, *169*, 111629. [CrossRef]
45. Santos, M.P.L.D.; Santos, O.V.D.; Conceição, L.R.V.D.; Teixeira-Costa, B.E.; Lourenço, L.D.F.H.; Sousa, C.L.L.D. Characterization of Lipid Extracts from Different Colors of Peach Palm Fruits—Red, Yellow, Green, and White—Obtained through Ultrasound-Assisted Green Extraction. *Foods* **2024**, *13*, 1475. [CrossRef]
46. Senna, C.; Soares, L.; Egea, M.B.; Fernandes, S.S. The Techno-Functionality of Chia Seed and Its Fractions as Ingredients for Meat Analogs. *Molecules* **2024**, *29*, 440. [CrossRef]
47. Lima, G.G.; Schoenherr, Z.C.P.; Magalhães, W.L.E.; Tavares, L.B.B.; Helm, C.V. Enzymatic Activities and Analysis of a Mycelium-Based Composite Formation Using Peach Palm (*Bactris gasipaes*) Residues on Lentinula Edodes. *Bioresour. Bioprocess.* **2020**, *7*, 58. [CrossRef]
48. Vieira, T.F.; Corrêa, R.C.G.; de Fatima Peralta Muniz Moreira, R.; Peralta, R.A.; de Lima, E.A.; Helm, C.V.; Garcia, J.A.A.; Bracht, A.; Peralta, R.M. Valorization of Peach Palm (*Bactris gasipaes* Kunth) Waste: Production of Antioxidant Xylooligosaccharides. *Waste Biomass Valorization* **2021**, *12*, 6727–6740. [CrossRef]
49. Martins, M.P.; Dagostin, J.L.A.; Franco, T.S.; de Muñiz, G.I.B.; Masson, M.L. Application of Cellulose Nanofibrils Isolated from an Agroindustrial Residue of Peach Palm in Cassava Starch Films. *Food Biophys.* **2020**, *15*, 323–334. [CrossRef]
50. Valério, T.P.; Szeremeta, L.A.; Pacheco, J.T.M.R.; Barros, B.C.B.; Sydney, E.B.; Danesi, E.D.G. Production of Oyster Mushroom (*Pleurotus ostreatus*) in Peach Palm By-Products: Effects on Composition and Maximization of Antioxidants Extraction. *Braz. Arch. Biol. Technol.* **2024**, *67*, e24230467. [CrossRef]
51. Nirmal, N.; Khanashyam, A.; Mundanat, A.; Shah, K.; Babu, K.; Thorakkattu, P.; Al-Asmari, F.; Pandiselvam, R. Valorization of Fruit Waste for Bioactive Compounds and Their Applications in the Food Industry. *Foods* **2023**, *12*, 556. [CrossRef] [PubMed]
52. Martínez-Girón, J.; Morón-Ortiz, Á.; Osorio, C.; Ordoñez-Santos, L.E.; Mapelli-Brahm, P. Coconut Beverage Enriched with Carotenoids Extracted from Peach Palm (*Bactris gasipaes*) by-Product: Impact on Colour, Sensory Attributes, and Bioaccessibility. *Int. J. Food Sci. Technol.* **2025**, *60*, vvaf164. [CrossRef]

53. Mohd Basri, M.S.; Abdul Karim Shah, N.N.; Sulaiman, A.; Mohamed Amin Tawakkal, I.S.; Mohd Nor, M.Z.; Ariffin, S.H.; Abdul Ghani, N.H.; Mohd Salleh, F.S. Progress in the Valorization of Fruit and Vegetable Wastes: Active Packaging, Biocomposites, By-Products, and Innovative Technologies Used for Bioactive Compound Extraction. *Polymers* **2021**, *13*, 3503. [CrossRef] [PubMed]
54. Santos, C.A.; Morais, M.A.B.; Mandelli, F.; Lima, E.A.; Miyamoto, R.Y.; Higasi, P.M.R.; Araujo, E.A.; Paixão, D.A.A.; Junior, J.M.; Motta, M.L.; et al. A Metagenomic 'Dark Matter' Enzyme Catalyses Oxidative Cellulose Conversion. *Nature* **2025**, *639*, 1076–1083. [CrossRef]
55. Camilleri, E.; Narayan, S.; Lingam, D.; Blundell, R. Mycelium-Based Composites: An Updated Comprehensive Overview. *Biotechnol. Adv.* **2025**, *79*, 108517. [CrossRef]
56. Chaurasia, P.K.; Bharati, S.L.; Singh, S.; Sivalingam, A.M.; Shankar, S.; Mani, A. Fungal Pretreatment Methods for Organic Wastes: Advances and Challenges in Biomass Valorization. *RSC Sustain.* **2025**, *3*, 1234–1266. [CrossRef]
57. Silva, A.C.G.; Modesto Junior, E.N.; de Medeiros, H.H.B.R.; Pena, R.D.S.; Chisté, R.C. Microcapsules of Carotenoids from Peels of Peach Palm Fruits (*Bactris gasipaes* Kunth): Optimization, Carotenoid Composition, and Stability over Storage. *ACS Food Sci. Technol. J.* **2025**, *5*, 274–283. [CrossRef]
58. Ferrari, M.H.; Souza Costa, M.; Villas Boas, F.; Lopes Leivas, C.; Maria Landi Franco, C.; Michielon de Souza, S.; Pedrosa Silva Clerici, M.T.; Mach Côrtes Cordeiro, L. Characterization and Technological Properties of Peach Palm (*Bactris gasipaes* Var. Gasipaes) Fruit Starch. *Food Res. Int.* **2020**, *136*, 109569. [CrossRef]
59. dos Santos, M.A.S.; Protázio, D.C.; da Costa, G.P.; Rebello, F.K.; Martins, C.M.; Bezerra, A.S.; da Silva Nogueira, A. Profile of Peach Palm Fruit Consumers in the Metropolitan Region of Belém, Pará, Brazilian Amazon. *Int. J. Innov. Educ. Res.* **2021**, *9*, 550–560. [CrossRef]
60. Martínez, J.M.; Moreno-Caicedo, L.P.; Loaiza-Loaiza, O.A. Sensory Dimensions of Peach-Palm Fruit (*Bactris gasipaes*) and Implications for Future Genetics. *Agron. Mesoam.* **2021**, *32*, 77–92. [CrossRef]
61. Hasan, M.M.; Islam, M.R.; Haque, A.R.; Kabir, M.R.; Khushe, K.J.; Hasan, S.M.K. Trends and Challenges of Fruit By-Products Utilization: Insights into Safety, Sensory, and Benefits of the Use for the Development of Innovative Healthy Food: A Review. *Bioresour. Bioprocess.* **2024**, *11*, 10. [CrossRef]
62. Kramer, Y.V.; Clement, C.R.; de Carvalho, J.C.; Fernandes, A.V.; da Silva, C.V.A.; Koolen, H.H.F.; Aguiar, J.P.L.; Nunes-Nesi, A.; Ramos, M.V.; Araújo, W.L.; et al. Understanding the Technical-Scientific Gaps of Underutilized Tropical Species: The Case of *Bactris gasipaes* Kunth. *Plants* **2023**, *12*, 337. [CrossRef]
63. Arantes, M.S.T.; Marques, G.S.; Hansel, F.A.; Zanoni, P.R.S.; Magalhães, W.L.E.; Silva, V.R.D.; Helm, C.V. Composition and Potential Utilization Strategies of By-Products from the Brazilian Peach Palm Industry. *Ciência Agrotecnologia* **2024**, *48*, 006224. [CrossRef]

Disclaimer/Publisher's Note: The statements, opinions and data contained in all publications are solely those of the individual author(s) and contributor(s) and not of MDPI and/or the editor(s). MDPI and/or the editor(s) disclaim responsibility for any injury to people or property resulting from any ideas, methods, instructions or products referred to in the content.

https://doi.org/10.3390/foods15040736

foods

MDPI

Review

Physical Processing-Assisted pH Shifting for Food Protein Modification: A Comprehensive Review

Ruiqi Long [1,2], Yuanyuan Huang [1,2], Mokhtar Dabbour [3], Benjamin Kumah Mintah [4,5], Jiayin Pan [1,2], Minquan Wu [1,2], Shengqi Zhang [1,2], Zhou Qin [1], Ronghai He [1,2,*] and Haile Ma [1,2]

[1] School of Food and Biological Engineering, Jiangsu University, 301 Xuefu Road, Zhenjiang 212013, China; 2212318004@stmail.ujs.edu.cn (R.L.); 2112318160@stmail.ujs.edu.cn (Y.H.); pjiayin@163.com (J.P.); 19710512320@163.com (M.W.); zhangshengqi0418@163.com (S.Z.); 19599960979@163.com (Z.Q.); mhl@ujs.edu.cn (H.M.)
[2] Institute of Food Physical Processing, Jiangsu University, 301 Xuefu Road, Zhenjiang 212013, China
[3] Department of Agricultural and Biosystems Engineering, Faculty of Agriculture, Benha University, Moshtohor, Qaluobia P.O. Box 13736, Egypt; mokhtar.dabbour@fagr.bu.edu.eg
[4] CSIR—Food Research Institute, Accra P.O. Box M20, Ghana; b.minta20@gmail.com
[5] Department of Agro-Processing Technology and Food Bio-Sciences, CSIR College of Science and Technology (CCST), Accra P. O. Box M32, Ghana
* Correspondence: heronghai1971@126.com; Tel.: +86-(511)-8878-0174

Abstract: The increasing demand for sustainable protein sources has intensified interest in improving the processing efficiency of traditional proteins and developing novel alternatives, particularly those derived from plants and algae. Among various processing technologies, pH shifting has attracted attention due to its simplicity, low cost, and capacity to effectively alter protein structure and functionality. However, employing pH shifting alone requires extremely acidic or alkaline conditions, which can lead to protein denaturation and the generation of undesirable by-products. To address these limitations, this review explores the integration of pH shifting with physical processing techniques such as ultrasound, high-pressure processing, pulsed electric fields, and thermal treatments. Moreover, this review highlights the effects of these combined treatments on protein conformational transitions and the resulting improvements in functional properties such as solubility, emulsification, foaming capacity, and thermal stability. Importantly, they reduce reliance on extreme chemical conditions, providing greater sustainability in industrial applications, particularly in food product development where milder processing conditions help preserve nutritional quality and functional properties. In that sense, this combined treatment approach provides a promising and eco-efficient protein modification strategy, and bridges technological innovation with sustainable resource utilization.

Keywords: pH shifting; ultrasonication; solubility; conformational attributes; plant protein

1. Introduction

With the rapid growth of global population and shifts in dietary preferences, the demand for high-quality protein has significantly increased, placing pressure on existing protein supply chains [1]. Traditional protein sources are currently in supply–demand imbalance, and as a consequence the selection of protein for various industrial applications is undergoing a significant process of diversification. Animal-derived proteins, including whey, casein, and meat proteins, have extensively been used in providing nutrition due to their comprehensive amino acid profiles [2,3]. However, they are limited by low resource conversion efficiency (15–40%) and a high carbon footprint (accounting for 14.5%

of total emissions). Moreover, plant-based proteins (e.g., soy, pea, and rice) are receiving much attention due to their bioavailability and sustainable production, yet challenges remain in terms of inconsistent extraction yields (40–68%) and difficulties in functional property modification [4,5]. Meanwhile, emerging biomass proteins such as algal proteins (e.g., spirulina, chlorella) and insect proteins (e.g., black soldier fly, beetle larvae) offer high reproduction efficiency (yield per unit area is up to 150 times greater than soybeans) and unique nutritional profiles. Despite these advantages, large-scale application and consumer acceptance continue to pose significant research challenges [4,6–8].

Recent studies have shown that the individual or combined use of physical, chemical, and biological modification methods can effectively improve the functional properties of proteins from various biological sources. These methods alter protein structures in targeted ways and enhance key functions such as solubility, emulsifying ability, and gel formation, which are crucial for both food and industrial applications [7,9–12]. Among various physical, chemical, and biological approaches for protein modification, the pH shifting method has recently emerged as a research theme of interest. Compared with traditional methods (including heat treatment, fermentation, etc.), it offers advantages such as simplicity, stability, and high efficiency. Moreover, its non-thermal mechanism, controllability, and environmentally friendly nature distinguish it from other methods.

The pH shifting technique has received considerable attention as a useful strategy in protein extraction and modification. By adjusting the acidity or alkalinity of solution, the pH shifting method effectively modifies protein structure, resulting in the augmentation of functional properties, thereby enhancing its application in industry [10,13,14]. In addition, the pH shifting method can enhance emulsifying properties, gelling ability, and thermal stability of proteins in a way that can increase the texture and sensory characteristics of food products [15,16]. Nevertheless, the use of the pH shifting technique presents certain practical limitations. Extreme pH conditions may cause protein denaturation, impair biological activity, and potentially lead to the formation of harmful by-products [12,17]. Also, the use of pH shifting alone for protein treatment often fails to yield the desired outcomes [18]. Therefore, to overcome this limitation, the combination of pH shifting techniques with physical modification methods, such as ultrasound, ultra-high pressure, and pulsed electric fields, has emerged as a critical research direction in recent years. The synergistic interactions resulting from these integrated methods can notably improve protein extraction efficiency and functional properties. Moreover, recent advancements in novel physical extraction techniques, including ultrasound-assisted extraction, ultra-high pressure processing, and pulsed electric field treatment, have significantly enhanced both extraction efficiency and protein purity [19]. Earlier studies show that protein structure can be modified using physical methods, thereby enhancing functional properties such as solubility and emulsification ability, and reducing allergenicity [20–23]. Compared to individual treatments, the combination of pH shifting with physical modification techniques often results in significantly enhanced functional properties of proteins, suggesting the presence of synergistic interactions. Notably, the use of physical methods in combination avoids the introduction of exogenous chemical agents. Moreover, the mechanisms involved in such combined treatments are relatively straightforward, making them more accessible for mechanistic interpretation [13,14,17].

Moreover, Momen et al. [24] presented a comprehensive review on the application of alkaline-mediated treatments for both the extraction and functional modification of proteins from plant and animal sources, and reported that alkaline conditions irreversibly unfolded protein structure, thereby improving solubility, emulsification, gelation, and bioactive compound binding properties. Additionally, Lou et al. [25] discussed the molten globule state of proteins induced by pH shifting. Their work focused on the structural characteristics of this intermediate state and its relevance to food processing applications, emphasizing the potential of partially unfolded proteins in improving emulsification, flavor retention, and gelation. Recently, Sultan et al. [26] provided an extensive review on the pH shifting technique for the extraction of plant-based proteins from various sources such as legumes, cereals, and oilseeds, and noticed that such a method had the ability to produce protein isolates with improved solubility, emulsification, foaming, and gelation properties. Concurrently, they systematically summarized the effects of extraction and precipitation pH on both protein yield and functionality, highlighting optimal pH ranges for various protein types. Furthermore, this review briefly mentioned the effect of pH shift combined with ultrasound on the structure and functionality of proteins, but did not provide a detailed explanation.

Although these studies have laid a foundation for understanding protein behavior under alkaline or pH-induced conditions, several limitations persist. including negligible functional improvement and extreme operational conditions. Notably, previous reports have largely overlooked the synergistic effects of combined pH shifting and physical processing techniques, such as high-intensity ultrasound (HIU), pulsed electric field (PEF), and high-pressure treatment. These emerging hybrid strategies may enable more precise control over protein unfolding, aggregation, or reconfiguration, thereby augmenting their functional characteristics. While pH shifting significantly enhances functional properties, such as solubility and emulsifying activity via structural changes (e.g., β-sheet dissociation, hydrophobic group exposure), its potential limitations should not be overlooked. Extreme pH conditions may induce chemical modifications of amino acid residues (e.g., serine, lysine), stimulating reduced nutritional quality and the formation of toxic compounds like lysinoalanine (LAL). While alkaline extraction can temporarily enhance the solubility of rice protein, overexposure to high pH levels (pH > 12) may result in irreversible deamidation and the loss of essential amino acids [5]. Therefore, precise control of the pH range and the development of multiscale synergistic modification techniques have become critical research directions.

Therefore, this review aims to provide an updated and comprehensive overview of the synergistic application of pH shifting and physical modification techniques in the structural and functional transformation of food proteins, from both plant and animal sources. The review systematically elucidates the physicochemical mechanisms governing protein conformational changes under extreme pH conditions, and elucidates their effects on secondary, tertiary, and quaternary structures. The review further explores how the integration of pH shifting and other physical techniques such as ultrasonication, microwave treatment, and pulsed electric field enhance key functional properties, including solubility, emulsification, gelling, foaming capacity, and bioactivity. Additionally, it discusses current and potential applications of these combined strategies in food processing. While critically addressing existing technological constraints, research gaps and future directions of combined pH-physical strategies for protein valorization in the food and agricultural industries were discussed.

2. Effect of pH Shifts on Proteins

The pH shifting treatment is a simple but effective chemical modification method. It operates by exposing proteins to highly acidic (pH < 2.5) or highly alkaline (pH > 10.5) conditions, which increases the surface charge of amino acids on protein molecules. This leads to enhanced electrostatic repulsion between protein molecules and promotes the unfolding of protein chains. Under ionic and dipole interactions, the binding between charged amino acids and water molecules increases, thereby improving protein solubility [27]. The treatment typically involves two stages. First, the protein solution is adjusted to an extreme pH (acidic or alkaline) and left to stand, allowing the elevated surface charge to induce molecular unfolding. Then, the pH is adjusted back to neutral (pH 7.0) and the solution is allowed to stand again. At this stage, the net surface charge decreases, weakening protein–water interactions, and hydrophobic forces drive the refolding and aggregation of protein molecules, often resulting in precipitation. During refolding, misfolding of protein chains can occur, altering the molecular structure and leading to functional modifications. The greater the degree of misfolding, the more pronounced the modification effect. This process induces various chemical reactions within protein molecules, ultimately altering their structure and enhancing functional properties. The pH shifting method is valued for its simplicity, precision, low cost, and significant modification efficiency. Only by adjusting the acidity or alkalinity of solution can one significantly improve protein solubility and functional properties such as emulsification, foaming, and thermal stability, while largely preserving the original structure and bioactivity.

Changes in pH significantly affect the structural stability and functional properties of proteins by modulating non-covalent interactions such as hydrogen bonds, ionic bonds, and hydrophobic interactions. At the molecular level, pH fluctuations induce the protonation or deprotonation of amino acid residues (e.g., His, Asp, Glu), directly altering hydrogen bond networks and ionic interactions. Under highly acidic conditions (e.g., pH < 3), α-helix structures become destabilized, leading to an increased content of β-sheet and random coil. Simultaneously, electrostatic repulsion is reduced due to charge shielding effects. Contrarily, alkaline pH may disrupt the stability of the hydrophobic core, resulting in the exposure of buried residues (Figure 1). This may cause partial unfolding of the tertiary structure, creating intermediate states with exposed thiol and hydrophobic groups, which provide a structural basis for subsequent aggregation or functional modifications. Figure 1 illustrates the changes in primary, secondary, and tertiary protein structures, as well as inter-molecular interactions, in response to the pH shifting process [28]. Under normal conditions, the primary structure of proteins remains stable after a pH shift, and the order of amino acids remains basically unchanged. However, under extreme pH shift conditions, the secondary structure usually shows that both the α-helix and β-fold undergo structural deconvolution and destruction of hydrogen bonding. There is also structural loosening and rearrangement, as well as an increase in random coils. Significant changes occur to the tertiary structure of proteins under pH shift conditions. The tertiary structure of proteins undergoes significant changes under pH shift treatment, mainly manifested by the destruction of salt bridges within the protein molecule and the exposure of hydrophobic groups and sulfhydryl (-SH) groups, which can form disulfide bonds (-S-S-) [29].

Given the critical role of pH shifting in modulating protein structure and function, relevant literature is summarized in Figure 2 to clearly illustrate the effects exerted by varying pH conditions on protein conformation. The phase behavior of ovalbumin gradually evolves with the decrease in pH, leading to formation of amorphous aggregates instead of gel bead-like aggregates, and spherulites instead of needle-like crystals [30]. For example, ovalbumin tends to form spherical aggregates rather than crystals at pH 4.7. However, moderate deviations from the isoelectric point (e.g., pH 3 or pH 8) enhance surface charge density, thereby improving protein solubility and dispersion. This property is of significant value in bio-separation processes. For instance, treating protein complexes under acidic conditions (pH 3–4) can increase solubility by 1.5–2 times compared to neutral pH, significantly enhancing extraction efficiency [31]. Moreover, plant storage proteins undergo quaternary structural destruction under extreme pH conditions (pH 1.5/12), generating soluble monomers and aggregates. The proteins after conformational changes exhibit excellent emulsifying and foaming properties due to side chain rearrangement. pH-induced structural transitions have successfully been utilized in food texture modulation and biomaterials engineering [32]. Notably, it was found that under extreme alkaline conditions (pH 11 and 11.5), protein solubility and recovery rates of rainbow trout by-products were significantly enhanced. Structural alterations included reduced particle size, decreased α-helix content, increased β-sheet and other secondary structures, and improved emulsification and foaming properties. These effects were further amplified under combined pH shifting and ultrasonication treatments [33]. Variation in pH conditions effectively alters protein structures, enhancing their functional properties such as solubility and emulsification, and exhibits synergistic potential when integrated with physical techniques like ultrasonication.

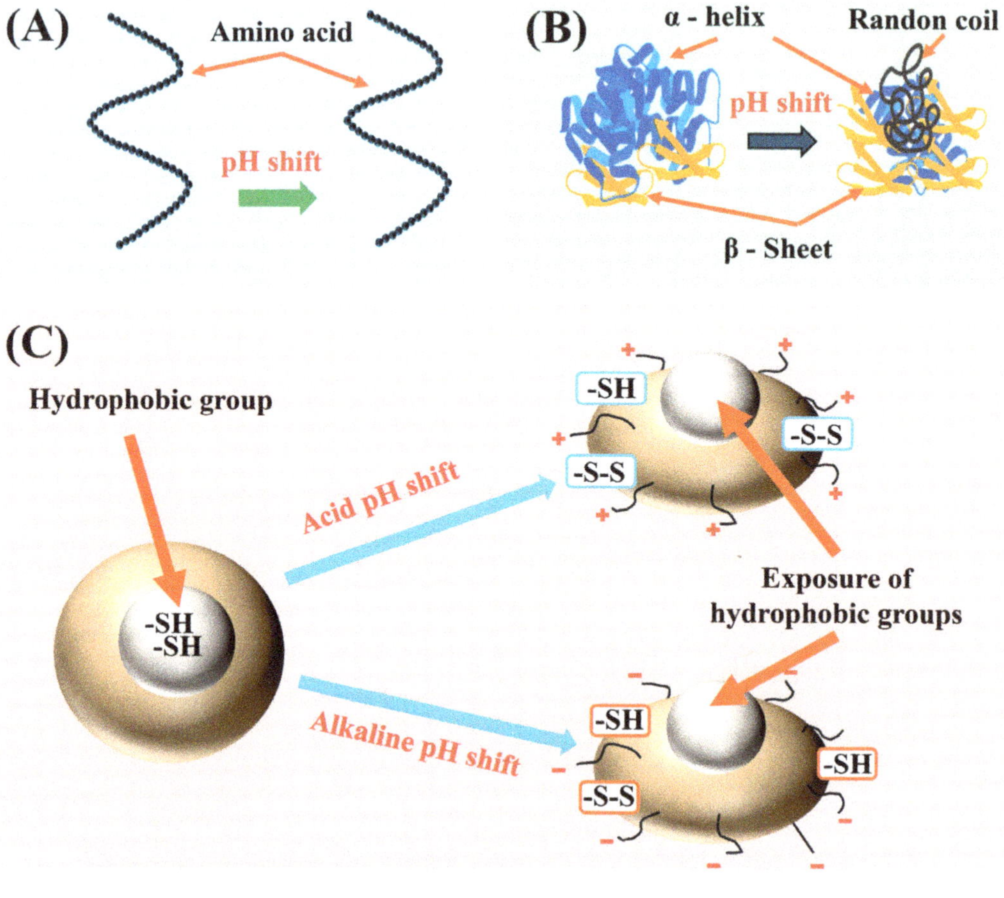

Figure 1. Effect of pH shifting on primary (**A**), secondary (**B**), and tertiary (**C**) structures of protein.

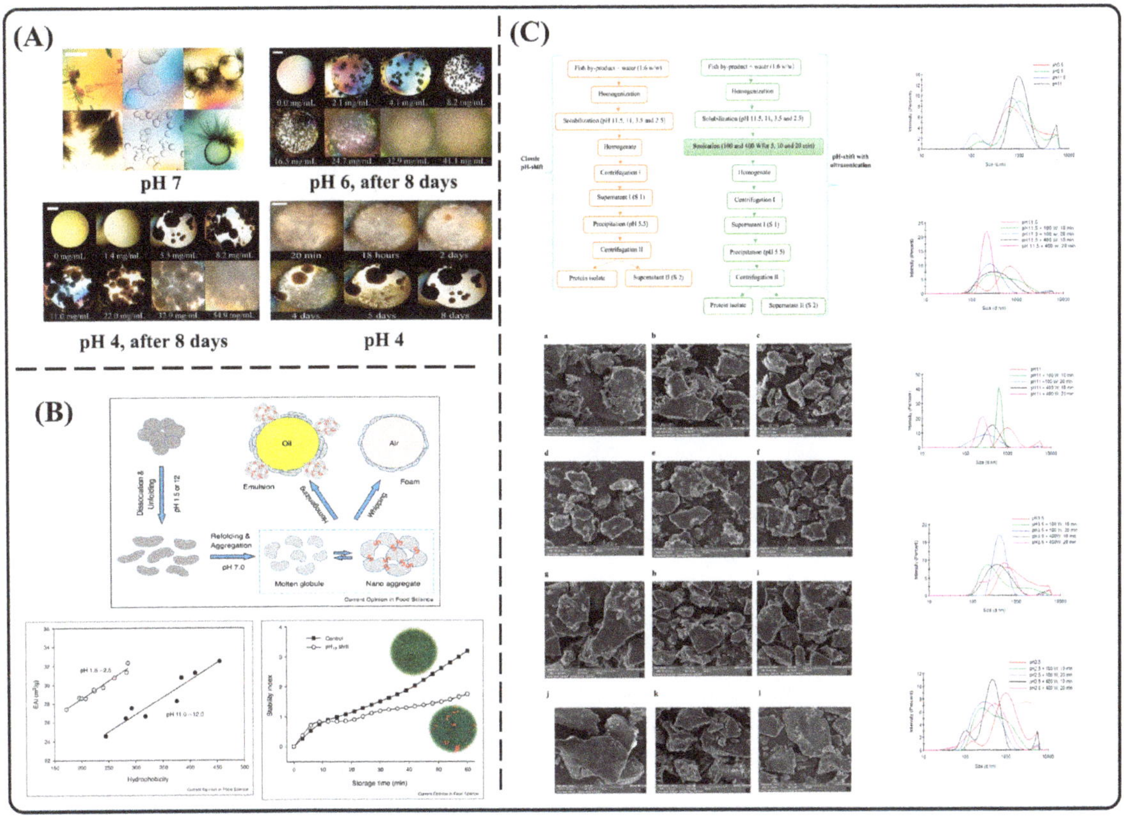

Figure 2. Phase behavior and spherulite growth of ovalbumin under different solution conditions (**A**) [30]; effects of pH changes on protein surface properties, emulsification performance, and emulsion stability (**B**) [32]; effects of ultrasound-assisted pH shift process on protein separation from fish by-products and its particle size distribution and surface microstructure (**C**) [33].

Recent studies have demonstrated that combining pH shifting with physical processing technologies (such as ultrasound, high-pressure, and pulsed electric fields) can overcome the limitations of standalone treatment. Ultrasound-induced cavitation can suppress excessive aggregation under extreme pH while enhancing interfacial adsorption and solubility (by over 50%). High-pressure treatment, by applying mechanical forces, cooperates with pH-induced conformational rearrangement to reduce disulfide bonds (with >90% retention of free thiol groups), thereby enhancing functionality and reducing LAL formation (by 60%). These synergistic strategies not only mitigate the side effects of the pH shifting process but also enable directed structural reconstruction through energy barrier modulation, offering innovative solutions for food (e.g., texture optimization of plant-based meat), pharmaceuticals (e.g., targeted delivery systems), and nutrition (e.g., development of high-bioavailability proteins). Therefore, a deeper exploration of such hybrid strategies holds great promise for advancing the precision and applicability of protein engineering in food systems [33].

3. Combination of pH Shifting and Physical Processing Techniques

3.1. Heat Treatment-Assisted pH Shifting

Temperature plays a crucial role in protein extraction and modification, with thermal treatment being one of the most commonly employed physical methods. By increasing temperature, thermal treatment enhances the interaction between protein and solvent molecules and increases the thermal kinetic energy of proteins. This disrupts non-covalent bonds within proteins including hydrogen bonds, hydrophobic interactions, and van der Waals forces to induce structural unfolding and rearrangement. Such controlled, heat-induced structural changes can significantly improve the functional properties of proteins, including solubility, emulsifying stability, and gelation ability.

When proteins are subjected to acidic or alkaline conditions far from their isoelectric point, the effects of heat-induced structural changes are further intensified. Under these conditions, significant alterations in surface charge distribution increase electrostatic repulsion, resulting in a looser protein structure. The thermal energy generated also enhances protein molecular mobility, accelerating the disruption of intramolecular non-covalent bonds and leading to further unfolding and exposure of functional groups. Consequently, many studies have combined controlled thermal treatment with pH shifting techniques to synergistically improve protein functionality and broaden their applicability in the food and agricultural sectors.

3.1.1. Effect of Heat Treatment-Assisted pH Shifting on Protein Structure

Many reports have indicated that an extreme pH shift combined with heating induces substantial protein unfolding and restructuring of secondary/tertiary structures [34]. Under thermal-alkaline conditions, globular proteins tend to lose their compact tertiary structure as strong electrostatic repulsion and thermal energy cause the molecules to partially unfold. A common outcome is the cleavage of stabilizing disulfide bonds, evidenced by a significant increase in free sulfhydryl groups and dissociation of high-molecular-weight subunits [12,35]. These unfolded proteins undergo notable secondary-structure transitions including a considerable decrease in α-helix content accompanied by an increase in β-sheet (or β-turn) structures after pH shifting and heat treatment, indicating that some helical regions refold into β-sheet conformations or other element structures [36]. This transition coupled with the exposure of hydrophobic amino acid residues and reactive sites reflects a more flexible, partially denatured conformation of protein. Notably, the newly exposed functional groups (e.g., thiols and hydrophobic groups) can intensify the intermolecular interactions. Sun et al. [37] reported that alkaline heating induced unfolding of the tertiary structure of soy protein, resulting in the exposure of previously buried reactive sites. This structural change enhanced enzymatic cross-linking mediated by transglutaminase, ultimately leading to the formation of stronger protein gels. Such findings indicate that pH shifting combined with thermal treatment considerably altered protein structure, resulting in the collapse of disulfide linkages. These changes may have converted α-helix into β-sheet structure, forming a less compact conformation with more reactive protein state conducive to aggregation and improved functional properties [12]. pH shifting induces conformational changes by altering the surface charge of proteins, while thermal treatment provides kinetic energy that disrupts non-covalent interactions, facilitating molecular unfolding and stimulating the exposure of functional groups [38]. Under alkaline conditions combined with heat, globular proteins lose their compact tertiary structures due to strong electrostatic repulsion and thermal energy, leading to partial unfolding of protein [39]. This process frequently involves the cleavage of stabilizing disulfide bonds, resulting in increased free sulfhydryl groups and dissociation of high-molecular-weight protein subunits [40]. Additionally, proteins undergo secondary-structure transitions, specifically a decrease in α-helix content and an increase in β-sheet structures, suggesting a refolding from helical to extended conformation. Such structural changes expose hydrophobic amino acid residues and reactive sites, significantly enhancing protein functionality [41].

3.1.2. Effect of Heat Treatment-Assisted pH Shifting on Protein Functionality

The combination of pH shifting and thermal treatment has recently emerged as a promising synergistic strategy for enhancing the functional properties of both plant and animal proteins. The enhanced extraction efficiency of protein is primarily attributed to improved solubility, which is directly caused by protein unfolding induced through pH shifting and heating. The micromorphology of Silkworm Pupa Protein Isolates (SPPI) was reported by Xu et al. [12] to be altered by alkaline pH shifting (pH 12.5) and heat treatment (80 °C) for 60 min, with the disulfide bonds between macromolecular subunits (72 and 95 kDa) being destroyed, resulting in reduced particle size and increased zeta potential and free sulfhydryl content of the isolates. The fluorescence spectra analysis showed red shifts with increased pH and fluorescence intensity improved with temperature (40–90 °C), implying the alterations in the tertiary structure of protein. This improvement may reduce the aggregation and increase the molecular flexibility following the cleavage of disulfide bonds and decrease in α-helix content, thereby augmenting protein solubility.

pH shifting combined with heat treatment can also improve the emulsifying properties of protein. Sun et al. [42] demonstrated that heat-assisted pH shifting treatment at pH 12 and 70 °C for 2 h significantly enhanced the emulsifying stability of pumpkin seed protein isolate (PSPI), achieving an internal oil phase volume fraction of up to 80%, and maintaining emulsion stability even after centrifugation at 10,000 g for 60 min and storage for 30 days. The data from sodium dodecyl sulfate-polyacrylamide gel electrophoresis (SDS-PAGE), UV–visible, intrinsic fluorescence, and Fourier transform infrared spectroscopy (FTIR) indicated that the primary, secondary, and tertiary structures of PSPI were disrupted, resulting in changes in interfacial tension, wettability, particle size, and dispersibility of protein. This may have promoted the adsorption of modified PSPI nanoparticles to the surface of oil droplets and improved the emulsion stability. In addition to emulsification, the combination of pH shifting and heat effectively enhances foaming properties, linked closely to protein unfolding and structural flexibility [42]. Chang et al. [36] demonstrated that combining pH shifting (pH 12, maintained for 1 h at room temperature) with controlled heating (70 °C for 30 min) significantly enhanced the foaming capacity of pea protein isolates. This improvement was mainly attributed to the formation of soluble protein aggregates, structural transformation from β-sheet to α-helix, and increased surface hydrophobicity, resulting in a more flexible protein conformation at the air–water interface. Protein unfolding increases flexibility and availability of hydrophobic sites that facilitate rapid adsorption at air–water interfaces, thereby stabilizing foam. The enhancement of gel properties is also driven by structural changes. Functional groups such as thiol groups are exposed through structural unfolding, promoting stronger intermolecular interactions and covalent cross-linking, thereby strengthening the protein network within the gel. Wang et al. [43] reported the changes in the structure and gel properties of peanut protein isolate (PPI1) under the synergistic effect of temperature and pH shifting. The breaking force and water-holding capacity of pH 10-treated PPI1 10–40 (at 40 °C) gel were 2.2 times and 2.15 times higher than that of the pH 7-adjusted sample at 25 °C. Moreover, the solubility, free sulfhydryl content, and surface hydrophobicity of PPI1 10–40 were 1.3 times, 1.8 times, and 1.6 times that of the pH 7-adjusted sample, respectively, which resulted in enhanced covalent or non-covalent interactions between proteins.

The combination of pH shifting and thermal treatment can significantly enhance functional properties of protein, including emulsifying ability, thermal stability, foaming capacity, and gelation performance. The existing literature suggests that this synergistic method not only improves protein functionality but also offers broad application potential in food, pharmaceutical, and related fields. Yang et al. [44] employed a combined pH shifting (pH 11) and thermal treatment (70 °C for 20 min) strategy to improve the functionality of whey protein isolate (WPI), successfully developing a novel method for preparing whey protein isolate-tryptophan (WPI-Trp) nanoparticles. Similarly, Sun et al. [42] successfully developed a food-grade Pickering emulsifier by employing a heat-assisted pH shifting treatment, specifically adjusting the pH to 12 and heating at 70 °C for 2 h followed by neutralization to pH 7, significantly enhancing the emulsifying properties of pumpkin seed protein isolate (PSPI) nanoparticles. Also, Zhong et al. [45] simulated gastrointestinal digestion and noticed that Pickering emulsions prepared using LP-Res nanoparticles under pH 11 and 60 °C (pH 11, 60 °C-LP-Res) effectively protected resveratrol (Res) and vitamin D3 from degradation and precipitation. Furthermore, Nisov et al. [46] reported that increasing the pH of raw plant protein materials, such as pea protein concentrate (PPC), pea protein isolate (PPI2), rice protein isolate (RP), and isolated wheat gluten (WG), to pH 7 prior to freeze-drying and extrusion processing at temperatures between 115 and 160 °C effectively enhanced the structural alignment and mechanical strength of the extrudates, thereby promoting their potential use as attractive plant-based meat analogues. Similarly, Zhu et al. [47] modified protein and starch mixtures by adjusting the pH to 8.0 and applying thermal treatment at 90 °C for 3 h. This treatment improved the emulsifying properties of rice starch and whey protein isolate conjugates, likely as a result of Maillard-type interactions and enhanced interfacial functionality.

The combined application of pH shifting and thermal treatment has been shown to significantly enhance protein functionality, including solubility, emulsifying capacity, foaming ability, and gelation. These improvements are primarily attributed to heat-induced protein unfolding and the exposure of reactive groups under extreme pH conditions, which promote intermolecular interactions and structural rearrangements. This strategy holds considerable potential for use in food formulation, functional delivery systems, and the development of plant-based protein products. Despite these advantages, several challenges persist. Most existing studies are conducted under controlled laboratory conditions, limiting their applicability to real food systems. Additionally, the outcomes are highly protein-specific and sensitive to processing parameters such as pH, temperature, and holding time. Excessive thermal treatment may also lead to undesirable protein aggregation, loss of nutritional value, and off-flavor formation. Therefore, optimizing processing conditions and elucidating the underlying molecular mechanisms are essential for broader industrial implementation. To address these limitations, researchers are investigating non-thermal approaches such as ultrasound-assisted pH shifting, which enables effective protein modification while minimizing thermal degradation.

3.2. *Ultrasound-Assisted pH Shifting*

Ultrasound treatment, a green and non-thermal physical processing technology, has widely been used in the extraction and modification of proteins [48]. Ultrasound, as a mechanical wave typically operating at frequencies above 20 kHz, has gained considerable attention across the food, medical, and industrial sectors due to its strong penetrability, absence of chemical contamination, and high controllability. It significantly improves protein extraction efficiency by disrupting the cellular matrix. Beyond facilitating extraction, ultrasound primarily alters the physical, structural, and functional properties of proteins through shear forces generated by cavitation. These forces disrupt non-covalent interactions such as hydrogen bonds and hydrophobic interactions, promoting protein unfolding or partial denaturation, and subsequently altering secondary and tertiary structures of protein [49]. As a result, protein solubility and reactivity are enhanced.

Moreover, ultrasound influences the surface properties of proteins by exposing hydrophobic regions, thereby enhancing their functionalities [50]. It can also disintegrate protein aggregates or precipitates, restoring proteins to their native states or improving their dispersibility, which further contributes to improved solubility and functional performance in food and other applications. However, ultrasound may also induce protein oxidation, aggregation, and cross-linking, which can negatively affect hydrophobicity, solubility, emulsification, and foaming properties [51]. Furthermore, ultrasound can alter various physical properties such as particle size, rheological behavior, conductivity, and ζ-potential, directly influencing the functional and nutritional quality of proteins [52].

The growing global demand for plant-based proteins has driven the need for efficient extraction techniques, making it a key research area in food science. Among these, ultrasound-assisted extraction (UAE) has gained significant attention due to its high efficiency, energy savings, and environmentally friendly characteristics. Figure 3 shows the effects of ultrasound and ultrasound assistance on the protein extraction rate and related structures. Studies have shown that UAE promotes the destruction of cell walls through cavitation, high shear force and mechanical energy transfer, allowing solvents to penetrate cells more easily, thereby enhancing the solubility of proteins and improving extraction efficiency [53]. For instance, in ultrasound-assisted alkaline extraction of pea protein isolate (PPI2), optimized conditions (solid-to-liquid ratio of 1:11.5 g/mL, pH 9.6, extraction time of 13.5 min, ultrasound amplitude of 33.7%) achieved a maximum PPI2 yield of 82.6%, significantly outperforming traditional alkaline extraction [11]. Additionally, UAE induced secondary and tertiary structural changes, causing partial unfolding and exposing hydrophobic groups, thereby improving solubility, emulsification, foaming capacity, and gel formation ability. For instance, in pecan protein extraction, an optimized ultrasound-assisted enzymatic method (400 W, 20 kHz, 5 s/3 s) improved protein yield to 25.51% by increasing substrate solubility and enzyme–protein interactions, accelerating chemical reactions and boosting protein recovery. This method also altered the secondary and tertiary structures of the protein, exposing hydrophobic groups and sulfhydryl (-SH) residues, which significantly enhanced solubility (70.77%), emulsification activity (120.56 m^2/g), and dispersibility (0.305) [54]. In beer spent grain (BSG) protein extraction, UAE further demonstrated its potential for protein recovery and functional improvement. Under optimized conditions (250 W, 20 min, duty cycle 60%), BSG protein yield increased from 45.71% (traditional method) to 86.16% (UAE method). Structural analysis revealed that ultrasound treatment increased β-sheet content by 5.83% while reducing α-helix, β-turn, and random coil contents by 4.76%, 0.33%, and 0.74%, respectively, indicating conformational modifications [55]. In summary, UAE exhibits tremendous potential in plant protein extraction and functional modification, significantly improving yield, structural properties, and bioactivity. Future research should further investigate the underlying mechanisms of UAE and explore its synergy with other auxiliary techniques, such as high-pressure homogenization, electric field treatments, and enzymatic hydrolysis, to achieve more efficient protein extraction and modification.

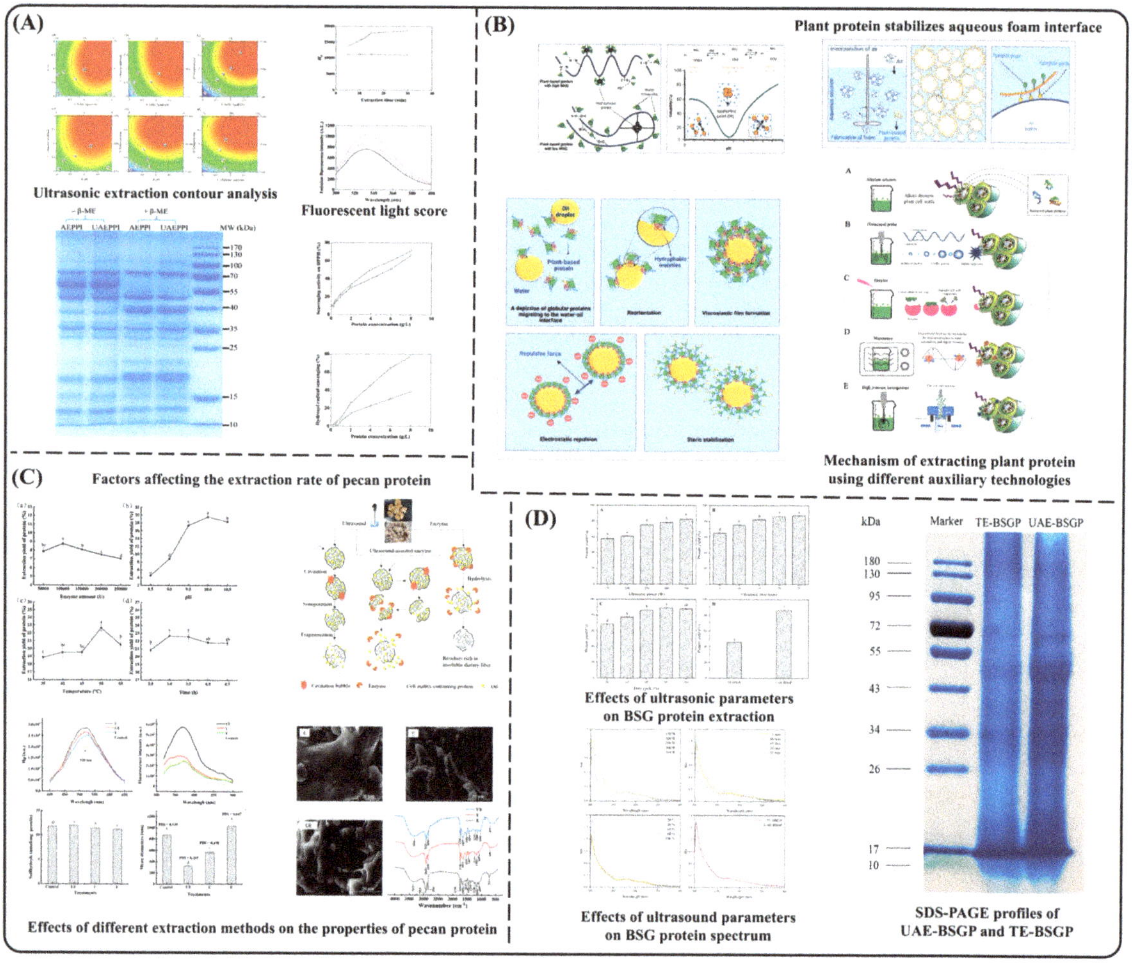

Figure 3. Effects of ultrasonic-assisted extraction on the properties and functions of pea protein (**A**) [11]; enhanced alkaline extraction techniques for plant-based proteins (**B**) [53]; effects of different extraction methods on the extraction rate, structure, and physicochemical properties of pecan protein (**C**) [54]; effects of ultrasonic parameters on BSG protein extraction yield, spectral characteristics, and molecular structure (**D**) [55].

Lately, the combined use of ultrasound and pH shifting has increasingly attracted interest in the field of protein modification. The synergistic mechanism of ultrasound-assisted pH shifting primarily relies on the complementary effects of both techniques: ultrasound-induced cavitation intensifies the structural alterations and reactivity of protein molecules, while pH shifting alters surface charge distribution, inducing further unfolding or rearrangement of protein structures. When proteins are exposed to acidic or alkaline environments far from their isoelectric point, surface charge and electrostatic repulsion increase, promoting chain relaxation and molecular unfolding. These structural changes are further amplified by ultrasound treatment, where the high shear forces generated by cavitation effectively disrupt non-covalent interactions within the protein, increasing structural unfolding and exposing more functional sites. Accordingly, the combined ultrasound and pH shifting treatment can significantly enhance protein functionalities, thus expanding their potential application in the food industry.

3.2.1. Effect of Ultrasonication-Assisted pH Shifting on Protein Structure

The combined effects of ultrasound (US) and pH shifting on protein structure have (in recent years) emerged as an important research direction, particularly in relation to particle size, secondary and tertiary structures, surface properties, and intermolecular interactions. The structural alterations induced by such treatments depend significantly on the pH shifting that occurs under acidic or alkaline conditions. Ultrasound-assisted pH shifting has shown notable synergistic effects in protein modification, with the underlying mechanisms summarized as follows.

Firstly, electrostatic repulsion induced by pH shifting promotes protein unfolding, leading to the exposure of internal hydrophobic groups and reactive sites. Simultaneously, cavitation-generated microjets from ultrasound enhance chain extension and improve solvent penetration [56]. Secondly, under extreme pH conditions, hydroxyl radicals (·OH) produced by ultrasound selectively oxidize sulfur-containing amino acid residues, enabling controlled disulfide bonds reformation and covalent cross-linking [57]. Thirdly, when the pH is readjusted to near the isoelectric point, ultrasound suppresses irreversible aggregation and, through mechanical shear, regulates the refolding process, resulting in functional particles with uniform size and optimized surface properties.

Previous study displayed that pH shifting and sonication significantly increased the emulsifying activity index (by 62.8%) and thermal stability of soy protein isolate ($p < 0.05$) [58]. Research further indicates that ultrasound-assisted pH shifting disrupts both non-covalent interactions (e.g., hydrogen bonds and hydrophobic interactions) and covalent bonds (e.g., disulfide linkages), resulting in a more relaxed and unfolded protein conformation. Changes in secondary structure are particularly evident, with a general reduction in α-helix and β-sheet content, accompanied by increases in β-turn and random coil. Zheng et al. [59] demonstrated that the combination of ultrasound treatment at an intensity of 400 W for 15 min with pH shifting (adjusted to pH 12 and subsequently neutralized) at 25 °C significantly altered the secondary structure of soy protein isolate. Specifically, the contents of α-helix and β-sheet were markedly reduced, while β-turn and random coil structures were increased, indicating a transition toward a more disordered and flexible conformation. Similarly, Dabbour et al. [60] found that ultrasound combined with pH shifting significantly decreased β-sheet content and increased random coil and β-turn content in cottonseed meal protein, implying unfolding in secondary structure. These structural changes are often associated with increased exposure of sulfhydryl groups, cleavage of disulfide bonds, enhanced surface hydrophobicity and reactivity, and reduced particle size [61].

3.2.2. Effect of Ultrasonication-Assisted pH Shifting on Functional Properties of Protein

Traditional physical and chemical methods have some limitations such as low efficiency, poor selectivity, and functional degradation. The application of ultrasound and pH shifting technologies has displayed significant advantages in enhancing extractability and functionality of protein. pH shifting and UAE have been indicated to effectively improve the extraction efficiency of both plant-based and marine by-product proteins, including pea protein isolate (PPI2), shrimp by-product protein, fish processing by-product protein, and oil body wastewater protein [62–64]. The cavitation effect and high shear forces generated by ultrasound facilitate the disruption of cell walls and the unfolding of protein structures, leading to an increase in solubility, thereby enhancing extraction efficiency [65]. For example, Yang et al. [66] reported that the ultrasound-assisted pH shifting treatment, in which the pH of pea protein isolate was adjusted to 12 followed by ultrasound application at 400 W and 20 kHz for 10 min at 25 °C, significantly improved protein solubility to over 90 percent. This was substantially higher than the solubility observed in samples treated with ultrasound alone or in the untreated control.

Moreover, the partial unfolding of proteins and the exposure of sulfhydryl and hydrophobic groups not only improve solubility but also enhance emulsifying activity, foaming capacity, gel-forming ability, and bioactivity [58,67]. For instance, proteins extracted from brewery spent grain using optimized UAE conditions exhibited significantly improved emulsifying, foaming, and fat-binding properties, attributable to conformational modifications in the secondary structure [8]. Table 1 shows the effect of pH shifting and ultrasound on functional properties of protein.

Table 1. The effects of pH shifting combined with ultrasound on protein functionality.

Sample	Condition	Functional Improvements	Reference
Peanut protein	pH shifting (pH 12) combined with ultrasound (400 W, 20 kHz, 10 min, 25 °C)	The solubility was 111.4% higher than that of the control (d-ppi), respectively.	[68]
Soybean protein isolate (SPI), potato protein isolate (PPI4) and soybean/potato protein complex (SPI/PPI4 complex)	pH shifting (pH 12) combined with ultrasound (360 W, 20 kHz, 30 min, 25 °C)	Increased protein–water interactions, resulting in enhanced solubility. The gel properties were significantly improved after treatment, and the hardness, elasticity, and water-holding capacity (WHC) were significantly improved.	[65]
Pleurotus ostreatus	pH shifting (pH 12) combined with ultrasound (45 kHz, 64 min, 35 °C)	Enhanced emulsifying properties and foam stability of protein concentrate.	[62]
Yeast protein (YP)	pH shifting (pH 12, 2 h) combined with ultrasound (390 W, 25 kHz, 20 min, 25 °C)	The combined treatment greatly reduced the particle size of the protein and increased the solubility, foaming performance, and surface hydrophobicity (H_0) value	[69]
Soy protein isolate (SPI)	Alkaline pH shifting (pH 9 1 h) combined with ultrasonic treatment (200 W, 300 W, 400 W, 10 min)	The viscosity of SPI decreased from 98.97 mPa. s to 22.83 mPa. s. These structural changes endow SPI with higher solubility (increasing from 81.13% to 91.53%), as well as better emulsifying and foaming properties.	[70]
Cottonseed meal protein (CSMP)	pH shifting (pH 1.5/3.5/9.5/11.5) and sonication (320 W, 20/60 kHz, 21 min)	Combined treatment at alkaline conditions increased absolute surface charge, solubility, foamability, and oil binding efficacy over control, and pH shifting alone ($p < 0.05$).	[60]
Pea protein isolate	pH shifting (pH 12, 1 h) and sonication (500 W, 20 kHz, 10 min, 25 °C)	PPI2 modified by ultrasound and pH change can successfully stabilize solid high internal phase lotion HIPEs with strong viscoelasticity and high stability.	[71]

Table 1. *Cont.*

Sample	Condition	Functional Improvements	Reference
Flaxseed protein isolate	pH shifting (pH 10, 2 h) and sonication (504 W, 20 kHz, 20 min, 25 °C)	UFPI-10 (FPI treated by ultrasound coupled with pH 10 cycling) possessed higher emulsification stability (ESI similar to 308.20 min), increasing by 1.74 times.	[72]
Chicken wooden breast myofibrillar protein (WBMP)	pH shifting (pH 11, 10 min, 4 °C) and sonication (400 W, 20 kHz)	WBMP emulsion more uniform, the gel strength and water-holding capacity of the protein gel increased.	[73]
Pine kernel protein (PKP)	Acidic pH shifting (pH 2, 30 min) and ultrasound (500 W, 20 kHz, 10 min, 30 °C)	Protein lotion shows higher viscoelasticity and stronger protein interaction, and lotion stability is enhanced.	[74]
Perilla protein isolate (PPI3)	pH shifting (pH 10/12, 30 min) and sonication (400 W, 20 kHz, 15 min)	the emulsifying and foaming properties of PPI3 could evidently enhance.	[66]
Chickpea protein isolate (CPI)	pH shifting (pH 2/12, 60 min) and sonication (300 W, 20 kHz, 20 min)	The foaming performance of CPI has significantly improved.	[75]
Coconut milk protein	pH shifting (pH 1/12, 60 min) and sonication (53 kHz, 40 W/L, 20 min, 25 °C)	Ultrasound can amplify the effect of pH shift on increasing the thermal stability of coconut milk by modifying functional properties and structures of coconut milk protein.	[76]
Shrimp proteins	alkaline pH shift (pH 12.5) combined with ultrasonication (300 W, 20 min)	The combined treatment of shrimp protein has a significant impact on its foaming and emulsifying properties.	[63]
Whey protein isolate	pH shifting (pH 2, 0–3 h) combined with ultrasound (600 W, 20 kHz, 30 min)	pH 2-shifting treatment combined with ultrasound could improve emulsion stability of WPI.	[77]
Native amaranth protein (APN)	pH shifting (pH 2/12, 1 h) combined with ultrasound (50% amplitude for 10 min)	Foaming capacity and stability were significantly increased in all treatments.	[78]
Faba bean protein isolate	pH shifting (pH 11) combined with ultrasound (1200 W, 20 kHz 10/20 min)	A striking enhancement in foaming capacity (from 93% to 306–386%) and stability (from 10 s to 473–974 s) was achieved by the combined treatment.	[79]
Barley protein isolate	pH shifting (pH 9) combined with ultrasound (20 kHz)	Ultrasound treatment improved both protein solubility and colloidal stability of the protein-enriched fraction at alkaline pH.	[80]

Ultrasound-assisted pH shifting protein modification holds great potential applications in functional ingredient delivery systems, enhancement of plant protein functionalities, and the design of novel food products. In the construction of delivery systems, this technique significantly improves the encapsulation efficiency of bioactive compounds by promoting the exposure of hydrophobic regions and the unfolding of protein structure. In a recent study, Liu et al. [81] demonstrated that combining pH 11 and 300 W ultrasound treatment for 20 min significantly enhanced the co-encapsulation of VE and QU in SLP matrices. This approach improved encapsulation efficiency, solubility, and antioxidant activity compared to untreated SLP, highlighting the potential of pH and ultrasound for modulating protein delivery systems. Similarly, Fang et al. [82] demonstrated that a combination of 540 W ultrasound-assisted pH shifting treatment at pH 12 for 5 min, followed by adjustment to pH 7.0, significantly enhanced the encapsulation efficiency of resveratrol in soy protein isolate (SPI) to 91.4 ± 4.3%. This result was substantially higher than the encapsulation efficiency achieved with ultrasound treatment alone (83 ± 3%) and significantly improved the functionality of SPI as a nanocarrier for hydrophobic compounds. In another study, Zhang et al. [71] fabricated solid high internal phase emulsions (HIPEs) with excellent elasticity and stability by modifying PPI2 with ultrasound (500 W, 10 min) at pH 12, followed by adjusting the pH to 7 to obtain modified PPI2 (MPPI2). MPPI2 and chitosan particles were used as emulsifier and co-stabilizer, respectively, to construct HIPEs with an enhanced interfacial adsorption and network structure, resulting in improved stability. These findings confirm the significant potential of ultrasound-pH synergistic modification in improving protein carrier performance, enhancing functional properties, and enabling the design of novel food formulations.

In summary, the combination of pH shifting and ultrasound alters protein conformation, enhancing solubility, emulsifying properties, and encapsulation capacity, thereby expanding their potential applications in bioactive delivery systems and emulsion-based food formulations. This aligns with previous discussions on the functional improvements of proteins induced by ultrasound-assisted pH modification.

3.3. *High Pressure-Assisted pH Shifting*

High-pressure (HP: 100–600 MPa) and ultra-high-pressure (UHP: >600 MPa) processing are non-thermal technologies that apply extreme mechanical forces to materials via hydrostatic pressure. HP treatment induces conformational alterations in protein by disrupting intermolecular non-covalent interactions such as hydrogen bonds and hydrophobic interactions [83]. Contrarily, UHP, due to its higher energy density, can additionally cause the disruption and reformation of covalent bonds, including disulfide bridges. Both techniques enable controlled structural modification of proteins at ambient or low temperatures while avoiding degradation of thermosensitive compounds. Consequently, they are widely applied in food, pharmaceutical, and other sectors for protein functionality enhancement and bioactivity preservation.

The combination of high-pressure and pH shifting is an effective strategy in protein modification and has extensively been used to improve functional properties of protein. Previous studies have exhibited that this combined approach effectively modulated protein conformation and enhanced its potential for delivering bioactive compounds [84–86]. High-pressure homogenization (HPH) can break disulfide-linked subunits into lower molecular weight fragments and convert highly ordered secondary structure into random coil, thereby facilitating the unfolding of protein [87]. In this regard, UHP has widely been adopted in the food processing industry, particularly for modifying biological macromolecules such as protein and starch. Under such conditions, UHP not only alters the higher-order structure of protein via physical interactions such as reducing the number of hydrogen bonds stabilizing protein structure, but also modulates their physicochemical properties, including solubility, particle size, and hydrophobicity [88]. These modifications ultimately enhance the functional performance of proteins in the food system by altering their structure [89].

3.3.1. Effect of High Pressure-Assisted pH Shifting on Protein Structure

Presently, the synergistic application of pH shifting and high-pressure treatment, particularly HPH, has attracted significant attention in the field of protein modification. This combined approach modulates protein structure at multiple levels by adjusting the pH environment and applying mechanical pressure [90,91]. These structural modifications provide a molecular basis for enhancing the functional properties of protein. For example, the synergistic effect of HPH and pH shifting results in significant alterations in the secondary and tertiary structure of protein, inducing the release of free hydrogen ions, and thus increasing the surface positive charge density of protein [84]. The combination of pH shifting and HPH promoted the depolymerization of protein aggregates, reduced particle size, and decreased surface hydrophobicity [85]. Additionally, HPH-assisted pH shifting enhanced the structural flexibility of hemp protein isolate, as evidenced by the increased content of random coil. Literature shows that, following HPH and pH shifting, the intrinsic fluorescence intensity significantly decreased and exhibited a red shift, while surface hydrophobicity increased, implying extensive unfolding in protein conformation [92]. When the pH approached the isoelectric point, the α-helix structure was converted into β-sheet and β-turn by up to 20%; additionally, HPH alone also promoted the transformation of α-helix into β-sheet, with a conversion rate of 15% observed under extreme acid and alkaline conditions [93]. Moreover, Laguna et al. reported that pea protein at pH 6.2 demonstrated a higher degree and rate of digestion, whereas high-pressure treatment masked this pH-dependent effect, which was consistent with the observed structural changes [94].

Moreover, under optimal high-pressure conditions (600 MPa), the emulsifying activity of protein isolate was significantly enhanced, primarily due to protein unfolding and the increased exposure of hydrophobic groups [95]. In this context, pH variation serves a regulatory role in the high-pressure modification of proteins [92]. In the case of soy protein isolate, this treatment has been noticed to increase the content of free sulfhydryl groups and surface hydrophobicity, reduce droplet size, and enhance zeta potential [96]. For hemp protein isolate, high-pressure treatment resulted in an increase in random coil content and a decrease accompanied by a red-shift in intrinsic fluorescence intensity, indicating a significant enhancement in structural flexibility [92]. In addition, the HPH-assisted pH-shifting treatment enhanced protein penetration and rearrangement at the oil–water interface, which in turn facilitated the formation of a more stable interfacial layer [92]. pH shifting and high-pressure processing (HPP), particularly high-pressure homogenization (HPH), holds great potential in protein modification; this combined approach induces multi-scale structural transformations by altering the pH environment and applying mechanical stress. As a result, significant changes in protein secondary and tertiary structures have been observed, including increased random coil content, α-helix to β-sheet transitions, and enhanced structural flexibility. These conformational modifications lead to structure changes such as reduced particle size, altered surface charge, and increased hydrophobic group exposure, which collectively provide the molecular foundation for improved protein functionalities.

3.3.2. Effect of High Pressure-Assisted pH Shifting on Functional Properties of Protein

This synergistic approach (e.g., pH shifting and high-pressure treatment, particularly HPH) has demonstrated significant advantages in enhancing solubility, emulsifying capacity, foaming properties, and interfacial activity. Zhu et al. [47] reported that a combination of pH shifting and high-pressure homogenization (HPH) treatment at a pressure of 500 MPa for five cycles at pH 3 increased protein solubility by 34.75%, and enhanced its binding affinity with vitamin B12, primarily through hydrogen bonding and hydrophobic interactions. Wang et al. [85] found that treatment of rice dreg protein with pH 12 and

20–100 MPa HPH increased solubility from 0.86% to 63.5%. Similarly, Yildiz and Yıldız et al. [97] observed significant improvements in solubility (from 7.85% to 78.97%) and soluble protein content in quinoa protein isolate by applying a sequential pH shifting treatment at alkaline pH 12 for 1 h, followed by high-pressure homogenization (250 MPa, single-pass) at ambient temperature. This combined approach effectively reduced particle sizes to 54 nm while enhancing surface hydrophobicity (198.0 ± 0.6) and antioxidant activity (9.8 ± 0.03%). Wang et al. [92] further observed that the solubility of hemp protein isolate reached a maximum of 62.8% following combined treatment with sequential pH shifting at alkaline pH 12.0 for 1 h, followed by high-pressure homogenization at 120 MPa for two consecutive passes at 25.0 ± 1.0 °C. This synergistic approach induced structural unfolding evidenced by a 2.2-fold increase in random coil content (from 44.2% to 98.3%) and particle size reduction from 1612 nm to 231 nm ($p < 0.05$), with treatment efficacy showing pressure-dependent optimization ($R^2 = 0.972$). Wang et al. [85] achieved a 126.7-fold increase in emulsifying activity and a 930.7-fold increase in foaming capacity of rice dreg protein through sequential pH shifting (pH 12.0, 25 °C, 4 h) combined with three cycles of high-pressure homogenization under 100 MPa at 25 °C. Tan et al. [96] demonstrated that the synergistic effect of pH shifting (pH 11.0, 25 °C, 30 min) and high hydrostatic pressure (400 MPa, 10 min) significantly enhanced the emulsion stability of soy protein isolate (ESI from 21.8 to 35.1 min), accompanied by 51.5% reduction in droplet size and 132% increase in emulsifying activity index (EAI from 12.7 to 29.6 m^2/g). Likewise, Ahmed et al. [95] also reported that under optimal conditions (600 MPa), high-pressure treatment alone markedly enhanced protein emulsification, with pH shifting playing a regulatory role in this process. Additionally, the synergistic application of pH shifting and high-pressure treatment substantially influenced the crystallization behavior and crystal morphology of protein, providing a molecular foundation for structural control and texture design in protein-based products [98]. Such modification has indicated broad applicability and promising potential across various plant proteins, including those derived from peas, rice dregs, soybeans, hemp seeds, and quinoa [53,99].

The application of pH shifting combined with high-pressure treatment in protein modification is primarily focused on areas such as bioactive compound delivery, novel food development, hypoallergenic food production, and overall improvement of food quality. For example, increased pH and high-pressure conditions significantly influence protein separation performance, particularly by affecting membrane permeability and salt rejection, thereby enabling the effective delivery of bioactive substances [100]. Khan et al. [101] found that emulsions stabilized by sweet potato protein under different pH levels combined with HPH have various applications in the food industry. HP treatment has also been shown to enhance the viscoelastic behavior of soy protein concentrate, contributing to improved protein content and textural properties of food products [89]. Interestingly, Teixeira et al. [102] demonstrated that pressurized liquid extraction (PLE), when used in conjunction with alkaline solvents (pH 9.0), effectively enhances protein recovery, yielding a high purity of 86.7% and improved thermal stability.

Moreover, HP treatment can effectively be combined with other modification techniques. Wang and Moraru [103] demonstrated optimal microstructural modification in milk protein concentrate (MPC) using pH 5.1 36 mg/g Ca combined with high-pressure homogenization (600 MPa, 5 °C, 3 min), achieving 58% porosity increase and 3.2-fold aggregate size enhancement. These findings highlight the potential use of HPH in the modification of MPC and open new avenues for the development of gel-based high-protein foods, such as puddings and portable protein bars. Figure 4 lists some applications of pH shifting combined with high pressure on protein.

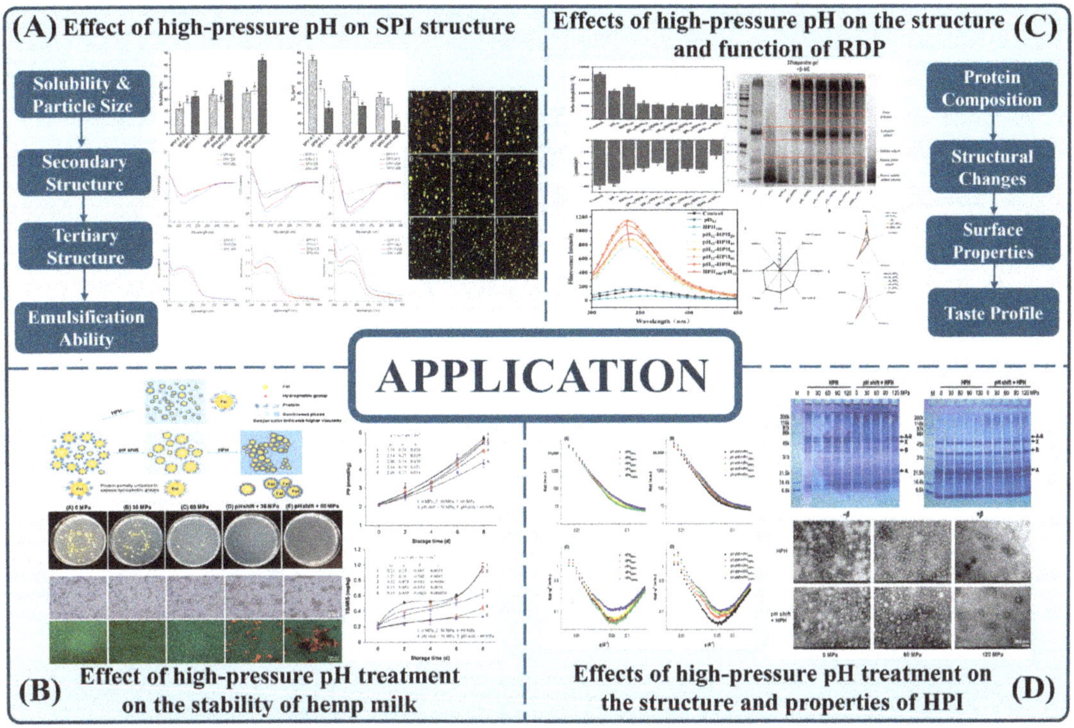

Figure 4. Effect of high pressure and pH treatment on the functionality and structure of various plant proteins. Soybean protein isolate (**A**) [96]; hemp milk (**B**) [104]; rice dreg protein (**C**) [85]; hempseed protein isolate (**D**) [92].

These studies collectively highlight the effectiveness of HPH and pH shifting in improving the functional properties of plant proteins. By optimizing processing conditions such as pH and pressure levels, different functional modifications can be achieved for different protein sources [92]. Alternatively, variations in protein response to these treatments necessitate further optimization for each specific protein type. Although these physical treatments offer substantial benefits in enhancing protein functionality, practical applications must also consider economic feasibility, sensory attributes, and nutritional implications. Future research should further investigate the applicability of these methods in real food systems and assess their long-term effects on human health.

3.4. Other Physical Methods-Assisted pH Shift Treatments

In addition to the extensive research on combining pH shifting with physical methods such as heat treatment, ultrasound, and high pressure for protein processing, other physical techniques can also be integrated with pH shifting to modify proteins. Examples include pulsed electric field, microwaves, and cold ions. Although limited research has been conducted on combining these methods with pH shifting for protein modification, each technique has effectively been used in protein processing and modification, making them promising alternatives. The following section will discuss existing studies and potential applications of these methods in combination with pH shifting for protein modification.

3.4.1. Pulsed Electric Field-Assisted pH Shifting

Pulsed electric field (PEF) is a non-thermal physical processing technology that applies a high-intensity electric field (typically 5–50 kV/cm) for a very short duration, inducing reversible or irreversible electroporation of biological cell membranes and thereby altering cell permeability [105]. This technology is widely used in food sterilization, ingredient extraction, and texture modification due to its efficiency, energy-saving properties, and ability to preserve food nutrients and sensorial characteristics.

Earlier studies show that PEF treatment (2.1 kV/cm) for 5–30 min at 25 °C significantly altered the secondary and tertiary structures of porcine myofibrillar proteins, and emulsification properties (EAI: 7.10 m^2/g, ESI: 179.67%) [106]. By applying high-voltage pulses at 1.00–1.25 kV/cm electric field strength, 20 μs pulse width, and 50 Hz frequency, PEF induced partial unfolding of bovine muscle proteins by disrupting Z-disk and I-band junctions, exposing internal hydrolysis sites (e.g., troponin T and myosin heavy chain), thereby improving in vitro protein digestibility by 18–31% ($p < 0.05$) after 180 min of simulated gastro-small intestinal digestion. [107]. Compared with traditional thermal and high-pressure treatments, PEF can trigger conformational changes under milder conditions, facilitating protein digestion in the gastrointestinal tract [108]. Moreover, PEF improved protein hydrophilicity and dispersibility, while exposing additional reactive sites on amino acid residues, which facilitated interactions with other macromolecules such as polysaccharides and polyphenols [109]. Notably, the synergistic effects of pulsed electric field (PEF) are significantly enhanced when applied in combination with a pH shifting treatment. Under natural pH conditions (5.49–5.66) close to the isoelectric point of myofibrillar proteins, pH shifting initiates conformational loosening by altering protein charge distribution. When coupled with constant-current PEF (12–40 mA, 1000 Hz, needle-spring electrodes) or constant-voltage PEF (20 V/cm, 1000 Hz), the electric field further promotes protein unfolding and formation of smaller aggregates through suppression of disulfide bond formation and carbonyl compound accumulation (reducing oxidation by 18.3–28.7% compared to conventional thawing) [110]. This cooperative modification provides a structural basis for subsequent improvements in protein functionality.

Although comprehensive investigations into the structural effects of PEF and pH shifting remain limited, primarily due to the complex and dynamic nature of PEF-induced mechanisms, accumulating evidence continues to support its potential in modulating protein molecular conformation [111]. Building upon these structural alterations, the combined application of PEF and pH shifting has been shown to significantly improve key functional attributes of proteins. Specifically, this integrated strategy enhances solubility, reduces aggregation, and increases the exposure of reactive functional groups, thereby contributing to improved digestibility, emulsifying activity, foaming capacity, and other interfacial functionalities. For example, Wang et al. [112] reported that combined treatment with pH shifting (pH 11) and moderate PEF strength (10 kV/cm) synergistically increased the solubility of soy protein isolate from 26.06% to 70.34%. Corresponding enhancements in emulsifying activity and foaming ability were attributed to conformational unfolding, reduced particle size, and elevated surface hydrophobicity. These structural modifications optimized molecular flexibility and interfacial adsorption, thereby enhancing emulsifying and foaming behaviors. Similarly, the combination of PEF (1.4–1.8 kV/cm, 653–695 kJ/kg) and pH shifting, particularly at pH 4, effectively enhanced protein digestibility by improving the enzymatic hydrolysis of egg-white protein while reducing protein aggregation, thereby increasing its functionality [111]. Additionally, PEF pre-treatment (1.1 kV/cm, 1.53 kJ/kg) combined with pH shifting to 5.0 during protein precipitation significantly increased the extraction yield of grass-clover juice by 25% and crude protein release by 31% [113]. In summary, the combination of PEF and pH shifting offers a novel processing strategy for the functional enhancement of both plant and animal proteins, providing a theoretical and practical basis for the development of high-performance foods and nutritional delivery systems.

Recent advances suggest that the combination of PEF and pH shifting also holds considerable potential in the delivery of bioactive compounds, molecular encapsulation, selective recovery and processing of target biomolecules, improvement of protein product quality, and non-thermal microbial inactivation. In particular, this combined treatment has been shown to modulate transmembrane protein transport during membrane separation processes by altering charge distribution and the electrochemical potential gradient across both protein surfaces and membrane interfaces, thereby enhancing separation selectivity and increasing membrane flux [114]. For example, Wang et al. [115] successfully modified soy protein isolate using a combination of pulsed electric field (10 kV/cm) and pH shifting to pH 11 to develop SPI-based nanoparticles for efficient lutein encapsulation. In addition, PEF is also recognized as an effective processing technology which, when combined with appropriate pH adjustment, not only improves protein extraction yields but also reduces energy consumption, making it suitable for large-scale applications [116].

In conclusion, the use of PEF in plant protein processing and bioactive peptide preparation has gained increasing attention. However, the impact of PEF alone under neutral pH is relatively mild, and its ability to enhance purification efficiency remains limited. This is primarily due to the thermal dependency of PEF, and the partial structural protection of proteins under these conditions, which prevents excessive denaturation or degradation. To overcome these challenges, combining PEF with pH shifting has emerged as an effective strategy in protein processing.

3.4.2. Microwave-Assisted pH Shifting

Microwave-assisted processing is a low-energy, low-pollution, non-thermal physical treatment technique that utilizes microwave radiation for rapid and uniform heating. Compared to conventional thermal methods, microwave treatment offers advantages such as faster heating rates, lower energy consumption, and improved uniformity. Because microwaves interact directly with polar groups within molecules, inducing molecular rotation and vibration, they thereby rapidly increase the internal temperature of the material and achieve uniform heating [117].

The combination of pH shifting and microwave treatment has been shown to significantly alter protein structure. Microwave irradiation promotes the rotation and vibration of polar moieties within protein molecules through rapid, directional heating, which induces unfolding and refolding of molecular structures within a short time frame. When coupled with pH shifting, the denaturation effects under extreme pH conditions are synergistically enhanced by microwave energy. This typically results in a decrease in α-helix content and an increase in β-sheet and random coil structures. Additionally, increases in surface hydrophobicity and free sulfhydryl content indicate molecular unfolding and the exposure of functional groups.

These conformational changes not only improve protein solubility and extraction efficiency but also enhance key functional properties such as emulsification, foaming, and antioxidant capacity. Consequently, this synergistic modification approach provides a promising strategy for the high-value utilization of proteins. pH shifting, as a widely used technique for improving protein solubility, when integrated with microwave treatment, has been demonstrated in numerous studies to synergistically enhance protein functionality. At the functional level, many studies have reported that the combination of microwave treatment and pH shifting significantly improved protein solubility, dispersibility, emulsifying properties, and thermal stability. For example, Han et al. [118] reported that the application of pH shifting (pH 12) and microwave heating (450 W, 90 °C, 2/4/6 min) to potato protein increased its solubility from 24.0 to 89.0%, surface hydrophobicity from 125 to 207, and reduced particle size from 249.95 μm to 90.37 nm, while simultaneously enhancing its dispersibility, emulsification, and thermal stability. Similarly, Das et al. [119] found that microwave-assisted extraction (600 W, 32 s) combined with pH modification (pH 8) optimized the yield and purity of soybean meal protein isolate, highlighting the synergistic effect of these techniques in enhancing extraction efficiency and quality control. Jahan et al. [120] further noticed that microwave (power 800, time 2 min) and pH shifting (pH 12) significantly improved the functional properties of mustard seed meal protein, thereby expanding its potential in plant protein processing.

In summary, the integration of microwave and pH shifting treatments not only induces substantial changes in protein structure but also effectively enhances their industrial application potential, particularly in the extraction, functional improvement, and sustainable utilization of plant-based proteins.

3.5. Multiple Methods-Assisted pH Shifting

With the advancement of protein processing research, the use of multi-method strategies for protein modification has emerged as a prominent area of focus. Modern processing approaches are no longer limited to the combination of two techniques; rather, when synergistic enhancements in performance and efficiency are observed, multiple processing technologies are increasingly applied in tandem. For proteins with inherently poor functional properties, specific and often complex processing conditions are required to achieve desirable modifications. Researchers have thus employed various techniques either to enhance specific functional attributes or to develop composite systems for targeted applications.

Building on this trend, an increasing number of studies have demonstrated the functional advantages of integrating physical, chemical, and enzymatic techniques to tailor protein properties for specific applications. Zhang et al. [121] optimized the extraction of proteins from herring by-products through the integration of radial discharge high-shear homogenization and ultrasound-assisted pH shifting treatment. The homogenization was conducted at 20,000 rpm for 3 min to disrupt cellular structures, followed by alkaline solubilization at pH 11.5 and subsequent isoelectric precipitation at pH 5.5. Ultrasound was applied at a power of 400 W for 10 min during the pH shifting process to enhance mass transfer and protein unfolding. This combined methodology not only improved protein recovery efficiency but also effectively retained the antioxidant activity of the extracts. These findings support the synergistic potential of mechanical and chemical treatments in modulating protein structure and functionality. Similarly, Igartúa et al. [18] investigated the structural and functional modification of rice protein isolates using a sequential treatment comprising pH shifting, high-intensity ultrasound, and thermal processing. The proteins were first solubilized under alkaline conditions at pH 11.0 to promote unfolding and dispersion. Subsequently, ultrasound treatment was applied using a probe-type sonicator (20 kHz) at full amplitude (100%) for 5 min to enhance mass transfer and facilitate molecular rearrangement. This was followed by thermal treatment at 85 °C for 30 min to stabilize the induced conformational changes. The combined treatment resulted in a marked increase in protein solubility and surface hydrophobicity, along with the formation of loosely associated aggregates. These structural alterations collectively contributed to improved plasticity and functional adaptability of the rice proteins. Furthermore, Sun et al. [37] examined the gelation behavior of soy protein isolate (SPI) subjected to enzymatic modification with transglutaminase (TGase) under controlled alkaline and thermal conditions. The treatment was performed at pH 8.0 with TGase added at a concentration of 20 U/g protein, followed by incubation at 50 °C for 60 min to facilitate cross-linking. This process resulted in a significant enhancement of gel hardness and water-holding capacity, which was attributed to the formation of covalent bonds between glutamine and lysine residues, thereby strengthening the protein network structure. Similarly, Karabulut et al. [122] explored the functional modification of hemp protein isolates through a combination of thermal ultrasound, HPH, and pH shifting. Alkaline solubilization was conducted at pH 11.0 for 60 min to promote protein unfolding, followed by isoelectric precipitation at pH 4.5. Ultrasound treatment was applied using a 25 kHz probe at 400 W for 30 min under controlled heating at 60 °C. Subsequently, the protein dispersion underwent HPH at 100 MPa for three cycles to induce further particle size reduction and structural disruption. This multi-step treatment significantly improved the solubility and emulsifying properties of hemp protein and contributed to a more relaxed and accessible protein structure, thereby enhancing its potential functionality in food formulations. Notably, the combination of multiple techniques is becoming a cutting-edge strategy in protein modification.

Like the protein-polysaccharide conjugates produced via pH shifting treatment described in Section 2, physical field-assisted pH regulation techniques can also promote covalent conjugation between proteins and polysaccharides. Jiang et al. found that ultrasound-assisted pH shifting accelerated the Maillard reaction of pea protein isolate and inulin and produced conjugates with a high degree of grafting [123]. Compared to the individual protein components, these conjugates exhibited significantly improved thermal stability, antioxidant activity, and interfacial properties. In a related study, Ding et al. [124] demonstrated that ultrasound-assisted pH shifting treatment effectively improved the interfacial adsorption behavior of whey protein isolate (WPI)-carboxymethyl cellulose (CMC) complexes. The treatment involved adjusting the system to an alkaline pH of 11.0 to induce protein unfolding, followed by acidification to pH 5.0 to facilitate complex formation near the isoelectric point of WPI. Ultrasound was applied at a power of 450 W and a frequency of 20 kHz for 10 min to enhance molecular interactions and structural rearrangements. This approach significantly increased the adsorption capacity at the oil–water interface, likely due to improved electrostatic and hydrogen bonding interactions between the protein and polysaccharide components. These findings underscore the potential of ultrasound-assisted pH shifting to modulate protein–polysaccharide interactions for emulsification applications.

These studies collectively demonstrate that the integration of pH shifting with physical treatments such as ultrasound, homogenization, and thermal processing can systematically optimize protein functionality by modulating conformational structure, aggregation behavior, and molecular interactions. This multi-technique exhibits synergistic effects, often yielding results greater than the sum of individual methods. However, despite the promising outcomes, the underlying mechanisms of such synergistic treatments remain complex. Current analyses rely primarily on macroscopic characterization, while the precise molecular mechanisms require further investigation, indicating a broad scope for future research.

4. Conclusions and Perspectives

Protein extraction and modification are critical to enhancing the value of agricultural products and advancing alternative protein sources. Among emerging strategies, the integration of pH shifting with physical treatments, such as thermal processing, ultrasound, high-pressure, and pulsed electric fields, has shown significant promise in improving protein structure and functionality. This combined approach enhances key functional properties, including solubility, emulsifying and foaming capacity, and thermal stability, while mitigating the limitations of single-modality treatments. Its application in emulsifiers, delivery systems, and functional foods underscores its potential for sustainable innovation in food science. Despite these advancements, several challenges hinder widespread industrial adoption. High equipment costs, particularly for high-pressure and pulsed electric field technologies, limit scalability. Additionally, extreme pH or intense physical treatments may cause partial protein degradation, emphasizing the need for optimized processing parameters. The synergistic mechanisms underlying these combined techniques remain insufficiently understood and warrant further molecular-level investigation. Moreover, comprehensive safety assessments and regulatory compliance evaluations are essential for ensuring consumer safety.

Future research should prioritize the development of cost-effective, environmentally sustainable, and efficient modification strategies that maintain protein integrity and functionality. Integrating insights from food science, biophysics, and computational modeling will be essential to optimize processes, clarify synergistic effects, and facilitate industrial translation. Addressing these challenges will enable the broader application of pH shifting combined with physical treatments, supporting the functional enhancement and sustainable utilization of protein resources in food and agricultural systems.

Author Contributions: Conceptualization, R.L. and R.H.; methodology, R.L.; validation, M.W., S.Z., and Z.Q.; investigation, J.P.; writing—original draft preparation, R.L.; writing—review and editing, M.D., B.K.M., Y.H., J.P., H.M., and R.H.; visualization, R.L.; supervision, Y.H.; funding acquisition, R.H. All authors have read and agreed to the published version of the manuscript.

Funding: This study received financial support from the National Key R&D Program of China under grant number 2024YFD21011039 (within program 2024YFD2101100), and was supported by the Primary Research and Development Plan of Shandong Province (2022CXGC010603), and sponsored by a project funded by the Priority Academic Program Development of Jiangsu Higher Education Institutions (PAPD).

Institutional Review Board Statement: Not applicable.

Informed Consent Statement: Not applicable.

Data Availability Statement: No new data were created or analyzed in this study. Data sharing is not applicable to this article.

Conflicts of Interest: The authors declare that they have no known competing financial interests or personal relationships that could have appeared to influence the work reported in this paper.

References

1. Gastaldello, A.; Giampieri, F.; De Giuseppe, R.; Grosso, G.; Baroni, L.; Battino, M. The Rise of Processed Meat Alternatives: A Narrative Review of the Manufacturing, Composition, Nutritional Profile and Health Effects of Newer Sources of Protein, and Their Place in Healthier Diets. *Trends Food Sci. Technol.* **2022**, *127*, 263–271. [CrossRef]
2. Karabulut, G.; Subasi, B.; Ivanova, P.; Goksen, G.; Chalova, V.; Capanoglu, E. Towards Sustainable and Nutritional-Based Plant Protein Sources: A Review on the Role of Rapeseed. *Food Res. Int.* **2025**, *202*, 115553. [CrossRef] [PubMed]
3. Wang, X.; Zhang, L.; Chen, L.; Wang, Y.; Okonkwo, C.; Yagoub, A.; Wahia, H.; Zhou, C. Application of Ultrasound and Its Real-Time Monitoring of the Acoustic Field during Processing of Tofu: Parameter Optimization, Protein Modification, and Potential Mechanism. *Compr. Rev. Food Sci. Food Saf.* **2023**, *22*, 2747–2772. [CrossRef]
4. Mintah, B.K.; He, R.; Agyekum, A.A.; Dabbour, M.; Golly, M.K.; Ma, H. Edible Insect Protein for Food Applications: Extraction, Composition, and Functional Properties. *J. Food Process Eng.* **2020**, *43*, e13362. [CrossRef]
5. Zhang, Z.; Wang, Y.; Li, Y.; Dai, C.; Ding, Q.; Hong, C.; He, Y.; He, R.; Ma, H. Effect of Alkali Concentration on Digestibility and Absorption Characteristics of Rice Residue Protein Isolates and Lysinoalanine. *Food Chem.* **2019**, *289*, 609–615. [CrossRef]
6. Gao, R.; Yu, Q.; Shen, Y.; Chu, Q.; Chen, G.; Fen, S.; Yang, M.; Yuan, L.; McClements, D.J.; Sun, Q. Production, Bioactive Properties, and Potential Applications of Fish Protein Hydrolysates: Developments and Challenges. *Trends Food Sci. Technol.* **2021**, *110*, 687–699. [CrossRef]
7. Pan, J.; Xu, H.; Cheng, Y.; Mintah, B.K.; Dabbour, M.; Yang, F.; Chen, W.; Zhang, Z.; Dai, C.; He, R.; et al. Recent Insight on Edible Insect Protein: Extraction, Functional Properties, Allergenicity, Bioactivity, and Applications. *Foods* **2022**, *11*, 2931. [CrossRef] [PubMed]

8. Wen, C.; Zhang, J.; Duan, Y.; Zhang, H.; Ma, H. A Mini-Review on Brewer's Spent Grain Protein: Isolation, Physicochemical Properties, Application of Protein, and Functional Properties of Hydrolysates. *J. Food Sci.* **2019**, *84*, 3330–3340. [CrossRef]
9. Mintah, B.K.; He, R.; Dabbour, M.; Agyekum, A.A.; Xing, Z.; Golly, M.K.; Ma, H. Sonochemical Action and Reaction of Edible Insect Protein: Influence on Enzymolysis Reaction-Kinetics, Free-Gibbs, Structure, and Antioxidant Capacity. *J. Food Biochem.* **2019**, *43*, e12982. [CrossRef]
10. Pan, J.; Xu, H.; Dabbour, M.; Mintah, B.K.; Chen, W.; Yang, F.; Zhang, Z.; Cheng, Y.; Dai, C.; He, R.; et al. Effect of Alkaline pH-Shifting Process on Extraction Rate, Structural, and Functional Properties of Black Soldier Fly (*Hermetia Illucens*) Larvae Protein. *LWT-Food Sci. Technol.* **2023**, *185*, 115180. [CrossRef]
11. Wang, F.; Zhang, Y.; Xu, L.; Ma, H. An Efficient Ultrasound-Assisted Extraction Method of Pea Protein and Its Effect on Protein Functional Properties and Biological Activities. *LWT-Food Sci. Technol.* **2020**, *127*, 109348. [CrossRef]
12. Xu, H.; Pan, J.; Dabbour, M.; Mintah, B.K.; Chen, W.; Yang, F.; Zhang, Z.; Cheng, Y.; Dai, C.; He, R.; et al. Synergistic Effects of pH Shift and Heat Treatment on Solubility, Physicochemical and Structural Properties, and Lysinoalanine Formation in Silkworm Pupa Protein Isolates. *Food Res. Int.* **2023**, *165*, 112554. [CrossRef] [PubMed]
13. Mao, C.; Wu, J.; Zhang, X.; Ma, F.; Cheng, Y. Improving the Solubility and Digestibility of Potato Protein with an Online Ultrasound-Assisted PH Shifting Treatment at Medium Temperature. *Foods* **2020**, *9*, 1908. [CrossRef]
14. Yang, J.; Peng, D.; Jin, W.; Geng, F.; Cheng, C.; Wang, L.; Zhang, H.; Duan, Y.; Deng, Q. Redesign of Air/Oil-Water Interface via Physical Fields Coupled with pH Shifting to Improve the Emulsification, Foaming, and Digestion Properties of Plant Proteins. *Crit. Rev. Food Sci. Nutr.* **2023**, *65*, 1093–1108. [CrossRef]
15. Tang, Q.; Roos, Y.H.; Miao, S. Structure, Gelation Mechanism of Plant Proteins versus Dairy Proteins and Evolving Modification Strategies. *Trends Food Sci. Technol.* **2024**, *147*, 104464. [CrossRef]
16. Yang, Z.; Xie, C.; Bao, Y.; Liu, F.; Wang, H.; Wang, Y. Oat: Current State and Challenges in Plant-Based Food Applications. *Trends Food Sci. Technol.* **2023**, *134*, 56–71. [CrossRef]
17. Li, Y.; Zhang, Z.; Ren, W.; Wang, Y.; Mintah, B.K.; Dabbour, M.; Hou, Y.; He, R.; Cheng, Y.; Ma, H. Inhibition Effect of Ultrasound on the Formation of Lysinoalanine in Rapeseed Protein Isolates during pH Shift Treatment. *J. Agric. Food Chem.* **2021**, *69*, 8536–8545. [CrossRef]
18. Igartúa, D.E.; Dichano, M.C.; Ferrari, S.B.; Palazolo, G.G.; Cabezas, D.M. Combination of pH-Shifting, Ultrasound, and Heat Treatments to Enhance Solubility and Emulsifying Stability of Rice Protein Isolate. *Food Chem.* **2024**, *433*, 137319. [CrossRef]
19. Dabbour, M.; He, R.; Mintah, B.; Ma, H. Antioxidant Activities of Sunflower Protein Hydrolysates Treated with Dual-Frequency Ultrasonic: Optimization Study. *J. Food Process Eng.* **2019**, *42*, e13084. [CrossRef]
20. Golly, M.K.; Ma, H.; Duan, Y.; Liu, D.; Quaisie, J.; Tuli, J.A.; Mintah, B.K.; Dzah, C.S.; Agordoh, P.D. Effect of Multi-Frequency Countercurrent Ultrasound Treatment on Extraction Optimization, Functional and Structural Properties of Protein Isolates from Walnut (*Juglans Regia* L.) Meal. *J. Food Biochem.* **2020**, *44*, e13210. [CrossRef]
21. Hou, F.; Ding, W.; Qu, W.; Oladejo, A.O.; Xiong, F.; Zhang, W.; He, R.; Ma, H. Alkali Solution Extraction of Rice Residue Protein Isolates: Influence of Alkali Concentration on Protein Functional, Structural Properties and Lysinoalanine Formation. *Food Chem.* **2017**, *218*, 207–215. [CrossRef] [PubMed]
22. Jin, J.; Ma, H.; Wang, W.; Luo, M.; Wang, B.; Qu, W.; He, R.; Owusu, J.; Li, Y. Effects and Mechanism of Ultrasound Pretreatment on Rapeseed Protein Enzymolysis. *J. Sci. Food Agric.* **2016**, *96*, 1159–1166. [CrossRef] [PubMed]
23. Pan, J.; Zhang, Z.; Mintah, B.K.; Xu, H.; Dabbour, M.; Cheng, Y.; Dai, C.; He, R.; Ma, H. Effects of Nonthermal Physical Processing Technologies on Functional, Structural Properties and Digestibility of Food Protein: A Review. *J. Food Process Eng.* **2022**, *45*, e14010. [CrossRef]
24. Momen, S.; Alavi, F.; Aider, M. Alkali-Mediated Treatments for Extraction and Functional Modification of Proteins: Critical and Application Review. *Trends Food Sci. Technol.* **2021**, *110*, 778–797. [CrossRef]
25. Lou, K.; Zheng, Y.; Wang, L.; Zhou, C.; Wang, J.; Pan, D.; Wu, Z.; Cao, J.; Zhang, H.; Xia, Q. Molten Globule-State Protein Structure: Perspectives from Food Processing Applications. *Food Res. Int.* **2024**, *198*, 115318. [CrossRef]
26. Sultan, Z.; Ashfaq, A.; Jahan, K.; Qadri, O.S.; Younis, K.; Yousuf, O. pH Shift Extraction Technique for Plant Proteins: A Promising Technique for Sustainable Development. *Energy Nexus* **2024**, *16*, 100329. [CrossRef]
27. Chen, Q.; Ji, H.; Wang, Z.; Wang, Y.; Wang, X.; Chen, Z. Effects of Different Charged Polysaccharides on the Gelation Properties and in Vitro Digestibility of Potato Protein Gel: Insight into Underlying Mechanisms. *Food Hydrocoll.* **2025**, *164*, 111187. [CrossRef]
28. Zhao, R.; Fu, W.; Li, D.; Dong, C.; Bao, Z.; Wang, C. Structure and Functionality of Whey Protein, Pea Protein, and Mixed Whey and Pea Proteins Treated by pH Shift or High-Intensity Ultrasound. *J. Dairy Sci.* **2024**, *107*, 726–741. [CrossRef]
29. Hu, Z.; Wang, Y.; Ma, Z.; Cheng, T.; Guo, Z.; Zhou, L.; Wang, Z. Impacts of Industrial Modification on the Structure and Gel Features of Soy Protein Isolate and Its Composite Gel with Myofibrillar Protein. *Foods* **2023**, *12*, 1982. [CrossRef]
30. Dumetz, A.C.; Chockla, A.M.; Kaler, E.W.; Lenhoff, A.M. Effects of pH on Protein–Protein Interactions and Implications for Protein Phase Behavior. *Biochim. Biophys. Acta BBA—Proteins Proteom.* **2008**, *1784*, 600–610. [CrossRef]

31. Han, Q.; Veríssimo, N.V.P.; Bryant, S.J.; Martin, A.V.; Huang, Y.; Pereira, J.F.B.; Santos-Ebinuma, V.C.; Zhai, J.; Bryant, G.; Drummond, C.J.; et al. Scattering Approaches to Unravel Protein Solution Behaviors in Ionic Liquids and Deep Eutectic Solvents: From Basic Principles to Recent Developments. *Adv. Colloid Interface Sci.* **2024**, *331*, 103242. [CrossRef]
32. Jiang, J.; Wang, Q.; Xiong, Y.L. A pH Shift Approach to the Improvement of Interfacial Properties of Plant Seed Proteins. *Curr. Opin. Food Sci.* **2018**, *19*, 50–56. [CrossRef]
33. Pezeshk, S.; Rezaei, M.; Hosseini, H.; Abdollahi, M. Impact of pH-Shift Processing Combined with Ultrasonication on Structural and Functional Properties of Proteins Isolated from Rainbow Trout by-Products. *Food Hydrocoll.* **2021**, *118*, 106768. [CrossRef]
34. Lu, X.; Qian, S.; Wu, X.; Lan, T.; Zhang, H.; Liu, J. Research Progress of Protein Complex Systems and Their Application in Food: A Review. *Int. J. Biol. Macromol.* **2024**, *265*, 130987. [CrossRef]
35. Yu, Y.; Guan, Y.; Wen, H.; Zhang, Y.; Liu, J.; Zhang, T. Mild Heating Assisted Alkaline pH Shifting Modify the Egg White Protein: The Mechanism and the Enhancement of Emulsifying Properties. *LWT* **2021**, *151*, 112094. [CrossRef]
36. Chang, L.; Chen, B.; Rao, J. Synergistic Effect of pH-Shift and Controlled Heating on Improving Foaming Properties of Pea Vicilin and Its Adsorption Behavior at the Air-Water Interface. *Food Hydrocoll.* **2023**, *145*, 109022. [CrossRef]
37. Sun, P.; Zhang, Q.; Zhao, Y.; Zhao, D.; Zhao, X.; Jiang, L.; Zhang, Y.; Wu, F.; Sui, X. Improving Gel Properties of Soy Protein Isolate through Alkaline pH-Shifting, Mild Heat Treatment, and TGase Cross-Linking. *Food Hydrocoll.* **2023**, *144*, 108924. [CrossRef]
38. Dabbour, M.; Jiang, H.; Mintah, B.; Wahia, H.; He, R. Ultrasonic-Assisted Protein Extraction from Sunflower Meal: Kinetic Modeling, Functional, and Structural Traits. *Innov. Food Sci. Emerg. Technol.* **2021**, *74*, 102824. [CrossRef]
39. Zhang, Y.; Wu, C.; Shen, X.; McClements, D.J.; Liu, X.; Liu, F. Effects of Combined Hot Alkaline and pH-Shift Treatments on Structure and Functionality of Legume Protein-EGCG Conjugates: Soybean-, Pea-, and Chickpea Protein-EGCG Systems. *Food Hydrocoll.* **2025**, *158*, 110424. [CrossRef]
40. Wang, Y.; Sun, S.; Shen, J.; Zou, B.; Zhang, L.; Xu, X.; Wu, C. Disulfide Bond Cleavage Combined with Critical pH Induced Unfolding and Assembly of Soy Protein and Its Encapsulation Effect on Curcumin. *Food Hydrocoll.* **2024**, *157*, 110358. [CrossRef]
41. Zhang, X.; Xu, J.; Sun, Y.; Zhang, H.; Guo, S. Alkaline-Heat Induced the Conformationally Flexible Regions of Soy Protein and Their Effect on Subunit Aggregation. *Food Chem.* **2025**, *477*, 143535. [CrossRef]
42. Sun, Y.; Wang, Q.; Li, C.; Wang, Y.; Mao, Y.; Yang, C.; Li, Y. Preparation of Pumpkin Seed Protein Isolate Nanoparticles by Heat-Assisted pH-Shifting: Enhanced Emulsification Performance and Dispersibility. *J. Food Eng.* **2024**, *377*, 112087. [CrossRef]
43. Wang, Y.; Yang, F.; Wu, M.; Li, J.; Bai, Y.; Xu, W.; Qiu, S. Synergistic Effect of pH Shifting and Mild Heating in Improving Heat Induced Gel Properties of Peanut Protein Isolate. *LWT* **2020**, *131*, 109812. [CrossRef]
44. Yang, L.; Dong, H.; Wang, J.; Dadmohammadi, Y.; Zhou, Y.; Lin, T.; Khongkomolsakul, W.; Meletharayil, G.; Kapoor, R.; Abbaspourrad, A. Fabrication and Characterization of Whey Protein Isolate-Tryptophan Nanoparticles by pH-Shifting Combined with Heat Treatment. *Food Res. Int.* **2024**, *196*, 115031. [CrossRef]
45. Zhong, M.; Sun, Y.; Sun, Y.; Fang, L.; Wang, Q.; Qi, B.; Li, Y. Soy Lipophilic Protein Self-Assembled by pH-Shift Combined with Heat Treatment: Structure, Hydrophobic Resveratrol Encapsulation, Emulsification, and Digestion. *Food Chem.* **2022**, *394*, 133514. [CrossRef]
46. Nisov, A.; Nikinmaa, M.; Nordlund, E.; Sozer, N. Effect of pH and Temperature on Fibrous Structure Formation of Plant Proteins during High-Moisture Extrusion Processing. *Food Res. Int.* **2022**, *156*, 111089. [CrossRef]
47. Zhu, P.; Du, X.; Liu, C.; Zhao, G.; Wang, M. Effects of pH during Dry-Heat Preparation on the Physicochemical and Emulsifying Properties of Rice Starch and Whey Protein Isolate Mixtures. *Food Hydrocoll.* **2023**, *140*, 108614. [CrossRef]
48. Ravindran, N.; Kumar Singh, S.; Singha, P. A Comprehensive Review on the Recent Trends in Extractions, Pretreatments and Modifications of Plant-Based Proteins. *Food Res. Int.* **2024**, *190*, 114575. [CrossRef]
49. Wang, F.; Li, J.; Qi, Q.; Mao, Y.; Yan, X.; Li, X.; Mu, Y.; Zhang, H.; Zhao, C.; Liu, J. Structural, Physicochemical and Digestive Properties of Non-Covalent and Covalent Complexes of Ultrasound Treated Soybean Protein Isolate with Soybean Isoflavone. *Food Res. Int.* **2024**, *189*, 114571. [CrossRef] [PubMed]
50. Su, J.; Cavaco-Paulo, A. Effect of Ultrasound on Protein Functionality. *Ultrason. Sonochem.* **2021**, *76*, 105653. [CrossRef]
51. Wang, Y.; Li, B.; Guo, Y.; Liu, C.; Liu, J.; Tan, B.; Guo, Z.; Wang, Z.; Jiang, L. Effects of Ultrasound on the Structural and Emulsifying Properties and Interfacial Properties of Oxidized Soybean Protein Aggregates. *Ultrason. Sonochem.* **2022**, *87*, 106046. [CrossRef]
52. Rahman, M.M.; Lamsal, B.P. Ultrasound-Assisted Extraction and Modification of Plant-Based Proteins: Impact on Physicochemical, Functional, and Nutritional Properties. *Compr. Rev. Food Sci. Food Saf.* **2021**, *20*, 1457–1480. [CrossRef]
53. Hadidi, M.; Aghababaei, F.; McClements, D.J. Enhanced Alkaline Extraction Techniques for Isolating and Modifying Plant-Based Proteins. *Food Hydrocoll.* **2023**, *145*, 109132. [CrossRef]
54. Wang, Q.; Wang, Y.; Huang, M.; Hayat, K.; Kurtz, N.C.; Wu, X.; Ahmad, M.; Zheng, F. Ultrasound-Assisted Alkaline Proteinase Extraction Enhances the Yield of Pecan Protein and Modifies Its Functional Properties. *Ultrason. Sonochem.* **2021**, *80*, 105789. [CrossRef]
55. Li, W.; Yang, H.; Coldea, T.E.; Zhao, H. Modification of Structural and Functional Characteristics of Brewer's Spent Grain Protein by Ultrasound Assisted Extraction. *LWT* **2021**, *139*, 110582. [CrossRef]

56. Zhang, X.; Guo, Q.; Shi, W. Ultrasound-Assisted Processing: Changes in Gel Properties, Water-Holding Capacity, and Protein Aggregation of Low-Salt *Hypophthalmichthys Molitrix* Surimi by Soy Protein Isolate. *Ultrason. Sonochem.* **2023**, *92*, 106258. [CrossRef]
57. Jiang, S.; Li, Q.; Wang, T.; Huang, Y.; Guo, Y.; Meng, X. Utilizing Ultrasound Combined with Quinoa Protein to Improve the Texture and Rheological Properties of Chinese Style Reduced-Salt Pork Meatballs (Lion's Head). *Ultrason. Sonochem.* **2024**, *109*, 106997. [CrossRef]
58. Wang, N.; Zhou, X.; Wang, W.; Wang, L.; Jiang, L.; Liu, T.; Yu, D. Effect of High Intensity Ultrasound on the Structure and Solubility of Soy Protein Isolate-Pectin Complex. *Ultrason. Sonochem.* **2021**, *80*, 105808. [CrossRef]
59. Zheng, X.; Zou, B.; Zhang, J.; Cai, W.; Na, X.; Du, M.; Zhu, B.; Wu, C. Recent Advances of Ultrasound-Assisted Technology on Aquatic Protein Processing: Extraction, Modification, and Freezing/Thawing-Induced Oxidation. *Trends Food Sci. Technol.* **2024**, *144*, 104309. [CrossRef]
60. Dabbour, M.; Hamoda, A.; Xu, H.; Mintah, B.K.; Wahia, H.; Betchem, G.; Yolandani; He, R.; Ma, H.; Fikry, M. pH-Shifting and Sonication Synergistically Altered Cottonseed Protein: Correlating the Conformational and Functional Characteristics. *Ind. Crops Prod.* **2024**, *222*, 120043. [CrossRef]
61. Jiang, S.; Ding, J.; Andrade, J.; Rababah, T.M.; Almajwal, A.; Abulmeaty, M.M.; Feng, H. Modifying the Physicochemical Properties of Pea Protein by pH-Shifting and Ultrasound Combined Treatments. *Ultrason. Sonochem.* **2017**, *38*, 835–842. [CrossRef]
62. Sai-Ut, S.; Watchasit, S.; Pongsetkul, J.; Kingwascharapong, P.; Suriyarak, S.; Grossmann, L.; Zhang, W.; Rawdkuen, S. Enhancing Protein Extraction from Pleurotus Ostreatus Using Synergistic pH-Shifting and Ultrasonic Technology: Optimization via RSM and 1H NMR-Based Metabolomic Profiling. *LWT* **2024**, *211*, 116895. [CrossRef]
63. Pezeshk, S.; Rezaei, M.; Hosseini, H.; Abdollahi, M. Ultrasound-Assisted Alkaline pH-Shift Process Effects on Structural and Interfacial Properties of Proteins Isolated from Shrimp by-Products. *Food Struct.* **2022**, *32*, 100273. [CrossRef]
64. Song, H.; Zhong, M.; Sun, Y.; Yue, Q.; Qi, B. Ultrasound-Assisted Alkali Removal of Proteins from Wastewater Generated during Oil Bodies Extraction. *Ultrason. Sonochem.* **2023**, *96*, 106436. [CrossRef] [PubMed]
65. Sun, Y.; Wang, L.; Wang, H.; Zhou, B.; Jiang, L.; Zhu, X. Effect of pH-Shifting and Ultrasound on Soy/Potato Protein Structure and Gelation. *Food Hydrocoll.* **2025**, *159*, 110672. [CrossRef]
66. Yang, J.; Duan, Y.; Geng, F.; Cheng, C.; Wang, L.; Ye, J.; Zhang, H.; Peng, D.; Deng, Q. Ultrasonic-Assisted pH Shift-Induced Interfacial Remodeling for Enhancing the Emulsifying and Foaming Properties of Perilla Protein Isolate. *Ultrason. Sonochem.* **2022**, *89*, 106108. [CrossRef]
67. Li, H.; Hu, Y.; Zhao, X.; Wan, W.; Du, X.; Kong, B.; Xia, X. Effects of Different Ultrasound Powers on the Structure and Stability of Protein from Sea Cucumber Gonad. *LWT* **2021**, *137*, 110403. [CrossRef]
68. Zhang, X.; Yan, D.; Qiu, W.; Chen, S.; Hu, Y.; Jin, J.; Udenigwe, C.C. Effects of Ultrasound Combined with pH Shift Modification on Functional and Structural Properties of Peanut Proteins. *Int. J. Biol. Macromol.* **2024**, *283*, 137874. [CrossRef] [PubMed]
69. Ma, C.; Wan, Q.; Song, J.; Hao, T.; Xia, S.; Shen, S.; Li, K.; Xue, C.; Jiang, X. Ultrasound-Assisted pH Shift-Induced Interfacial Remodeling for Enhancing Soluble Yeast Protein Content: Effects on Structure and Interfacial Properties of Proteins under Different Treatment Conditions. *Food Hydrocoll.* **2024**, *149*, 109521. [CrossRef]
70. Kang, Z.; Zhang, S.; Kong, Y.; Wu, Z.; Li, Y.; Liu, T.; Xie, F. Modification of Soybean Protein Isolate by pH-Shifting Combined with Ultrasonic Treatment: Structural, Viscosity, and Functional Properties. *Food Struct.* **2024**, *42*, 100383. [CrossRef]
71. Zhang, J.; Zhao, S.; Liu, Q.; Chen, Q.; Liu, H.; Kong, B. High Internal Phase Emulsions Stabilized by Pea Protein Isolate Modified by Ultrasound and pH-Shifting: Effect of Chitosan Self-Assembled Particles. *Food Hydrocoll.* **2023**, *141*, 108715. [CrossRef]
72. Yang, J.; Duan, Y.; Zhang, H.; Huang, F.; Wan, C.; Cheng, C.; Wang, L.; Peng, D.; Deng, Q. Ultrasound Coupled with Weak Alkali Cycling-Induced Exchange of Free Sulfhydryl-Disulfide Bond for Remodeling Interfacial Flexibility of Flaxseed Protein Isolates. *Food Hydrocoll.* **2023**, *140*, 108597. [CrossRef]
73. Wang, K.; Liu, H.; Sun, J. Improved Gelling and Emulsifying Properties of Chicken Wooden Breast Myofibrillar Protein by High-Intensity Ultrasound Combination with pH-Shifting. *Poult. Sci.* **2023**, *102*, 103063. [CrossRef]
74. Wang, J.; Wang, X.; Wang, W.; Zhang, L.; Zhao, Y. Functionalization of Pine Kernel Protein by pH-Shifting Combined with Ultrasound Treatments: Further Improvement with Increasing Acidity. *Int. J. Biol. Macromol.* **2023**, *248*, 125884. [CrossRef]
75. Wang, Y.; Wang, S.; Li, R.; Wang, Y.; Xiang, Q.; Li, K.; Bai, Y. Effects of Combined Treatment with Ultrasound and pH Shifting on Foaming Properties of Chickpea Protein Isolate. *Food Hydrocoll.* **2022**, *124*, 107351. [CrossRef]
76. Sun, Y.; Chen, H.; Chen, W.; Zhong, Q.; Shen, Y.; Zhang, M. Effect of Ultrasound on pH-Shift to Improve Thermal Stability of Coconut Milk by Modifying Physicochemical Properties of Coconut Milk Protein. *LWT* **2022**, *167*, 113861. [CrossRef]
77. Jiang, Z.; Gao, Y.; Li, J.; Wang, K.; Ma, C.; Sun, D.; Hussain, M.A.; Qayum, A.; Hou, J. Consecutive pH-Shift and Ultrasound Treatment Modify the Physicochemical Properties of Whey Protein Isolate. *Int. Dairy J.* **2022**, *127*, 105211. [CrossRef]
78. Figueroa-González, J.J.; Lobato-Calleros, C.; Vernon-Carter, E.J.; Aguirre-Mandujano, E.; Alvarez-Ramirez, J.; Martínez-Velasco, A. Modifying the Structure, Physicochemical Properties, and Foaming Ability of Amaranth Protein by Dual pH-Shifting and Ultrasound Treatments. *LWT* **2022**, *153*, 112561. [CrossRef]

79. Alavi, F.; Chen, L.; Emam-Djomeh, Z. Effect of Ultrasound-Assisted Alkaline Treatment on Functional Property Modifications of Faba Bean Protein. *Food Chem.* **2021**, *354*, 129494. [CrossRef] [PubMed]
80. Silventoinen, P.; Sozer, N. Impact of Ultrasound Treatment and pH-Shifting on Physicochemical Properties of Protein-Enriched Barley Fraction and Barley Protein Isolate. *Foods* **2020**, *9*, 1055. [CrossRef]
81. Liu, Y.; Tan, X.; Li, L.; Xie, T.; Teng, F. Co-Encapsulation of Vitamin E and Quercetin by Soybean Lipophilic Proteins Based on pH-Shifting and Ultrasonication: Focus on Interaction Mechanisms, Structural and Physicochemical Properties. *Food Chem.* **2024**, *460*, 140608. [CrossRef] [PubMed]
82. Fang, Z.; Cai, X.; Wu, J.; Zhang, L.; Fang, Y.; Wang, S. Effect of Simultaneous Treatment Combining Ultrasonication and pH-Shifting on SPI in the Formation of Nanoparticles and Encapsulating Resveratrol. *Food Hydrocoll.* **2021**, *111*, 106250. [CrossRef]
83. Baldelli, A.; Shi, J.; Singh, A.; Guo, Y.; Fathordoobady, F.; Amiri, A.; Pratap-Singh, A. Effect of High-Pressure on Protein Structure, Refolding, and Crystallization. *Food Chem. Adv.* **2024**, *5*, 100741. [CrossRef]
84. Gao, Y.; Gao, T.; Li, L.; Chi, H.; Teng, F. Modification of Soybean Lipophilic Protein Based on pH-Shifting and High-Pressure Homogenization: Focus on Structure, Physicochemical Properties and Delivery Vehicle. *Food Chem.* **2025**, *463*, 141001. [CrossRef]
85. Wang, Y.; Liu, X.; Luo, S.; Zhong, C.; Ye, J.; Liu, C. The Impact of pH Shifting Combined High-Pressure Homogenization on Structural and Functional Properties of Rice Dreg Protein. *Innov. Food Sci. Emerg. Technol.* **2024**, *91*, 103520. [CrossRef]
86. Yildiz, G. Effects of High-Intensity Ultrasound, High-Pressure Processing, and Their Combination with pH-Shifting on the Techno-Functionality and Digestibility of Melon Seed Protein Isolate. *Food Res. Int.* **2025**, *208*, 116219. [CrossRef]
87. Parlak, M.E.; Sarıcaoglu, F.T.; Yilmaz, M.T. Application of High-Pressure Homogenization-Assisted pH-Shift to Enhance Techno-Functional and Interfacial Properties of Lentil Protein Isolate. *Food Hydrocoll.* **2024**, *157*, 110425. [CrossRef]
88. Mune Mune, M.A.; Stănciuc, N.; Grigore-Gurgu, L.; Aprodu, I.; Borda, D. Structural Changes Induced by High Pressure Processing in Bambara Bean Proteins at Different pH. *LWT* **2020**, *124*, 109187. [CrossRef]
89. Alvarez, P.A.; Ramaswamy, H.S.; Ismail, A.A. High Pressure Gelation of Soy Proteins: Effect of Concentration, pH and Additives. *J. Food Eng.* **2008**, *88*, 331–340. [CrossRef]
90. Hall, A.E.; Moraru, C.I. Structure and Function of Pea, Lentil and Faba Bean Proteins Treated by High Pressure Processing and Heat Treatment. *LWT* **2021**, *152*, 112349. [CrossRef]
91. Mulla, M.Z.; Subramanian, P.; Dar, B.N. Functionalization of Legume Proteins Using High Pressure Processing: Effect on Technofunctional Properties and Digestibility of Legume Proteins. *LWT* **2022**, *158*, 113106. [CrossRef]
92. Wang, Q.; Tang, Z.; Cao, Y.; Ming, Y.; Wu, M. Improving the Solubility and Interfacial Absorption of Hempseed Protein via a Novel High Pressure Homogenization-Assisted pH-Shift Strategy. *Food Chem.* **2024**, *442*, 138447. [CrossRef]
93. Liu, Q.; Chen, A.; Hong, P.; Zhou, C.; Li, X.; Xie, M. pH-Induced Interface Protein Structure Changes to Adjust the Stability of Tilapia Protein Isolate Emulsion Prepared by High-Pressure Homogenization. *Food Chem. X* **2024**, *24*, 101841. [CrossRef]
94. Laguna, L.; Picouet, P.; Guàrdia, M.D.; Renard, C.M.G.C.; Sarkar, A. In Vitro Gastrointestinal Digestion of Pea Protein Isolate as a Function of pH, Food Matrices, Autoclaving, High-Pressure and Re-Heat Treatments. *LWT* **2017**, *84*, 511–519. [CrossRef]
95. Ahmed, J.; Al-Ruwaih, N.; Mulla, M.; Rahman, M.H. Effect of High Pressure Treatment on Functional, Rheological and Structural Properties of Kidney Bean Protein Isolate. *LWT* **2018**, *91*, 191–197. [CrossRef]
96. Tan, M.; Xu, J.; Gao, H.; Yu, Z.; Liang, J.; Mu, D.; Li, X.; Zhong, X.; Luo, S.; Zhao, Y.; et al. Effects of Combined High Hydrostatic Pressure and pH-Shifting Pretreatment on the Structure and Emulsifying Properties of Soy Protein Isolates. *J. Food Eng.* **2021**, *306*, 110622. [CrossRef]
97. Yildiz, G.; Yıldız, G. A New Approach to Enhance Quinoa Protein Nano-Aggregates: Combined pH Shifting—High Pressure Homogenization. *Food Chem.* **2023**, *415*, 135800. [CrossRef] [PubMed]
98. Kadri, A.; Jenner, G.; Damak, M.; Lorber, B.; Giegé, R. Crystallogenesis Studies of Proteins in Agarose Gel—Combined Effect of High Hydrostatic Pressure and pH. *J. Cryst. Growth* **2003**, *257*, 390–402. [CrossRef]
99. Research Progress in Soybean Lipophilic Protein (LP): Extraction, Structural, Techno-Functional Properties, and High-Performance Food Applications. *Trends Food Sci. Technol.* **2024**, *147*, 104440. [CrossRef]
100. Zulkali, M.M.D.; Ahmad, A.L.; Derek, C.J.C. Membrane Application in Proteomic Studies: Preliminary Studies on the Effect of pH, Ionic Strength and Pressure on Protein Fractionation. *Membr. Drink. Ind. Water Prod.* **2005**, *179*, 381–390. [CrossRef]
101. Khan, N.M.; Mu, T.-H.; Zhang, M.; Arogundade, L.A. The Effects of pH and High Hydrostatic Pressure on the Physicochemical Properties of a Sweet Potato Protein Emulsion. *Food Hydrocoll.* **2014**, *35*, 209–216. [CrossRef]
102. Teixeira, R.F.; Araujo, T.R.; Oliveira, D.d.; Zielinski, A.A.F. Unveiling the Potential of Pressurized Liquid Extraction for Recovering Protein Fractions from Broken Black Beans: Insights into Thermal and Structural Properties. *Food Hydrocoll.* **2024**, *149*, 109649. [CrossRef]
103. Wang, L.; Moraru, C.I. High-Pressure Structuring of Milk Protein Concentrate: Effect of pH and Calcium. *J. Dairy Sci.* **2021**, *104*, 4074–4083. [CrossRef]
104. Wang, Q.; Jiang, J.; Xiong, Y.L. High Pressure Homogenization Combined with pH Shift Treatment: A Process to Produce Physically and Oxidatively Stable Hemp Milk. *Food Res. Int.* **2018**, *106*, 487–494. [CrossRef]

105. Malik, M.A.; Sheikh, M.A.; Mir, N.A. A Review on Pulsed Electric Field Modification of Proteins: Effect on the Functional and Structural Properties. *Food Biosci.* **2024**, *61*, 104636. [CrossRef]
106. Kim, Y.J.; Shin, D.-M.; Oh, E.-J.; Chun, Y.G.; Shin, J.-K.; Choi, Y.-S.; Kim, B.-K. Mechanisms Underlying the Changes in the Structural, Physicochemical, and Emulsification Properties of Porcine Myofibrillar Proteins Induced by Prolonged Pulsed Electric Field Treatment. *Food Chem.* **2024**, *456*, 140024. [CrossRef] [PubMed]
107. Chian, F.M.; Kaur, L.; Oey, I.; Astruc, T.; Hodgkinson, S.; Boland, M. Effect of Pulsed Electric Fields (PEF) on the Ultrastructure and in Vitro Protein Digestibility of Bovine Longissimus Thoracis. *LWT* **2019**, *103*, 253–259. [CrossRef]
108. Farjami, T.; Babaei, J.; Nau, F.; Dupont, D.; Madadlou, A. Effects of Thermal, Non-Thermal and Emulsification Processes on the Gastrointestinal Digestibility of Egg White Proteins. *Trends Food Sci. Technol.* **2021**, *107*, 45–56. [CrossRef]
109. Xu, F.-Y.; Wen, Q.-H.; Wang, R.; Li, J.; Chen, B.-R.; Zeng, X.-A. Enhanced Synthesis of Succinylated Whey Protein Isolate by Pulsed Electric Field Pretreatment. *Food Chem.* **2021**, *363*, 129892. [CrossRef] [PubMed]
110. Yang, N.; Yao, H.; Zhang, A.; Jin, Y.; Zhang, X.; Xu, X. Effect of Constant-Current Pulsed Electric Field Thawing on Proteins and Water-Holding Capacity of Frozen Porcine Longissimus Muscle. *Food Chem.* **2024**, *454*, 139784. [CrossRef] [PubMed]
111. Liu, Y.-F.; Oey, I.; Bremer, P.; Silcock, P.; Carne, A. Proteolytic Pattern, Protein Breakdown and Peptide Production of Ovomucin-Depleted Egg White Processed with Heat or Pulsed Electric Fields at Different pH. *Food Res. Int.* **2018**, *108*, 465–474. [CrossRef]
112. Wang, R.; Wang, L.-H.; Wen, Q.-H.; He, F.; Xu, F.-Y.; Chen, B.-R.; Zeng, X.-A. Combination of Pulsed Electric Field and pH Shifting Improves the Solubility, Emulsifying, Foaming of Commercial Soy Protein Isolate. *Food Hydrocoll.* **2023**, *134*, 108049. [CrossRef]
113. Guo, X.; Aganovic, K.; Bindrich, U.; Juadjur, A.; Hertel, C.; Ebert, E.; Macke, J.g.; Geil, C.; Heinz, V. Extraction of Protein from Juice Blend of Grass and Clover Pressed by a Pilot Pressing Facility Combined with a Pulsed Electric Field Treatment. *Future Foods* **2022**, *6*, 100173. [CrossRef]
114. Saxena, A.; Shahi, V.K. pH Controlled Selective Transport of Proteins through Charged Ultrafilter Membranes under Coupled Driving Forces: An Efficient Process for Protein Separation. *J. Membr. Sci.* **2007**, *299*, 211–221. [CrossRef]
115. Wang, R.; Zeng, M.-Q.; Wu, Y.-W.; Teng, Y.-X.; Wang, L.-H.; Li, J.; Xu, F.-Y.; Chen, B.-R.; Han, Z.; Zeng, X.-A. Enhanced Encapsulation of Lutein Using Soy Protein Isolate Nanoparticles Prepared by Pulsed Electric Field and pH Shifting Treatment. *Food Chem.* **2023**, *424*, 136386. [CrossRef] [PubMed]
116. Akaberi, S.; Krust, D.; Müller, G.; Frey, W.; Gusbeth, C. Impact of Incubation Conditions on Protein and C-Phycocyanin Recovery from Arthrospira Platensis Post- Pulsed Electric Field Treatment. *Bioresour. Technol.* **2020**, *306*, 123099. [CrossRef] [PubMed]
117. Liang, Y.; Wu, K.; He, D.; Ou, C.; Lin, J.; Chai, X.; Xiang, Y.; Duan, X.; Cha, Q.; Zhang, X.; et al. Physicochemical and Functional Properties of Cinnamon Essential Oil Emulsions Stabilized by Galactomannan-Rich Aqueous Extract from Gleditsia Sinensis Seeds and Soy Protein Isolate. *Int. J. Biol. Macromol.* **2025**, *295*, 139601. [CrossRef]
118. Han, W.; Shi, W.; Gong, D.; Zhang, G. Improvement of Solubility, Emulsification Property and Stability of Potato Protein by pH-Shifting Combined with Microwave Treatment and Interaction with Pectin. *Food Biosci.* **2023**, *56*, 103301. [CrossRef]
119. Das, D.; Panesar, P.S.; Saini, C.S. Effect of pH Shifting on Different Properties of Microwave-Extracted Soybean Meal Protein Isolate. *Food Bioprocess Technol.* **2024**, *17*, 640–655. [CrossRef]
120. Jahan, K.; Sultan, Z.; Younis, K.; Mir, S.S.; Yousuf, O. pH-Shift Extraction Followed by Microwave and Ultrasound Modified Functional Properties of Mustard Meal Protein. *Biocatal. Agric. Biotechnol.* **2024**, *60*, 103295. [CrossRef]
121. Zhang, J.; Ström, A.; Bordes, R.; Alminger, M.; Undeland, I.; Abdollahi, M. Radial Discharge High Shear Homogenization and Ultrasonication Assisted pH-Shift Processing of Herring Co-Products with Antioxidant-Rich Materials for Maximum Protein Yield and Functionality. *Food Chem.* **2023**, *400*, 133986. [CrossRef]
122. Karabulut, G.; Kapoor, R.; Yemis, O.; Feng, H. Manothermosonication, High-Pressure Homogenization, and Their Combinations with pH-Shifting Improve the Techno-Functionality and Digestibility of Hemp Protein. *Food Hydrocoll.* **2024**, *150*, 109661. [CrossRef]
123. Jiang, W.; Zhang, Y.; Julian McClements, D.; Liu, F.; Liu, X. Impact of Pea Protein-Inulin Conjugates Prepared via the Maillard Reaction Using a Combination of Ultrasound and pH-Shift Treatments on Physical and Oxidative Stability of Algae Oil Emulsions. *Food Res. Int.* **2022**, *156*, 111161. [CrossRef] [PubMed]
124. Ding, S.; Ye, X.; Qu, L.; Mu, J.; Huang, L.; Dai, C. Modification of Whey Protein Isolate by Ultrasound-Assisted pH Shift for Complexation with Carboxymethylcellulose: Structure and Interfacial Properties. *Int. J. Biol. Macromol.* **2023**, *252*, 126479. [CrossRef] [PubMed]

Disclaimer/Publisher's Note: The statements, opinions and data contained in all publications are solely those of the individual author(s) and contributor(s) and not of MDPI and/or the editor(s). MDPI and/or the editor(s) disclaim responsibility for any injury to people or property resulting from any ideas, methods, instructions or products referred to in the content.

MDPI AG
Grosspeteranlage 5
4052 Basel
Switzerland
Tel.: +41 61 683 77 34

Foods Editorial Office
E-mail: foods@mdpi.com
www.mdpi.com/journal/foods

Disclaimer/Publisher's Note: The title and front matter of this reprint are at the discretion of the Guest Editors. The publisher is not responsible for their content or any associated concerns. The statements, opinions and data contained in all individual articles are solely those of the individual Editors and contributors and not of MDPI. MDPI disclaims responsibility for any injury to people or property resulting from any ideas, methods, instructions or products referred to in the content.

www.ingramcontent.com/pod-product-compliance
Lightning Source LLC
Chambersburg PA
CBHW041948260926
51850CB00018B/2008